T0226225

Communications
in Computer and Information Science　　　871

Commenced Publication in 2007
Founding and Former Series Editors:
Alfredo Cuzzocrea, Xiaoyong Du, Orhun Kara, Ting Liu, Dominik Ślęzak,
and Xiaokang Yang

Editorial Board

More information about this series at http://www.springer.com/series/7899

Anton Eremeev · Michael Khachay
Yury Kochetov · Panos Pardalos (Eds.)

Optimization Problems and Their Applications

7th International Conference, OPTA 2018
Omsk, Russia, July 8–14, 2018
Revised Selected Papers

 Springer

Editors
Anton Eremeev (ID)
Sobolev Institute of Mathematics SB RAS
Omsk
Russia

Michael Khachay (ID)
Krasovsky Institute of Mathematics
  and Mechanics
Ekaterinburg
Russia

Yury Kochetov (ID)
Sobolev Institute of Mathematics SB RAS
Novosibirsk
Russia

Panos Pardalos (ID)
Industrial and Systems Engineering
University of Florida
Gainesville, FL
USA

and

Laboratory of Algorithms and Technologies
  for Network Analysis (LATNA)
Higher School of Economics
Moscow
Russia

ISSN 1865-0929   ISSN 1865-0937   (electronic)
Communications in Computer and Information Science
ISBN 978-3-319-93799-1   ISBN 978-3-319-93800-4   (eBook)
https://doi.org/10.1007/978-3-319-93800-4

Library of Congress Control Number: 2018947312

Printed on acid-free paper

This Springer imprint is published by the registered company Springer International Publishing AG
part of Springer Nature
The registered company address is: Gewerbestrasse 11, 6330 Cham, Switzerland

Preface

This volume contains the proceedings of the 7th International Conference on Optimization Problems and Their Applications (OPTA 2018), held in Omsk, Russia, July 8–14, 2018. The conference was organized by the Russian Operational Research Society, Sobolev Institute of Mathematics, Krasovsky Institute of Mathematics and Mechanics, Novosibirsk State University, Dostoevsky Omsk State University, the Higher School of Economics in Nizhny Novgorod, and several other institutions. This year, the conference started in the city of Omsk and continued countryside in a picturesque place Chernoluchye, located in a pine forest in the vicinity of Omsk. Concurrently with the OPTA 2018 conference, an eponymous school seminar, OPTA-SCL 2018, was held, where invited speakers gave plenary lectures on the subjects of their individual expertise.

The previous conferences of this series, named "Optimization Problems and Economical Applications," were organized in 1997, 2003, 2006, 2009, 2012, and 2015 and chaired by Professor Alexander Kolokolov. OPTA 2018 was devoted to the memory of Alexander Kolokolov, who passed away in 2017.

The OPTA conference belongs to a group of international conferences on optimization and operations research in the Siberian and Ural areas of Russia, covering a wide range of topics in operations research, mathematical programming, discrete optimization, and their applications. This group also involves the International Conference on Discrete Optimization and Operations Research (DOOR), Baikal International Triennial School Seminar Methods of Optimization and Their Applications (BITSS), and the conference Mathematical Programming and Applications (MPA). The main purpose of these events is to provide a forum where researchers can exchange ideas, identify promising directions for theoretical studies and application domains, and foster new collaborations.

In response to the call for papers, we received 73 submissions. The papers included in this volume were carefully selected by the Program Committee on the basis of reports from two or more reviewers. Only 27 submissions were selected for inclusion in this volume. The abstracts of invited talks made by eminent speakers are also included here. We thank the Program Committee members and the external reviewers for their helpful remarks and fair evaluation of the submissions. We also thank the Organizing Committee members for their input.

Finally, we thank our sponsors, the Russian Foundation for Basic Research, the Novosibirsk State University, the Laboratory of Algorithms and Technologies for Networks Analysis (LATNA), the Higher School of Economics in Nizhny Novgorod, and Dostoevsky Omsk State University for their support.

April 2018

Anton Eremeev
Michael Khachay
Yury Kochetov
Panos Pardalos

Organization

Organizing Committee

Alexander Adelshin	Sobolev Institute of Mathematics SB RAS, Russia
Pavel Borisovsky	Sobolev Institute of Mathematics SB RAS, Russia
Ksenia Chernykh	Sobolev Institute of Mathematics SB RAS, Russia
Anton Eremeev (Chair)	Sobolev Institute of Mathematics SB RAS, Russia
Adil Erzin	Sobolev Institute of Mathematics SB RAS, Russia
Alexander Gnusarev	Sobolev Institute of Mathematics SB RAS, Russia
Victor Il'ev	Dostoevsky Omsk State University, Russia
Natalia Ivanova (Secretary)	Sobolev Institute of Mathematics SB RAS, Russia
Yury Kochetov	Sobolev Institute of Mathematics SB RAS, Russia
Nina Kochetova	Sobolev Institute of Mathematics SB RAS, Russia
Julia Kovalenko	Sobolev Institute of Mathematics SB RAS, Russia
Pavel Kuznetsov	Dostoevsky Omsk State University, Russia
Tatyana Levanova	Sobolev Institute of Mathematics SB RAS, Russia
Timur Medvedev	Higher School of Economics, Russia
Alexander Morshinin	Sobolev Institute of Mathematics SB RAS, Russia
Leonid Popov	Krasovsky Institute of Mathematics and Mechanics UB RAS, Russia
Artem Pyatkin	Sobolev Institute of Mathematics SB RAS, Russia
Anna Romanova	Dostoevsky Omsk State University, Russia
Tatyana Sergienko	Dostoevsky Omsk State University, Russia
Vladimir Servakh	Sobolev Institute of Mathematics SB RAS, Russia
Anatoly Solovyov	Siberian State Automobile and Highway University, Russia
Nikolay Tyunin	Sobolev Institute of Mathematics SB RAS, Russia
Gennady Zabudsky	Sobolev Institute of Mathematics SB RAS, Russia
Lidia Zaozerskaya	Sobolev Institute of Mathematics SB RAS, Russia
Igor Ziegler	Sobolev Institute of Mathematics SB RAS, Russia

Program Committee

Ekaterina Alekseeva	Colisweb Company, France
Edilkhan Amirgaliev	Institute of Information and Computational Technologies MES RK, Republic of Kazakhstan
Adil Bagirov	Federation University Australia, Australia
Evripidis Bampis	Universite Pierre et Marie Curie, France
Olga Battaia	ISAE-Supaero, France
Sergey Belim	Dostoevsky Omsk State University, Russia
Vladimir Beresnev	Sobolev Institute of Mathematics SB RAS, Russia

René van Bevern	Novosibirsk State University, Russia
Valentina Bykova	Siberian Federal University, Russia
Duc-Cuong Dang	University of Nottingham, UK
Tatjana Davidović	Mathematical Institute SASA, Serbia
Stephan Dempe	TU Freiberg University of Mining and Technology, Germany
Alexandre Dolgui	IMT Atlantique, France
Anton Eremeev (Co-chair)	Sobolev Institute of Mathematics SB RAS, Russia
Stefka Fidanova	Institute of Information and Communication Technologies BAS, Bulgaria
Edward Gimadi	Sobolev Institute of Mathematics SB RAS, Russia
Alexander Grigoriev	Maastricht University, The Netherlands
Evgeny Gurevsky	University of Nantes, France
Klaus Jansen	University of Kiel, Germany
Vyacheslav Kalashnikov	Tecnologico de Monterrey, Mexico
Valeriy Kalyagin	Higher School of Economics, Russia
Valery Karpov	Institute of Economics and Industrial Engineering SB RAS, Russia
Alexander Kelmanov	Sobolev Institute of Mathematics SB RAS, Russia
Mikhail Khachay (Co-chair)	Krasovsky Institute of Mathematics and Mechanics UB RAS, Russia
Oleg Khamisov	Melentiev Energy Systems Institute SB RAS, Russia
Yury Kochetov (Co-chair)	Sobolev Institute of Mathematics SB RAS, Russia
Vladimir Kotov	Belarusian State University, Belarus
Mikhail Kovalyov	United Institute of Informatics Problems NASB, Belarus
Per Kristian Lehre	University of Birmingham, UK
Bertrand Lin	National Chiao Tung University, Taiwan
Vittorio Maniezzo	University of Bologna, Italy
Vladimir Mazalov	Institute of Applied Mathematical Research of KRC RAS, Russia
Nenad Mladenović	Mathematical Institute SASA, Serbia
Frank Neumann	The University of Adelaide, Australia
Rolf Niedermeier	Technische Universität Berlin, Germany
Dmitry Novikov	V.A. Trapeznikov Institute of Control Sciences RAS, Russia
Evgeni Nurminski	Far Eastern Federal University, Russia
Panos Pardalos (Co-chair)	Higher School of Economics, Russia and University of Florida, USA
Soumyendu Raha	Indian Institute of Science, India
Colin Reeves	Coventry University, Coventry, UK
Konstantin Rudakov	Federal Research Center Informatics and Control RAS, Russia
Natalia Shakhlevich	University of Leeds, UK
Angelo Sifaleras	University of Macedonia, Greece

Alexander Strekalovsky   Matrosov Institute for System Dynamics and Control
                          Theory SB RAS, Russia
Vitaly Strusevich   University of Greenwich, UK
Maxim Sviridenko   Yahoo Research, USA
Ider Tseveendorj   Universite de Versailles-Saint Quentin en Yvelines, France
Yury Tsoy   Solidware, Republic of Korea
Aida Valeeva   Ufa State Aviation Technical University, Russia
Valery Vasil'ev   Sobolev Institute of Mathematics SB RAS, Russia
Xin Yao   Southern University of Science and Technology, China
Igor Zabotin   Kazan Federal University, Russia

Additional Reviewers

Abiad, Aida
Adelshin, Alexander
Antsyz, Sergey
Bagirov, Adil
Baklanov, Artem
Berestovskiy, Valeriy
Berger, Andre
Berikov, Vladimir
Berndt, Sebastian
Borisovsky, Pavel
Buzdalov, Maxim
Čvokić, Dimitrije
Davydov, Ivan
Erzin, Adil
Fedor, Stonyakin
Filimonov, Viacheslav
Golak, Julian
Gribanova, Irina
Il'ev, Victor
Kazakov, Alexander

Kazakovtsev, Lev
Kelmanov, Alexander
Khamisov, Oleg
Khandeev, Vladimir
Klokov, Sergey
Kobylkin, Konstantin
Kononov, Alexander
Kononova, Polina
Kovalenko, Julia
Kuznetsov, Pavel
Lagutaeva, Daria
Langelaar, Matthijs
Levanova, Tatyana
Maack, Marten
Maksimov, Vyacheslav
Neznakhina, Katherine
Oda, Yoshiaki
Orlov, Andrei
Panin, Artem
Plyasunov, Alexander

Poberiy, Maria
Popov, Leonid
Romanova, Anna
Servakh, Vladimir
Shamray, Natalia
Shenmaier, Vladimir
Simanchev, Ruslan
Skarin, Vladimir
Strekalovsky, Alexander
Strusevich, Vitaly
Timmermans, Veerle
Tsidulko, Oxana
Ushakov, Anton
Vasilyev, Igor
Weiss, Christian
Zakharov, Aleksey
Zaozerskaya, Lidia
Zhadan, Vitaly
Zolotykh, Nikolai

Invited Talks

Combinatorial Techniques to Optimally Customize Machining/Assembly Lines

Alexandre Dolgui

Ecole des Mines de Nantes, Nantes, France
alexandre.dolgui@imt-atlantique.fr

Keywords: Machining lines · Assembly lines · Line design · Line balancing
Combinatorial optimization · MIP · Graph theory · Metaheuristics · Heuristics

Problems of combinatorial design of complex machining/assembly lines are considered. Operations are partitioned into groups which are performed either by a team of workers for manually lines or by a piece of equipment for automated lines (for example by a multi-spindge head). Constraints related to the design of worker teams or spindle heads, working position and line configurations, as well as precedence constraints related to operations, are given. Such problems consist in minimizing the estimated cost of the corresponding machining/assembly line, while reaching a given cycle time and satisfying all constraints. A decision support system is developed. Several optimisation methods were implemented: MIP formulation, a constrained shortest path approach, Branch and Bound techniques, metaheurstics and heuristics. The developed decision-aid software tool is presented. Industrial examples are reported for different types of lines.

PQSQ Potentials and Tropic Methods
in Machine Learning

Alexander Gorban

University of Leicester, Leicester, UK
ag153@le.ac.uk

Keywords: Machine learning · Theory of PQSQ potentials

We develop a new machine learning framework (theory and application) allowing one to deal with arbitrary error potentials of not-faster than quadratic growth, imitated by piece-wise quadratic function of subquadratic growth (PQSQ error potential). This universal framework is able to deal with a large family of error potentials. We exploit the fact that finding a minimum of a piece-wise quadratic function, or, in other words, a function which is the minorant of a set of quadratic functionals, can be almost as computationally efficient as optimizing the standard quadratic potential. The theory of PQSQ potentials uses min,+ algebras and can be considered as a part of tropical mathematics.

New Approaches for Multiprocessor Scheduling Problem with Incomplete Information

Vladimir Kotov

Belarusian State University, Minsk, Belarus
kotovvm@yandex.by

Keywords: Combinatorial optimization · Multiprocessor scheduling

Combinatorial optimization problems come with various paradigms on how an instance is revealed to a solving algorithm. The very common offline paradigm assumes that the entire instance is known in advance. On the opposite end, one can deal with the pure online scheme, where the instance is revealed part by part, unpredictable to the algorithm, and no further knowledge on these parts is assumed. In between these two extremes, and also highly relevant for many practical applications, are semi-online paradigms, where at least some characteristics of the instance in general are assumed to be known, for example, the total instance size or distributions of some internal values. The well-known classical multiprocessor scheduling problem is a fundamental and well-investigated scheduling problem both in the offline and the online setting. A set of n independent jobs is to be processed on m parallel machines in order to minimize the makespan. We present some approaches such as bunch techniques, a dynamic discrete lower bound and other priority rules. These approaches allow to design online and semi online algorithms with the best known worst-case performances.

Optimization, Modeling, and Data Sciences for Sustainable Energy Systems

Panos Pardalos

University of Florida, USA
pardalos@ufl.edu

Keywords: Energy systems · Optimization

For decades, power systems have been playing an important role in humanity. Industrialization has made energy consumption an inevitable part of daily life. Due to our dependence on fuel sources and our large demand for energy, power systems have become interdependent networks rather than remaining independent energy producers. This talk will focus on the problems arising in energy systems as well as recent advances in optimization, modeling, and data sciences techniques to address these problems. Among the topics to be discussed are emission constrained hydrothermal scheduling, electricity and gas networks expansion, as well as reliability analysis of power grid.

References

1. Bjorndal, E., Bjorndal, M., Pardalos, P.M., Ronnqvist, M.: Energy, Natural Resources and Environmental Economics. Springer, Heidelberg (2010)
2. Zheng, Q.P., Rebennack, S., Pardalos, P.M., Iliadis, N., Pereira, M.: Handbook of CO2 in Power Systems. Springer, Heidelberg (2012)
3. Rassia, S.T., Pardalos, P.M.: Cities for Smart Environmental and Energy Futures. Springer, Heidelberg (2013)
4. Pardalos, P.M., Rebennack, S., Pereira, M.V.F., Iliadis, N.A., Pappu V.: Handbook of Wind Power Systems. Springer, Berlin (2014)
5. Eksioglu, S., Rebennack, S., Pardalos, P.M.: Handbook of Bioenergy: Bioenergy Supply Chain – Models and Applications. Springer, Berlin (2015)
6. Rebennack, S., Pereira, M.V.F., Pardalos, P.M.: Stochastic hydro-thermal scheduling under CO2 emissions constraints. IEEE Trans. Power Syst. **27**(1), 58–68 (2012)
7. Rahmani, M., Vinasco, G., Rider, J.M., Romero, R., Pardalos, P.M.: Multistage transmission expansion planning considering fixed series compensation allocation. IEEE Trans. Power Syst. **28**(4), 3795–3805 (2013)
8. Resener, M., Hanera, S., Pereira, L.A., Pardalos, P.M.: Mixed-integer LP model for volt/var control and energy losses minimization in distribution systems. Electr. Power Syst. Res. **140**, 895–905 (2016)
9. Boroojeni, K.G., Amini, M.H., Iyengar, S.S., Rahmani, M., Pardalos, P.M.: An economic dispatch algorithm for congestion management of smart power networks - an oblivious routing approach. Energy Syst. **8**(3), 643–667 (2017)

Contents

Location Problems

On Minimizing Supermodular Functions on Hereditary Systems   3
  Victor Il'ev and Svetlana Il'eva

A New Model of Competitive Location and Pricing with the Uniform
Split of the Demand   16
  Aleksandr V. Kononov, Artem A. Panin, and Aleksandr V. Plyasunov

Branch and Bound Method for the Weber Problem with Rectangular
Facilities on Lines in the Presence of Forbidden Gaps   29
  Gennady G. Zabudsky and Natalia S. Veremchuk

Scheduling and Routing Problems

Inapproximability Lower Bounds for Open Shop Problems
with Exact Delays.   45
  Alexander Ageev

Exact Solution of One Production Scheduling Problem   56
  Pavel Borisovsky

Towards Tractability of the Euclidean Generalized Traveling Salesman
Problem in Grid Clusters Defined by a Grid of Bounded Height   68
  Michael Khachay and Katherine Neznakhina

Worst-Case Analysis of a Modification of the Brucker-Garey-Johnson
Algorithm   78
  Julia Memar, Yakov Zinder, and Aleksandr V. Kononov

Reduction of the Pareto Set in Bicriteria Asymmetric Traveling
Salesman Problem.   93
  Aleksey O. Zakharov and Yulia V. Kovalenko

Optimization Problems in Data Analysis

Randomized Algorithms for Some Clustering Problems   109
  Alexander Kel'manov, Vladimir Khandeev, and Anna Panasenko

An Approximation Polynomial Algorithm for a Problem of Searching
for the Longest Subsequence in a Finite Sequence of Points
in Euclidean Space   120
    Alexander Kel'manov, Artem Pyatkin, Sergey Khamidullin,
    Vladimir Khandeev, Yury V. Shamardin, and Vladimir Shenmaier

On Vector Summation Problem in the Euclidean Space   131
    Edward Kh. Gimadi, Ivan A. Rykov, and Yury V. Shamardin

Fast Numerical Evaluation of Periodic Solutions for a Class of Nonlinear
Systems and Its Applications for Parameter Estimation Problems   137
    Ivan Y. Tyukin, Jehan Mohammed Al-Ameri, Alexander N. Gorban,
    Jeremy Levesley, and Valery A. Terekhov

Mathematical Programming

On Vertices of the Simple Boolean Quadric Polytope Extension   155
    Andrei V. Nikolaev

On Accuracy Estimates for One Regularization Method
in Linear Programming   170
    Leonid D. Popov

Binary Solutions to Some Systems of Linear Equations   183
    Alexandr V. Seliverstov

Variant of the Cutting Plane Method with Approximation of the Set
of Constraints and Auxiliary Functions Epigraphs   193
    I. Ya. Zabotin and K. E. Kazaeva

Game Theory and Economical Applications

Sorger Game Under Uncertainty: Discrete Case   207
    Natalia V. Adukova and Konstantin N. Kudryavtsev

Public-Private Partnership Models with Tax Incentives: Numerical
Analysis of Solutions.   220
    Sergey Lavlinskii, Artem A. Panin, and Aleksandr V. Plyasunov

Fuzzy Core Allocations in a Mixed Economy of Arrow-Debreu Type   235
    Valery A. Vasil'ev

Applied Optimization Problems and Metaheuristics

Convergence Analysis of Swarm Intelligence Metaheuristic Methods.   251
    Tatjana Davidović and Tatjana Jakšić Krüger

An Optimization Model for Empty Tank Cars Movement at Railway
Petroleum Logistics Market   267
   Ivan A. Davydov

Complexity of Bi-objective Buffer Allocation Problem in Systems
with Simple Structure   278
   Alexandre B. Dolgui, Anton V. Eremeev, Mikhail Y. Kovalyov,
   and Vyacheslav S. Sigaev

Inventory Policies in Dual Sourcing Systems with Uncertain Yield   288
   Adriana F. Gabor and Andrei Sleptchenko

A Genetic Algorithm for the Pooling-Inventory-Capacity Problem
in Spare Part Supply Systems.   296
   Hasan Hüseyin Turan, Andrei Sleptchenko, and Fuat Kosanoglu

A Core Heuristic and the Branch-and-Price Method for a Bin Packing
Problem with a Color Constraint.   309
   Artem Kondakov and Yury Kochetov

On Calculation and Estimation of Flow Transmission Probability
in a Communication Network.   321
   Alexey S. Rodionov, Olga A. Yadykina, and Denis A. Migov

Profit Maximization and Vehicle Fleet Planning for a Harbor
Logistics Company   331
   Natalia B. Shamray and Nina A. Kochetova

Author Index   343

Location Problems

On Minimizing Supermodular Functions on Hereditary Systems

Victor Il'ev[1,2(✉)] and Svetlana Il'eva[2]

[1] Omsk State Technical University, Omsk, Russia
`iljev@mail.ru`
[2] Dostoevsky Omsk State University, Omsk, Russia
`iljeva@mail.ru`

Abstract. The problem of minimizing a supermodular set function is considered. A special case of this problem is the well-known NP-hard minimization p-median problem. The main results of the paper are tight a priori and a posteriori bounds on worst-case behaviour of a "reverse" greedy (steepest descent) algorithm of minimizing a supermodular set function on comatroid. As a corollary, approximation guarantees of this algorithm for the general minimization p-median problem improving the known bounds are obtained.

Keywords: Supermodular function · Hereditary system · Comatroid
Greedy algorithm · Approximation guarantee

1 Introduction

Let I be a finite set, 2^I be a Boolean lattice of all subsets of I. A set function $f : 2^I \to \mathbb{R}_+$ is called *supermodular*, if for all $X, Y \subseteq I$

$$f(X \cup Y) + f(X \cap Y) \geq f(X) + f(Y)$$

and *submodular* if the inverse inequality holds. If for all $X, Y \subseteq I$ the inequality is satisfied with equality, then the set function f is said to be *modular*. Note that a nondecreasing set function f with $f(\emptyset) = 0$ is modular if and only if it is additive.

A *hereditary system* $\mathcal{H}$ on I can be defined as a Boolean lattice 2^I with a distinguished family $\mathcal{A} \subseteq 2^I$ that satisfies the following heredity axiom:

$$(A \in \mathcal{A}, \ A' \subseteq A) \Rightarrow A' \in \mathcal{A}.$$

Sometimes it is said that the family $\mathcal{A}$ is an *independence system* or a *hereditary family* [7]. The sets of the family $\mathcal{A}$ are called *independent*, all the other sets of the lattice 2^I are called *dependent sets* of the hereditary system $\mathcal{H}$.

Note that the family $\mathcal{D} = 2^I \setminus \mathcal{A}$ of all dependent sets satisfies the following "heredity up" axiom:

$$(D \in \mathcal{D}, \ D \subseteq D') \Rightarrow D' \in \mathcal{D}.$$

A. Eremeev et al. (Eds.): OPTA 2018, CCIS 871, pp. 3–15, 2018.
https://doi.org/10.1007/978-3-319-93800-4_1

Clearly, the families $\mathcal{A}$ and $\mathcal{D} = 2^I \setminus \mathcal{A}$ determine each other uniquely, therefore these families can be viewed as different sides of the hereditary system $\mathcal{H}$. We shall write $\mathcal{H} = (I, \mathcal{A})$ or $\mathcal{H} = (I, \mathcal{D})$ depending on which side of the hereditary system $\mathcal{H}$ is of our interest.

Bases of a hereditary system $\mathcal{H}$ are maximal (under inclusion) independent sets and *circuits* of $\mathcal{H}$ are minimal dependent sets. The families of all bases and all circuits of $\mathcal{H}$ are denoted by $\mathcal{B}$ and $\mathcal{C}$, respectively. A *base of a set* $X \subseteq I$ is a maximal independent set contained in X, a *circuit of a set* X is a minimal dependent set containing X.

Matroids and comatroids are the important special cases of hereditary systems. These objects can be defined as follows.

A hereditary system $\mathcal{H} = (I, \mathcal{A})$ is called a *matroid* if all bases of every set $X \subseteq I$ have the same cardinality. The cardinality r of any base of I is called the *rank* of a matroid. As an example, we consider the *p-uniform matroid* $(I, \mathcal{A}_p)$, where $\mathcal{A}_p = \{A \subseteq I : |A| \leq p\}$, $p < |I|$.

Given a hereditary system $\mathcal{H} = (I, \mathcal{A})$, we define the *complementary system* or the *cosystem* $\overline{\mathcal{H}} = (I, \overline{\mathcal{D}})$ by $\overline{\mathcal{D}} = \{I \setminus A : A \in \mathcal{A}\}$. Obviously, $\overline{\mathcal{A}} = \{I \setminus D : D \in \mathcal{D}\}$.

The hereditary system complementary to a matroid is said to be a *comatroid*. It is easy to see that a hereditary system is a comatroid if and only if all circuits of every set $X \subseteq I$ have the same cardinality. The cardinality p of any circuit of $\emptyset$ is called the *girth* of the comatroid. The comatroid $(I, \mathcal{D}_p)$ complementary to the $(n - p)$-uniform matroid on I is called the *p-uniform comatroid*. Clearly, $\mathcal{D}_p = \{D \subseteq I : |D| \geq p\}$.

Problems of minimizing supermodular functions on hereditary systems and lattices arise in different areas of discrete optimization, and the study of these problems has more than fifty-year history. Among first works in this direction we can mark the paper [4], where the problem of minimizing a supermodular function on a Boolean lattice was considered.

Many important problems such as the well-known minimization p-median problem [20] and some variants of the half-product problem [17] are the special cases of the problem of minimizing a supermodular function. The known graph correlation clustering problem [12] closely connected with studying systems of equations over graphs [14] can be reduced to this problem.

In the past decades, a great amount of optimization problems with concrete types of supermodular functions were studied and a lot of theoretical results were obtained. For example, in [18], a description of the problems of minimizing supermodular functions on the different types of lattices is contained and the previously obtained theoretical results, on the basis of which the problems of minimizing supermodular functions on these lattices have been solved, are shown. This allows the authors to consider the methods for solving the problems of minimizing supermodular functions as a new field of mathematical programming – supermodular programming [18].

Our aim is to solve the following combinatorial optimization problem approximately:

$$\min \{f(X) : X \in \mathcal{C}\}, \tag{1}$$

where $\mathcal{C}$ is the family of all circuits of a comatroid on I of girth p, $p < n = |I|$, and $f : 2^I \to \mathbb{R}_+$ is a nonincreasing supermodular function, $f(I) = 0$.

Problem (1) is NP-hard because it contains the well-known NP-hard minimization p-median problem as a special case [16]. The mathematical model of the *minimization p-median problem* can be formulated as follows:

$$\min \{f(X) : X \subseteq I, |X| = p\}, \tag{2}$$

where

$$f(X) = \sum_{j \in J} \min_{i \in X} a_{ij}, \tag{3}$$

$A = (a_{ij})$ is a nonnegative matrix with the row index set I and the column index set J. The problem is called *metric* if $I = J$, the matrix A is symmetric, and the assignment costs a_{ij} satisfy the triangle inequality: $a_{ij} + a_{jk} \geq a_{ik}$ for all $i, j, k \in I$. Note that the set function (3) is nonincreasing. One can see that after the extension

$$f(\emptyset) = \max_{\substack{X,Y \subseteq I, \\ X \cap Y = \emptyset}} \{f(X) + f(Y) - f(X \cup Y)\} \tag{4}$$

f becomes supermodular. Therefore the p-median problem can be viewed as problem (1) on the p-uniform comatroid.

We consider a "reverse" variant of the greedy heuristic, a discrete analogue of the steepest descent algorithm, to solve problem (1) of minimizing a supermodular function on a comatroid approximately. Approximation guarantees of the greedy reverse algorithm for problem (1) are obtained. This result improves and supplements the bounds obtained in [8,10,11]. A new technique presented in this paper makes it possible to get also a posteriori bounds. As a corollary we obtain bounds on the worst-case behaviour of the greedy reverse algorithm for the general minimization p-median problem, i.e., the problem which may not be metric, improving the results obtained in [13].

In Sect. 2, we describe the greedy reverse algorithm GR and cite known results. In Sect. 3, we show that Algorithm GR always finds a solution with the value at most $1 + s$ times the optimal value, where s is a curvature of a nonincreasing supermodular function f, a parameter which describes the rate of decrease of f. Section 4 contains a posteriori bounds on the worst-case behaviour of the algorithm. Finally, in Sect. 5, we apply our results to the general minimization p-median problem.

2   Known Results

The following maximization analogue of problem (1) was studied in the literature:

$$\max \{f(X) : X \in \mathcal{B}\}, \tag{5}$$

where $\mathcal{B}$ is the family of all bases of a matroid on I of rank p, $p < n = |I|$, $f : 2^I \to \mathbb{R}_+$ is a nondecreasing submodular function, $f(\emptyset) = 0$.

Problem (5) is NP-hard since it contains the well-known NP-hard maximization p-median problem as a special case.

The following simple algorithm, a discrete analogue of the steepest ascent algorithm, is usually applied to solve problem (5) approximately.

Algorithm GA (greedy algorithm)
Step 0: Set $X_0 = \emptyset$. Go to step 1.
Step i $(i \geq 1)$: Select $x_i \notin X_{i-1}$ such that

$$f(X_{i-1} \cup \{x_i\}) = \max_{\substack{x \notin X_{i-1}, \\ X_{i-1} \cup \{x\} \in \mathcal{A}}} f(X_{i-1} \cup \{x\}).$$

Set $X_i = X_{i-1} \cup \{x_i\}$. If $i < p$, then go to step $i+1$, else stop. Return $S_{GA} = X_p$.
End.

It is easy to see that Algorithm GA always finds a base of the matroid, i.e., S_{GA} is a feasible solution to problem (5).

In [6], the lower bound on the approximation guarantee of Algorithm GA for the maximization p-median problem was obtained:

$$\frac{f(S_{GA})}{f(S_O)} \geq 1 - \left(\frac{p-1}{p}\right)^p \geq \frac{e-1}{e} \approx 0,63,$$

where S_O is an optimal solution and S_{GA} is the solution retrieved by Algorithm GA.

In [21], this result was extended to the problem

$$\max\{f(X) : X \subseteq I, |X| = p\}, \tag{6}$$

where $f : 2^I \to \mathbb{R}_+$ is a nondecreasing submodular function, $f(\emptyset) = 0$.

In [5], the bound for problem (6) was refined by using the additional information on the objective function:

$$\frac{f(S_{GA})}{f(S_O)} \geq \frac{1}{c}\left(1 - \left(\frac{p-c}{p}\right)^p\right), \tag{7}$$

where c is the *curvature* of f, a parameter describing the rate of growth of a nondecreasing submodular function. It is defined as

$$c = \max_{\substack{x \in I, \\ f(\{x\}) > f(\emptyset)}} \frac{(f(\{x\}) - f(\emptyset)) - (f(I) - f(I \setminus \{x\}))}{f(\{x\}) - f(\emptyset)}.$$

Obviously, $c \in [0, 1]$ and $c = 0$ if and only if f is modular.

In [5], the authors also derived bound (7) for problem (5) and proved a simpler bound:

$$\frac{f(S_{GA})}{f(S_O)} \geq \frac{1}{1+c} \geq \frac{1}{2}. \tag{8}$$

Unfortunately, the situation is dramatically different in the case of minimizing a supermodular function. Constant-factor polynomial time approximation algorithms were given only for the metric minimization p-median problem [1–3,15,19]. Moreover, it was shown in [20] that existence of a constant-factor approximation polynomial time algorithm for the general minimization p-median problem implies $P = NP$. Note that a constant can be replaced here by any increasing function of n. Clearly, the same statement remains true for problem (1).

We consider the following "reverse" version of the greedy heuristic, a discrete analogue of the steepest descent algorithm, to solve problem (1) approximately.

Algorithm GR (greedy reverse)
Step 0: Set $X_0 = I$. Go to step 1.
Step i ($i \geq 1$): Select $x_i \in X_{i-1}$ such that

$$f(X_{i-1} \setminus \{x_i\}) = \min_{\substack{x \in X_{i-1}, \\ X_{i-1} \setminus \{x\} \in \mathcal{D}}} f(X_{i-1} \setminus \{x\}).$$

Set $X_i = X_{i-1} \setminus \{x_i\}$. If $i < n - p$, then go to step $i + 1$, else stop. Return $S_{GR} = X_{n-p}$.
End.

It is easy to see that Algorithm GR always finds a circuit of the comatroid, i.e., S_{GR} is a feasible solution to problem (1).

Note that Algorithm GR can yield an arbitrary bad solution to problem (1) even on the p-uniform comatroid. However an additional information on the objective function makes it possible to get approximation guarantees of Algorithm GR.

Consider the *curvature* of a nonincreasing supermodular function f:

$$c = \max_{\substack{x \in I, \\ f(\{x\}) < f(\emptyset)}} \frac{(f(\emptyset) - f(\{x\})) - (f(I \setminus \{x\}) - f(I))}{f(\emptyset) - f(\{x\})}. \qquad (9)$$

For set functions of curvature $c < 1$ we define the characteristic $s = c/(1 - c)$, wich we also will call the curvature. Both parameters c and s characterize the rate of decrease of a nonincreasing supermodular function. Note that $c \in [0, 1]$, $s \in [0, +\infty)$, and $c = s = 0$ if and only if f is modular.

In [8], the following special case of problem (1) with an arbitrary supermodular objective function $f : 2^I \to \mathbb{R}_+$ was considered:

$$\min\{f(X) : X \subseteq I, |X| = p\} \qquad (10)$$

and the upper bound on the worst-case behaviour of Algorithm GR for problem (10) was obtained:

$$\frac{f(S_{GR})}{f(S_O)} \leq \frac{1}{s} \left(\left(\frac{q + s}{q} \right)^q - 1 \right), \qquad (11)$$

where S_O is an optimal solution to problem (10), S_{GR} is the solution returned by Algorithm GR, and $q = n - p$. In [11], this result was extended to problem (1).

For objective functions with curvature $s < 1$ a simpler bound was obtained [10]:

$$\frac{f(S_{GR})}{f(S_O)} \leq \frac{1}{1-s} . \tag{12}$$

The main goal of this work is to obtain new approximation guarantees of Algorithm GR to problems (1) and (2), which improve and supplement bounds (11) and (12).

3   Approximation Guarantees of Algorithm GR

Now let us turn back to problem (1).

For $X \subseteq I$ and $x \in X$, we set $d_x(X) = f(X \setminus \{x\}) - f(X) \geq 0$. Evidently, the curvature (9) can now be rewritten as

$$c = \max_{\substack{x \in I, \\ d_x(\{x\}) > 0}} \frac{d_x(\{x\}) - d_x(I)}{d_x(\{x\})} .$$

Let us also introduce the following notation:

$$Y = I \setminus \{y_1, \ldots, y_k\}, \ \overline{Y} = I \setminus Y = \{y_1, \ldots, y_k\} \ \ (k \in \{1, \ldots, n\}),$$
$$Y_0 = I, \ Y_i = I \setminus \{y_1, \ldots, y_i\} \ \ (i = 1, \ldots, k).$$

In [8], the following properties of nonincreasing supermodular functions were proved.

Proposition 1. $f(Y) = \sum_{y_i \in \overline{Y}} d_{y_i}(Y_{i-1})$.

Proposition 2. $d_x(X) \geq d_x(Y)$ *for all* $X, Y, \ X \subseteq Y \subseteq I$, *and for every* $x \in X$.

In [9], an equivalent definition of a comatroid was given.

Proposition 3. *A hereditary system* $\mathcal{H} = (I, \mathcal{D})$ *is a comatroid if and only if the family* $\mathcal{D}$ *of its dependent sets satisfies the following axiom:*

$$(D, D' \in \mathcal{D}, |D| = |D'| + 1) \Rightarrow \exists x \in D \setminus D' : D \setminus \{x\} \in \mathcal{D}.$$

Algorithm GR consecutively finds the sets $X_0 = I, X_1, \ldots, X_q = S_{GR}$, where $q = n - p$, $X_i = X_{i-1} \setminus \{x_i\} = I \setminus \{x_1, \ldots, x_i\}$, and $x_i \in X_{i-1}$ is the element chosen at step i such that

$$d_{x_i}(X_{i-1}) = \min_{\substack{x \in X_{i-1}, \\ X_{i-1} \setminus \{x\} \in \mathcal{D}}} d_x(X_{i-1}) \ \ (i = 1, \ldots, q). \tag{13}$$

Lemma 1. *For any circuit* $C \in \mathcal{C}$ *of a comatroid* $\mathcal{H} = (I, \mathcal{D})$ *the elements of the set* $\overline{C} = I \setminus C = \{c_1, \ldots, c_q\}$ *can be ordered so that* $d_{c_i}(X_{i-1}) \geq d_{x_i}(X_{i-1})$, $i = 1, \ldots, q$. *Furthermore, if* $c_i \in \overline{C} \cap \overline{S}_{GR} = \{c_1, \ldots, c_q\} \cap \{x_1, \ldots, x_q\}$, *then* $c_i = x_i$.

Proof. Assume that the elements $c_q, c_{q-1} \ldots, c_{i+1}$ are already found. If $x_i \notin C_i = C \cup \{c_q, c_{q-1}, \ldots, c_{i+1}\}$, then we set $c_i = x_i$. If $x_i \in C_i$, then, by Proposition 3, there exists $x \in X_{i-1} \setminus C_i$ such that $X_{i-1} \setminus \{x\} \in \mathcal{D}$, and we set $c_i = x$. Since x_i is the element chosen by Algorithm GR, it follows from (13) that $d_{c_i}(X_{i-1}) \geq d_{x_i}(X_{i-1})$. The lemma is proved. □

Lemma 2. $(1 - c)d_x(X) \leq d_x(I)$ *for all $X \subseteq I$ and for any $x \in X$.*

Proof. It follows from the definition of curvature that

$$c \geq \frac{d_x(\{x\}) - d_x(I)}{d_x(\{x\})}$$

for every $x \in I$ such that $d_x(\{x\}) > 0$. Hence $cd_x(\{x\}) \geq d_x(\{x\}) - d_x(I)$, i.e., $(1 - c)d_x(\{x\}) \leq d_x(I)$. If $d_x(\{x\}) = 0$, then the last inequality evidently holds. By Proposition 2, $d_x(X) \leq d_x(\{x\})$. Thus we obtain $(1 - c)d_x(X) \leq (1 - c)d_x(\{x\}) \leq d_x(I)$. The lemma is proved. □

Theorem 1. *The following bound on the worst-case behaviour of Algorithm GR holds:*

$$(1 - c)f(S_{GR}) \leq f(S_O), \tag{14}$$

where S_O is an optimal solution to problem (1) and S_{GR} is the solution returned by Algorithm GR.

Proof. Let $\overline{S}_O = \{c_1, \ldots, c_q\}$, where the elements c_i are ordered according to Lemma 1, and let $C_i = I \setminus \{c_1, \ldots, c_i\}$, $i = 1, \ldots, q$. By Proposition 1, $f(S_O) = \sum_{c_i \in \overline{S}_O} d_{c_i}(C_{i-1})$. It follows from Proposition 2 that $d_{c_i}(C_{i-1}) \geq d_{c_i}(I)$, $i = 1, ..., q$. Thus,

$$f(S_O) = \sum_{c_i \in \overline{S}_O} d_{c_i}(C_{i-1}) \geq \sum_{c_i \in \overline{S}_O} d_{c_i}(I). \tag{15}$$

Using Proposition 1, we obtain

$$f(S_{GR}) = \sum_{x_i \in \overline{S}_{GR}} d_{x_i}(X_{i-1}) = \sum_{x_i \in \overline{S}_{GR} \cap \overline{S}_O} d_{x_i}(X_{i-1}) + \sum_{x_i \in \overline{S}_{GR} \setminus \overline{S}_O} d_{x_i}(X_{i-1}).$$

By Lemma 1, for $i = 1, \ldots, q$, $d_{x_i}(X_{i-1}) \leq d_{c_i}(X_{i-1})$ whence,

$$f(S_{GR}) \leq \sum_{x_i \in \overline{S}_{GR} \cap \overline{S}_O} d_{x_i}(X_{i-1}) + \sum_{c_i \in \overline{S}_O \setminus \overline{S}_{GR}} d_{c_i}(X_{i-1}). \tag{16}$$

By Lemma 2, for $x = x_i$ and $X = X_{i-1}$, we have $(1 - c)d_{x_i}(X_{i-1}) \leq d_{x_i}(I)$, $i = 1, \ldots, q$. Hence,

$$(1 - c) \sum_{x_i \in \overline{S}_{GR} \cap \overline{S}_O} d_{x_i}(X_{i-1}) \leq \sum_{x_i \in \overline{S}_{GR} \cap \overline{S}_O} d_{x_i}(I) = \sum_{c_i \in \overline{S}_{GR} \cap \overline{S}_O} d_{c_i}(I) \tag{17}$$

(the last equality follows from Lemma 1).

Further, note that $\overline{S}_O \setminus \overline{S}_{GR} = S_{GR} \setminus S_O$, therefore for any $c_i \in \overline{S}_O \setminus \overline{S}_{GR}$ we have $c_i \in S_{GR} \setminus S_O \subseteq S_{GR} = X_q \subseteq X_{i-1}$, $i = 1, \ldots, q$. By Lemma 2, for $x = c_i$ and $X = X_{i-1}$, we obtain $(1-c)d_{c_i}(X_{i-1}) \leq d_{c_i}(I)$. Thus,

$$(1-c) \sum_{c_i \in \overline{S}_O \setminus \overline{S}_{GR}} d_{c_i}(X_{i-1}) \leq \sum_{c_i \in \overline{S}_O \setminus \overline{S}_{GR}} d_{c_i}(I). \tag{18}$$

Combining (16), (17) and (18), we get

$$(1-c)f(S_{GR}) \leq \sum_{c_i \in \overline{S}_{GR} \cap \overline{S}_O} d_{c_i}(I) + \sum_{c_i \in \overline{S}_O \setminus \overline{S}_{GR}} d_{c_i}(I) = \sum_{c_i \in \overline{S}_O} d_{c_i}(I).$$

Finally, taking into account (15), we obtain bound (14):

$$(1-c)f(S_{GR}) \leq \sum_{c_i \in \overline{S}_O} d_{c_i}(I) \leq f(S_O).$$

This completes the proof of Theorem 1. $\qquad\square$

To obtain a bound on the performance guarantee of Algorithm GR in terms of the curvature $s = c/(1-c)$, we assume below that $c < 1$.

Theorem 2. *For any nondecreasing supermodular objective function of problem (1) with the curvature $c \in [0, 1)$, it holds that*

$$f(S_{GR}) \leq (1+s)f(S_O), \tag{19}$$

where S_O is an optimal solution to problem (1) and S_{GR} is the feasible solution found by Algorithm GR.

Proof. Taking into account the condition $c \in [0, 1)$, inequality (14) can be rewritten in the form

$$f(S_{GR}) \leq \frac{f(S_O)}{1-c}. \tag{20}$$

Recall, that $s = c/(1-c)$; therefore $c = s/(1+s)$. Substitute $s/(1+s)$ for c in (20) to obtain (19). This completes the proof of Theorem 2. $\qquad\square$

Remark 1. Bounds (14) and (19) are tight. For example, if the objective function of problem (1) is modular, then $c = s = 0$, and inequalities (14) and (19) hold with equalities.

To compare bounds (14) and (19) with bounds (11) and (12) we assume that $f(S_O) \neq 0$. In the case $f(S_O) = 0$, we have, due to (19), $f(S_{GR}) = 0$, i.e., the algorithm GR finds an optimal solution. If $f(S_O) \neq 0$, then bounds (14) and (19) can be written as

$$\frac{f(S_{GR})}{f(S_O)} \leq \frac{1}{1-c}. \tag{21}$$

$$\frac{f(S_{GR})}{f(S_O)} \leq 1+s. \tag{22}$$

Remark 2. Bounds (21) and (22) are stronger than bound (12).

Indeed, for $s < 1$, we have $\frac{1}{1-c} = 1 + s \leq \frac{1}{1-s}$.

Now, we compare bound (22) with bound (11). The right-hand side of inequality (11) is the polynomial of degree $q-1$ of the variable s with a positive leading coefficient. Hence $\frac{1}{s}\left(\left(\frac{q+s}{q}\right)^q - 1\right) < 1 + s$ only for small values of q or small values of s. More precisely,

Remark 3. Bound (11) is better than bound (22) when at least one of the following conditions holds: (1) $q \in \{1, 2\}$ and $s > 0$; (2) $s \in (0; 1, 79]$.

However in most cases $1 + s < \frac{1}{s}\left(\left(\frac{q+s}{q}\right)^q - 1\right)$:

Remark 4. For $q \geq 3$, there exists s_q such that, for any $s > s_q$, bound (22) is better than bound (11).

The values of the parameter s_q for some q are presented in Table 1.

Table 1. The values of the parameter s_q.

q	3	4	5	6	...	9	...	40	...	100	...
s_q	18	6.97	4.78	3.88	...	2.89	...	1.98	...	1.87	...

4   A Posteriori Bounds

Theorems 1 and 2 give a priori bounds on the quality of the solution to problem (1) returned by Algorithm GR. Now we obtain a posteriori bounds.

Let us introduce the parameter (*greedy curvature*)

$$\bar{c} = \max_{\substack{x \in I, \\ d_x(S_{GR} \cup \{x\}) > 0}} \frac{d_x(S_{GR} \cup \{x\}) - d_x(I)}{d_x(S_{GR} \cup \{x\})}.$$

Clearly, $\bar{c} \in [0, 1]$ and $\bar{c} \leq c$. Arguing as earlier, we shall assume that $\bar{c} < 1$ and define $\bar{s} = \bar{c}/(1 - \bar{c})$. Note that $\bar{s} \leq s$.

The following lemma is analogous to Lemma 2.

Lemma 3. $(1 - \bar{c})d_x(X) \leq d_x(I)$ for all $X \subseteq I$ such that $S_{GR} \subseteq X$ and for any $x \in X$.

Proof. It follows from the definition of $\bar{c}$ that for every $x \in I$ such that $d_x(S_{GR} \cup \{x\}) > 0$

$$\bar{c} \geq \frac{d_x(S_{GR} \cup \{x\}) - d_x(I)}{d_x(S_{GR} \cup \{x\})}.$$

Hence $\bar{c}d_x(S_{GR} \cup \{x\}) \geq d_x(S_{GR} \cup \{x\}) - d_x(I)$, i.e., $(1 - \bar{c})d_x(S_{GR} \cup \{x\}) \leq d_x(I)$. If $d_x(S_{GR} \cup \{x\}) = 0$, then the last inequality evidently holds.

Since $S_{GR} \subseteq X$ and $x \in X$, we have $S_{GR} \cup \{x\} \subseteq X$. By Proposition 2, $d_x(X) \leq d_x(S_{GR} \cup \{x\})$. Thus we obtain $(1 - \bar{c})d_x(X) \leq (1 - \bar{c})d_x(S_{GR} \cup \{x\}) \leq d_x(I)$. The lemma is proved. $\qquad\square$

Theorem 3. *The following bound on the worst-case behaviour of Algorithm GR holds:*

$$(1 - \bar{c})f(S_{GR}) \leq f(S_O), \tag{23}$$

where S_O is an optimal solution to problem (1) and S_{GR} is the solution returned by Algorithm GR.

As a corollary of Theorem 3 we obtain

Theorem 4. *For any nondecreasing objective function of problem (1) with the greedy curvature $\bar{c} \in [0, 1)$, it holds that*

$$f(S_{GR}) \leq (1 + \bar{s})f(S_O), \tag{24}$$

where S_O is an optimal solution to problem (1) and S_{GR} is the feasible solution found by Algorithm GR.

Proofs of Theorems 3 and 4 are similar to ones of Theorems 1 and 2, but we use Lemma 3 and inequality (23) instead of Lemma 2 and inequality (14).

Remark 5. Bounds (23) and (24) are tight. In particular, (24) is satisfied with equality for modular objective functions, since in this case $\bar{s} = s = 0$. But $\bar{s}$ can be equal to 0 even when f is not modular, as the following example demonstrates.

Example 1. Consider the minimization p-median problem with the matrix

$$A = \begin{pmatrix} 0\,2\,1\,1 \\ 2\,0\,1\,1 \\ 2\,2\,0\,2 \\ 2\,2\,2\,0 \end{pmatrix}.$$

Here, $I = J = \{1, 2, 3, 4\}$, $p = 2$. Define the set function f according to (3):
$f(\{1\}) = 4$, $f(\{2\}) = 4$, $f(\{3\}) = 6$, $f(\{4\}) = 6$,
$f(\{1, 2\}) = 2$, $f(\{1, 3\}) = 3$, $f(\{1, 4\}) = 3$, $f(\{2, 3\}) = 3$, $f(\{2, 4\}) = 3$,
$f(\{3, 4\}) = 4$,
$f(\{1, 2, 3\}) = 1$, $f(\{1, 2, 4\}) = 1$, $f(\{1, 3, 4\}) = 2$, $f(\{2, 3, 4\}) = 2$,
$f(\{1, 2, 3, 4\}) = f(I) = 0$,
and set $f(\emptyset) = 7$ to make f supermodular.
We see that $S_{GR} = S_O = \{1, 2\}$ and $\bar{s} = 0$ whereas $s = \frac{1}{2} > 0$.

5   Application to the General p-median Problem

Consider the minimization p-median problem (2), where the objective function f is defined by (3) and (4):

$$f(X) = \sum_{j \in J} \min_{i \in X} a_{ij}, \quad f(\emptyset) = \max_{\substack{X, Y \subseteq I, \\ X \cap Y = \emptyset}} \{f(X) + f(Y) - f(X \cup Y)\}.$$

Note that f is nonincreasing supermodular function.

We suppose without loss of generality that for any $j \in J$ there exists $i \in I$ such that $a_{ij} = 0$. Otherwise subtract $\min_{i \in I} a_{ij}$ from all elements of every column $j \in J$. Thus we have $f(I) = 0$.

In [13], we proved bound (19) on the worst-case behaviour of Algorithm GR for problem (2) and showed that the curvature s can be calculated as follows:

$$s = \max_{i \in I} \frac{f(\emptyset) - \sum\limits_{j \in J} a_{ij}}{\sum\limits_{j \in J} \min\limits_{k \neq i} a_{kj}} - 1, \tag{25}$$

$$f(\emptyset) = \max_{i,k \in I} \sum_{j \in J} \max \{a_{ij}, a_{kj}\}. \tag{26}$$

Notice that equalities (25) and (26) provide an efficient method of calculating the curvature s of the objective function of the minimization p-median problem (2) in terms of assignment costs.

Now, as a corollary of Theorem 4 we obtain the following bound.

Theorem 5. *The following bound on the worst-case behaviour of Algorithm GR holds:*

$$f(S_{GR}) \leq (1 + \overline{s}) f(S_O), \tag{27}$$

where S_O is an optimal solution to problem (2), S_{GR} is the solution returned by Algorithm GR, and

$$\overline{s} \leq \max_{i \in I} \frac{f(\emptyset) - \sum\limits_{j \in J} a_{ij}}{\sum\limits_{j \in J} \min\limits_{k \neq i} a_{kj}} - 1.$$

Remark 6. Note that $\overline{s} \leq s$. Moreover, bound (27) can be essentially better than a priori bound (19) as the following example shows.

Example 2. Consider the minimization p-median problem (2) with the matrix

$$A = \begin{pmatrix} 0 & a & 3 & 2 & 2 \\ a & 0 & 4 & 5 & a \\ 1 & 4 & 0 & 5 & 5 \\ 2 & 5 & 5 & 0 & a \\ 2 & a & 5 & a & 0 \end{pmatrix}, \qquad a > 5.$$

Here, $I = J = \{1, 2, 3, 4, 5\}$, $p = 3$. Define the set function f according to (3):
$f(\{1\}) = a + 7$, $f(\{2\}) = 2a + 9$, $f(\{3\}) = 15$, $f(\{4\}) = a + 12$, $f(\{5\}) = 2a + 7$,
$f(\{1, 2\}) = 7$, $f(\{1, 3\}) = 8$, $f(\{1, 4\}) = 10$, $f(\{1, 5\}) = a + 5$, $f(\{2, 3\}) = 11$,
$f(\{2, 4\}) = a + 6$, $f(\{2, 5\}) = 11$, $f(\{3, 4\}) = 10$, $f(\{3, 5\}) = 10$, $f(\{4, 5\}) = 12$,
$f(\{1, 2, 3\}) = 4$, $f(\{1, 2, 4\}) = 5$, $f(\{1, 2, 5\}) = 5$, $f(\{1, 3, 4\}) = 6$,
$f(\{1, 3, 5\}) = 6$, $f(\{1, 4, 5\}) = 8$, $f(\{2, 3, 4\}) = 6$, $f(\{2, 3, 5\}) = 6$, $f(\{2, 4, 5\}) = 6$, $f(\{3, 4, 5\}) = 5$,
$f(\{1, 2, 3, 4\}) = 2$, $f(\{1, 2, 3, 5\}) = 2$, $f(\{1, 2, 4, 5\}) = 3$, $f(\{1, 3, 4, 5\}) = 4$,
$f(\{2, 3, 4, 5\}) = 1$, $f(\{1, 2, 3, 4, 5\}) = f(I) = 0$.

After extension $f(\emptyset) = 4a + 5$ the nonincreasing set function f becomes supermodular.

Algorithm GR finds the solution $S_{GR} = \{3, 4, 5\}$. It is easy to check that

$$\bar{c} = \max_{\substack{x \in I, \\ d_x(S_{GR} \cup \{x\}) > 0}} \frac{d_x(S_{GR} \cup \{x\}) - d_x(I)}{d_x(S_{GR} \cup \{x\})} = \frac{3}{5}, \quad \bar{s} = \frac{\bar{c}}{1 - \bar{c}} = \frac{3}{2}$$

whereas

$$c = \max_{\substack{x \in I, \\ d_x(\{x\}) > 0}} \frac{d_x(\{x\}) - d_x(I)}{d_x(\{x\})} = \frac{3a - 3}{3a - 2}, \quad s = 3a - 3.$$

Thus we obtain the a posteriori bound $1 + \bar{s} = \frac{5}{2}$ and the a priori bound $1 + s = 3a - 2$. Note that the latter value grows infinitely as $a \to +\infty$.

6 Conclusion

We study and approximately solve the problem of minimizing a supermodular function whose special case is the well-known NP-hard minimization p-median problem. The main results of the paper are new a priori and a posteriori bounds on the worst-case behaviour of a "reverse" greedy (steepest descent) algorithm of minimizing a supermodular function on comatroid. A priori bound generalizes our earlier approximation guarantees of the steepest descent algorithm for the minimization p-median problem. A posteriori bound in many cases turns out to be considerably better than a priori bound.

Acknowledgements. The research of the first author was supported by the RSF grant 17-11-01117.

References

1. Arya, V., Garg, N., Khandekar, R., Meyerson, A., Munagala, K., Pandit, V.: Local search heuristics for k-median and facility location problems. In: Proceedings of the 33rd Annual ACM Symposium on Theory of Computing, pp. 21–29 (2001)
2. Charikar, M., Guha, S.: Improved combinatorial algorithms for the facility location and k-median problems. In: Proceedings of the 40th Annual IEEE Symposium on Foundations of Computer Science, pp. 378–388 (1999)
3. Charikar, M., Guha, S., Tardos, E., Shmoys, D.B.: A constant-factor approximation algorithm for the k-median problem. In: Proceedings of the 31st Annual ACM Symposium on Theory of Computing, pp. 1–10 (1999)
4. Cherenin, V.P.: Solution of some combinatorial problems in optimal planning by the method of successive calculations. In: Scientific and Tutorial Proceedings of the Mathematical Economics Seminar of the Laboratory of Economic-Mathematical Methods. USSR Academy of Sciences 2, Gipromez, Moscow (1962). (in Russian)
5. Conforti, M., Cornuéjols, G.: Submodular set functions, matroids and the greedy algorithm: tight worst-case bounds and some generalizations of the Rado-Edmonds theorem. Discrete Appl. Math. **7**(3), 251–274 (1984)

6. Cornuéjols, G., Fisher, M.L., Nemhauser, G.L.: Location of bank accounts to optimize float: an analytic study of exact and approximate algorithms. Manage. Sci. **23**(8), 789–810 (1977)
7. Grötshel, M., Lovász, L.: Combinatorial optimization. In: Graham, R.L., Grötshel, M., Lovász, L. (eds.) Handbook of Combinatorics, vol. 2, pp. 1541–1598. Elsevier Science B.V, Amsterdam (1995)
8. Il'ev, V.: An approximation guarantee of the greedy descent algorithm for minimizing a supermodular set function. Discrete Appl. Math. **114**, 131–146 (2001)
9. Il'ev, V.: Hereditary systems and greedy-type algorithms. Discrete Appl. Math. **132**, 137–148 (2003)
10. Il'ev, V., Linker, N.: On the problem of minimizing a supermodular set function on comatroid. Vestnik Omskogo Universiteta **1**, 16–18 (2002). (in Russian)
11. Il'ev, V., Linker, N.: Performance guarantees of a greedy algorithm for minimizing a supermodular set function on comatroid. Eur. J. Oper. Res. **171**, 648–660 (2006)
12. Il'ev, V., Il'eva, S., Kononov, A.: Short survey on graph correlation clustering with minimization criteria. In: Kochetov, Y., Khachay, M., Beresnev, V., Nurminski, E., Pardalos, P. (eds.) DOOR 2016. LNCS, vol. 9869, pp. 25–36. Springer, Cham (2016). https://doi.org/10.1007/978-3-319-44914-2_3
13. Il'ev, V., Il'eva, S., Navrotskaya, A.: Approximate solution of the p-median minimization problem. Comput. Math. Math. Phys. **56**(9), 1591–1597 (2016)
14. Ilev, A.V., Remeslennikov, V.N.: Study of the compatibility of systems of equations over graphs and finding their general solutions. Vestnik Omskogo Universiteta **4**(86), 26–32 (2017). (in Russian)
15. Jain, K., Vazirani, V.V.: Primal-dual approximation algorithms for metric facility location and k-median problems. In: 1999 Proceedings of the 40th Annual IEEE Symposium on Foundations of Computer Science, pp. 2–13 (1999)
16. Kariv, O., Hakimi, S.L.: An algorithmic approach to network location problems. II. The p-medians. SIAM J. Appl. Math. **37**, 539–560 (1979)
17. Kellerer, H., Strusevich, V.A.: Optimizing the half-product and related quadratic boolean functions: approximation and scheduling applications. Ann. Oper. Res. **240**, 39–94 (2016)
18. Khachaturov, V.R., Khachaturov, R.V.: Supermodular programming on finite lattices. Comput. Math. Math. Phys. **52**(6), 855–878 (2012)
19. Li, S., Svensson, O.: Approximating k-median via peeudo-approximation. SIAM J. Comput. **45**(2), 530–547 (2013)
20. Nemhauser, G.L., Wolsey, L.A.: Integer and Combinatorial Optimization. Wiley, New York (1988)
21. Nemhauser, G.L., Wolsey, L.A., Fisher, M.L.: An analysis of approximations for maximizing submodular set functions - I. Math. Program. **14**(13), 265–294 (1978)

A New Model of Competitive Location and Pricing with the Uniform Split of the Demand

Aleksandr V. Kononov[1,2(✉)], Artem A. Panin[1,2],
and Aleksandr V. Plyasunov[1,2]

[1] Sobolev Institute of Mathematics, Novosibirsk, Russia
{alvenko,apljas}@math.nsc.ru, arteam1897@gmail.com
[2] Novosibirsk State University, Novosibirsk, Russia

Abstract. In this paper, a new optimization model of competitive facility location and pricing is introduced. This model is an extension of the well-known $(r|p)$-centroid problem. In the model, two companies compete for the client's demand. Each client has a finite budget and a finite demand. First, a company-leader determines a location of p facilities. Taking into account the location of leader's facilities, the company-follower determines a location of its own r facilities. After that, each company assigns a price for each client. When buying a product, the client pays the price of the product and its transportation. A client buys everything from a company with lower total costs if their total costs do not exceed the budget of the client. If the cost of buying a product from both companies is the same, the demand of clients is distributed equally among them. The goal is to determine a location of leader's facilities and set the prices in which the total income of the leader is maximal. Results about the computational complexity of the model are presented. Several special cases are considered. These cases can be divided into three categories: (1) polynomially solvable problems; (2) NP-hard problems; (3) problems related to the second level of the polynomial hierarchy. Finally, the complexity of the maxmin-2-Sat problem is discussed.

Keywords: Competitive location · Pricing · Split demand
Computational complexity

1 Introduction

Research in the field of competitive location problems was initiated in [1], where the process of choosing the location of facilities and the choice of the policy of pricing by two competitors in a finite segment with a uniform distribution of buyers were considered. The last decades, more and more attention is drawn to

The research of the first author was supported by RFBR grant 17-07-00513 and the research of the second and third authors was supported by RFBR grant 16-07-00319.

A. Eremeev et al. (Eds.): OPTA 2018, CCIS 871, pp. 16–28, 2018.
https://doi.org/10.1007/978-3-319-93800-4_2

problems in which the decision to place and pricing is taken by players competing with each other [2–5]. To date, many relevant problems are being addressed in this area and many interesting results have been obtained. In this paper, we consider a model that is in some ways an extension of the model of competitive location and pricing from [6].

Let us describe in more detail the problem with its novelty and differences from previous models. The problem is based on the Stackelberg game of two players - the leader and the follower. The players select their locations and then set prices in order to maximize their profits. The leader makes the decision first, and then the follower makes his move. Players consistently place their facilities in the finite set of predetermined locations. When all facilities are placed, the players set prices for a homogeneous product. Here, a discriminatory pricing strategy is used, when the player assigns a price for each client at each facility. In [6], the well-known Bertrand model was used to determine prices and distribute customers by facilities. In this model, the client is monopolized by the facility, where the minimum cost of maintenance is achieved and the monopoly price is assigned. In this paper, a new situation is considered when players can share the demand of clients when it is profitable for them. Obviously, if a player at his facility assigns a price to a client that does not exceed the minimum cost of services at the follower's facilities, then a rational client will prefer to be serviced by the leader since the follower can not offer a lower price. On the other hand, players can agree among themselves to establish the prices at the level of the maximum purchasing power of the client, and divide the customer's demand among themselves. We suppose that the demand will be shared equally. In other words, the client will make purchases at the facility of the leader, in a half of cases, and at the facility of the follower, in other cases.

In the paper, the main emphasis is placed on the computational complexity of finding of exact and approximate solutions for different variants of the problem [7].

The paper is organized as follows. In the next section, we formulate the problem. The third section contains results on the computational complexity of the general problem. In the fourth section, we consider the special cases of the problem, their complexity, and algorithms for their solution.

2   The Competitive Location and Pricing Problem with the Uniform Split of the Demand

We introduce the following notation:

$I = \{1, ..., m\}$ is the set of locations for the facilities of the leader and the follower;

$J = \{1, ..., n\}$ is the set of clients;

p is the number of facilities placed by the leader;

r is the number of facilities placed by the follower;

t_i is the unit cost of production in location i;

b_j is the budget of the client of j;

d_j is the demand for the client of j;

c_{ij} is the unit cost of transportation of product from the facility i to the client j.

To identify the placement of facilities of the leader and follower, we use the following variables:

$$x_i = \begin{cases} 1, & \text{if the leader placed the facility at the point } i, \\ 0 & \text{otherwise;} \end{cases}$$

$$y_i = \begin{cases} 1, & \text{if the follower placed the facility at the point } i, \\ 0 & \text{otherwise.} \end{cases}$$

For each client and each facility, we can calculate the prime cost of service. Let vector x denote the leader's choice and vector y denote the follower's choice, then $c_j(x) = \min\{d_j(t_i + c_{ij}) | x_i = 1\}$ is the prime cost of service for the client j by the leader and $c_j(y) = \min\{d_j(t_i + c_{ij}) | y_i = 1\}$ is the prime cost of service by the follower. When the facilities have been chosen, the pricing process for each client is implemented based on the Bertrand price competition model. Companies compete by setting prices simultaneously and clients choose a company with a lower price [6, 8, 9]. A client prefers the leader if the costs of service by the leader and the follower are the same.

Let $x_i = 1$, $c_j(x) = d_j(t_i + c_{ij})$ and $y_k = 1$, $c_j(y) = d_j(t_k + c_{kj})$. Suppose, that $c_j(x) \leq c_j(y)$, i.e. for the client j, the leader is the winner. Note, that the leader can set the price at the income-making level of the follower. Denote this price as q^1_{ijk}. Then from the equation $d_j(p^1_{ijk} + c_{ij}) = d_j(t_k + c_{kj})$ we get the price

$$q^1_{ijk} = t_k + c_{kj} - c_{ij}$$

for the client j. Hence, the income of the leader at point i is $w^1_{ijk} = d_j(q^1_{ijk} - t_i)$.

On the other hand, the revenue at the ith service point from the jth client does not exceed $b_j - d_j(t_i + c_{ij})$. It gives one more way of formation of the price. Denote this price as q^2_{ij}. Set $b_j - d_j(c_{ij} + q^2_{ij}) = 0$. We get

$$q^2_{ij} = \frac{b_j}{d_j} - c_{ij}.$$

Therefore, the income of the leader in this case is $w^2_{ij} = d_j(q^2_{ij} - t_i) = b_j - d_j(t_i + c_{ij})$. It is easy to see that

$$w^1_{ijk} - w^2_{ij} = d_j(q^1_{ijk} - q^2_{ij}) = d_j(t_k + c_{kj}) - b_j.$$

Let $w^3_{ij} = (d_j/2)(q^2_{ij} - t_i)$ is the income of the leader from the division of the demand in half between players.

Let's analyze the possible cases.

1. Let $q^2_{ij} \geq q^1_{ijk}$ and $w^1_{ijk} \leq w^3_{ij}$. That is, the income from monopolization is less than the income from the division of the demand in half between players. Therefore in our model, in this case, players agree to share the client's demand among themselves.

2. If $q_{ij}^2 \geq q_{ijk}^1$ and $w_{ijk}^1 > w_{ij}^3$, then the leader has more income when using a monopoly price q_{ijk}^1.

3. If $q_{ij}^2 < q_{ijk}^1$, then the leader gets the maximum possible income, that is equal to $w_{ij}^2 = d_j(q_{ij}^2 - t_i) = b_j - d_j(t_i + c_{ij})$ using the price q_{ij}^2.

Case $c_j(x) > c_j(y)$ is analyzed in a similar way.

Now, as in [6], we replace non-Boolean variables q_{ijk}^1, q_{ij}^2 with boolean variables.

$$z_{ijk}^{Lc} = \begin{cases} 1, \text{ if the client } j \text{ is serviced by the leader's facility } i \\ \quad \text{ with the price } q_{ijk}^1, \\ 0 \text{ otherwise;} \end{cases}$$

$$z_{ijk}^{Lb} = \begin{cases} 1, \text{ if the client } j \text{ is serviced by the leader's facility } i \\ \quad \text{ with the price } q_{ij}^2, \\ 0 \text{ otherwise;} \end{cases}$$

$$z_{ijk}^{Fc} = \begin{cases} 1, \text{ if the client } j \text{ is serviced by the follower's facility } k \\ \quad \text{ with the price } q_{ijk}^1, \\ 0 \text{ otherwise;} \end{cases}$$

$$z_{ijk}^{Fb} = \begin{cases} 1, \text{ if the client } j \text{ is serviced by the follower's facility } k \\ \quad \text{ with the price } q_{ij}^2 \\ 0 \text{ otherwise;} \end{cases}$$

$$z_{ijk} = \begin{cases} 1, \text{ if the client } j \text{ is serviced by the leader from the point } i \text{ and} \\ \quad \text{ the follower from the point } k \text{ simultaneously,} \\ 0 \text{ otherwise;} \end{cases}$$

This approach allows us to limit ourselves to only Boolean variables in the proposed model. The set $I_j(x)$ consists of locations that the follower can use to capture the client j:

$$I_j(x) = \{i \in I : c_{ij} + t_i < \min_{k \in I : x_k = 1}(c_{kj} + t_k)\}.$$

The competitive location and pricing problem with the uniform split of the demand can be represented as the linear Boolean bi-level optimization program. We propose the following model for the leader

$$\sum_{i \in I} \sum_{j \in J} \sum_{k \in I} (d_j(c_{kj} + t_k - c_{ij} - t_i)z_{ijk}^{Lc} + (b_j - d_j(c_{ij} + t_i))z_{ijk}^{Lb} \tag{1}$$

$$+ 0.5(b_j - d_j(c_{ij} + t_i))z_{ijk}) \to \max_{x,y,z^{Lc},z^{Lb},z^{Fc},z^{Fb},z}$$

under constraints:

$$\sum_{i \in I} x_i = p; \tag{2}$$

$$(y, z^{Lc}, z^{Lb}, z^{Fc}, z^{Fb}, z) \in \mathcal{F}^*(x); \tag{3}$$

$$x_i \in \{0,1\}; i \in I. \tag{4}$$

The objective function (1) defines the income of the leader. Here the first term corresponds to the case when the leader monopolizes the client, reducing his price to the level of the cost price of service at the follower's facilities. That is, in terms of prices and revenues, it is the case 2: $q_{ij}^2 \geq q_{ijk}^1$ and $w_{ijk}^1 > w_{ij}^3$. If the price of q_{ij}^2, determined by the budget, is less than the monopoly price q_{ijk}^1, then the second term linking the price to the budget level is used. The third term corresponds to the case when it is advantageous for the leader to share the client's budget with the follower. That is, the income from monopolization is less than the income from the division of the demand in half between players. Constraint (2) means that the leader must open exactly p facilities. From constraint (3) it follows that the distribution of clients between players and the player's incomes are determined on the basis of the optimal solution of the follower. Due to this constraint, our model is a bilevel programming problem. The set $\mathcal{F}^*(x)$ is the set of optimal solutions for the follower's parametric problem. As the parameter, we consider here the set of facility locations chosen by the leader.

For the follower, we propose the following model:

$$\sum_{i \in I} \sum_{j \in J} \sum_{k \in I} (d_j(c_{ij} + t_i - c_{kj} - t_k)z_{ijk}^{Fc} + (b_j - d_j(c_{kj} + t_k))z_{ijk}^{Fb} \tag{5}$$

$$+ 0.5(b_j - d_j(c_{kj} + t_k))z_{ijk}) \rightarrow \max_{y, z^{Lc}, z^{Lb}, z^{Fc}, z^{Fb}, z}$$

under constraints:

$$\sum_{i \in I} y_i = r; \tag{6}$$

$$\sum_{i,k \in I} (z_{ijk}^{Fc} + z_{ijk}^{Fb}) \leq \sum_{i \in I_j(x)} y_i; j \in J; \tag{7}$$

$$x_i + y_i \leq 1; i \in I; \tag{8}$$

$$\sum_{i,k \in I} (z_{ijk}^{Lc} + z_{ijk}^{Lb} + z_{ijk}^{Fc} + z_{ijk}^{Fb} + z_{ijk}) \leq 1; j \in J; \tag{9}$$

$$\sum_{k \in I} (z_{ijk}^{Lc} + z_{ijk}^{Lb} + z_{ijk}^{Fc} + z_{ijk}^{Fb} + z_{ijk}) \leq x_i; i \in I, j \in J; \tag{10}$$

$$\sum_{i \in I} (z_{ijk}^{Lc} + z_{ijk}^{Lb} + z_{ijk}^{Fc} + z_{ijk}^{Fb} + z_{ijk}) \leq y_k; k \in I, j \in J; \tag{11}$$

$$(c_{ij} + t_i)(z_{ijk}^{Lc} + z_{ijk}^{Lb} + z_{ijk}^{Fc} + z_{ijk}^{Fb} + z_{ijk}) \leq (c_{i'j} + t_{i'})x_{i'} \tag{12}$$
$$+ (1 - x_{i'})\overline{C}; i, i', k \in I, j \in J;$$

$$(c_{kj} + t_k)(z_{ijk}^{Lc} + z_{ijk}^{Lb} + z_{ijk}^{Fc} + z_{ijk}^{Fb} + z_{ijk}) \leq (c_{k'j} + t_{k'})y_{k'} \tag{13}$$
$$+ (1 - y_{k'})\overline{C}; i, k, k' \in I, j \in J;$$

$$d_j(c_{kj} + t_k - c_{ij} - t_i) \leq 0.5(b_j - d_j(c_{ij} + t_i))z_{ijk} + (1 - z_{ijk})\overline{C}; i, k \in I, j \in J; \quad (14)$$

$$\sum_{i \in I} \sum_{k \in I} (d_j(c_{kj} + t_k) - b_j)z_{ijk}^{Lc} \leq 0; j \in J; \quad (15)$$

$$\sum_{i \in I} \sum_{k \in I} (d_j(c_{ij} + t_i) - b_j)z_{ijk}^{Fc} \leq 0; j \in J; \quad (16)$$

$$z_{ijk}^{Lc}, z_{ijk}^{Lb}, z_{ijk}^{Fc}, z_{ijk}^{Fb}, z_{ijk}, y_i \in \{0, 1\}; i, k \in I, j \in J. \quad (17)$$

The objective function (5) defines the income of the follower. The components of the objective function have the same meaning as the terms of the objective function of the leader. The constraint (6) means that the follower must open exactly r facilities. The constraints (7) and (9) implement the mechanism of distribution of clients between players. If $I_j(x) = \emptyset$ then the leader monopolize the client since he has the minimal servicing cost there. Otherwise, the client may belong to the follower if he chooses one of the points of the set $I_j(x)$ as the location for one of his facilities. The constraint (8) prohibits to the leader and follower to place facilities at the same point. From the constraints (10) and (11) it follows that the client cannot be serviced at a point where there are no open facilities. The constraints (9), (12) and (13) imply that we have selected a unique pair of open facilities for the client j, one for the leader (i) and another for the follower (k), and the chosen leader's facility achieves the smallest servicing cost for the client. Similarly, the smallest cost of service for client in follower's facilities is achieved at his chosen facility. Further assume that the client was monopolized by the leader and we consider the optimal solution of the bilevel problem. Then if the prices q_{ijk}^1 and q_{ij}^2 are nontrivial, then it follows from the restriction of (9) that one of the variables z_{ijk}^{Lc}, z_{ijk}^{Lb}, z_{ijk} is equal to 1. Let $d_j(t_k + c_{kj}) - b_j \leq 0$ and $w_{ijk}^1 \leq w_{ij}^3 = (d_j/2)(q_{ij}^2 - t_i)$, that is, case 1 is executed. Then the income from monopolization is less than the income from the division of the demand in half between players. So, $z_{ijk} = 1$ and the restriction (14) holds. Suppose, that $d_j(t_k + c_{kj}) - b_j \leq 0$ and $w_{ijk}^1 > w_{ij}^3$, then the leader has more income when using a monopoly price q_{ijk}^1. Then $z_{ijk}^{Lc} = 1$ and the restriction (15) holds. Finally, if $d_j(t_k + c_{kj}) - b_j > 0$, then the leader gets the maximum possible income equals to $w_{ij}^2 = d_j(q_{ij}^2 - t_i) = b_j - d_j(t_i + c_{ij})$ using the price q_{ij}^2. Constraints (14) and (16) are interpreted in a similar way for client j monopolized by the follower.

Further, we will assume that the initial data for the problem is rational.

3   The Computational Complexity

We recall the definition of the first level of the polynomial hierarchy of complexity classes of decision problems. The first level consists of classes P, NP and co-NP. The class P contains problems solvable in polynomial time on deterministic Turing machines. The class NP is defined as the class of problems solvable in polynomial time on nondeterministic Turing machines. The third basic class co-NP consists of decision problems whose complements belong to NP. These classes are also denoted as Δ_1^P, Σ_1^P, and Π_1^P, respectively. The second level of the polynomial hierarchy is defined by deterministic and nondeterministic Turing machines with oracle [7]. It is said that the decision problem belongs to class Δ_2^P if there exists a deterministic Turing machine with an oracle that recognizes its in polynomial time, using as oracle some language from class NP. Similarly, the decision problem belongs to class Σ_2^P if there exists a nondeterministic Turing machine with an oracle that recognizes its in polynomial time, using as oracle some language from class NP.

In order to proceed optimization problems, we use the concept of the standard decision problem corresponding to the optimization problem. We associate optimization problem L with the following decision problem $D(L)$. The input of this problem is the input of the problem L and an arbitrary rational number k. In the problem $D(L)$ it is necessary to decide whether a feasible solution exists with the objective function value large or equal to k. Class PO (correspondingly, $\Delta_2^P O$) includes optimization problems for which the standard decision problem lies in class P (correspondingly, Δ_2^P). Similarly, class NPO (correspondingly, $\Sigma_2^P O$) includes optimization problems for which the standard decision problem lies in class NP (correspondingly, Σ_2^P).

In this section, we analyze the complexity of the competitive location and pricing problem and its subproblems. We start from the following lemma.

Lemma 1. *The competitive location and pricing problem with the uniform split of the demand belongs to the class $\Sigma_2^P O$.*

Proof. In the standard decision problem, it is necessary to find a solution to the problem with the value of the objective function at least k. Such a problem can be solved by brute force enumeration of all locations of leader's facilities and by solving the parametric problem of the follower. In other words, we can guess the necessary location x of the leader's facilities and the corresponding optimal solution $(y, z^{Lc}, z^{Lb}, z^{Fc}, z^{Fb}, z)$ of the follower in a non-deterministic time and then check constraints (2)–(4) in polynomial time, using a suitable NP-oracle. The verification of constraints (2) and (4) is trivial. As an NP-oracle, let's take the standard decision problem for the follower's problem. Since the follower's objective function is limited, with the help of the oracle and binary search we will find the optimal value of the parametric problem of the follower for the given location x of the leader's facilities. If the variables $(y, z^{Lc}, z^{Lb}, z^{Fc}, z^{Fb}, z)$ satisfy the constraints of the follower's problem and the value of the objective function on this feasible solution coincides with the previously found optimal value, then the constraint (3) is satisfied. Since the number of calls to an oracle is limited by the logarithm of the length of the record of the initial data of the problem, then to verify constraints (2)–(4), a polynomial time is sufficient. Thus, the problem of the leader belongs to class $\Sigma_2^P O$.

Theorem 1. *The competitive location and pricing problem with the uniform split of the demand is Σ_2^P-hard.*

Proof. We reduce the following problem to the competitive location and pricing problem.

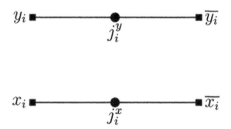

Fig. 1. Facilities and profitable clients that correspond to variables x_i and y_i.

Problem 1 ($\exists\forall 3, 4SAT$)
Input: We are given two vectors $x = (x_1, \ldots, x_p)$ and $y = (y_1, \ldots, y_r)$ of Boolean variables and a formula $\varphi(x, y)$ in the disjunctive normal form. Each conjunction contains exactly one variable from x and either two or three variables from y.
Question: Does there exist a truth assignment of x such that for all assignments of y the formula $\varphi(x, y)$ is satisfied?

As shown in [10] $\exists\forall 3, 4SAT$ is Σ_2^p-complete.

Given an instance of $\exists\forall 3, 4SAT$ and let k be the number of conjunction in $\varphi(x, y)$.

We construct the following instance of the competitive location and pricing problem. For each variable $x_i (y_i)$ we introduce two *profitable* facility locations x_i and $\overline{x_i}$ (y_i and $\overline{y_i}$) corresponding to literals x_i and $\overline{x_i}$ (y_i and $\overline{y_i}$), respectively. Between the facility locations x_i and $\overline{x_i}$ (y_i and $\overline{y_i}$) we insert a *profitable* client j_i^x (j_i^y) which is connected by arcs to both facility locations (see Fig. 1). The length of arcs that connect j_i^x (j_i^y) with x_i (y_i) and $\overline{x_i}$ ($\overline{y_i}$) is equal to k, i.e. $d(x_i, j_i^x) = d(j_i^x, \overline{x_i}) = d(y_i, j_i^y) = d(j_i^y, \overline{y_i}) = k$. The budget of the client j_i^x is equal to $16k^2 + k$ and the budget of the client j_i^y is equal to $12k^2 + k$. We will call locations x_i and $\overline{x_i}$ (y_i and $\overline{y_i}$) *alternative* facility locations.

For each conjunction κ_s, we introduce two facility locations alt_s, sin_s and four clients j_s^{con}, j_s^{alt}, j_s^{sin}, j_s^{ad}. The reduction of $x_i \wedge y_{i_1} \wedge y_{i_2} \wedge y_{i_3}$ is illustrated in Fig. 2. The client j_s^{con} is directly connected with five facility locations alt_s, x_i, $\overline{y_{i_1}}$, $\overline{y_{i_2}}$, and $\overline{y_{i_3}}$. We set $d(j_s^{con}, x_i) = 10k^2 + k$, $d(j_s^{con}, alt_s) = 10k^2$, and $d(j_s^{con}, \overline{y_i})$ is equal to $10k^2 + 1$ for all i. The budget of j_s^{con} is equal to $10k^2 + k + 1$. We will call the clients j_s^{con} *conflicting* clients. The client j_s^{alt} is directly connected with two facility locations x_i and alt_s, wherein alt_s is located between j_s^{alt} and j_s^{con}. We set $d(alt_s, j_s^{alt}) = 5k^2$ and $d(x_i, j_s^{alt}) = 16k^2 - k$. The budget of j_s^{alt} is equal to $16k^2$. The client j_s^{ad} is at a distance of 0 from the facility location $\overline{x_i}$ and his budget is equal to k. The client j_s^{sin} and the facility location sin_s are removed a sufficient distance from the other vertices and $d(sin_s, j_s^{sin}) = 0$. The budget of j_s^{sin} is equal to $11k^2 + \varepsilon$, where $0 < \varepsilon < 1$. The transportation cost between two vertices given by the shortest path. For example, a network for the formula $\varphi(x, y) = (\overline{x_1} \wedge y_1 \wedge \overline{y_2} \wedge y_3) \vee (x_2 \wedge y_2 \wedge \overline{y_3})$ is described in Fig. 3. The number of leader's facilities p coincides with the number of Boolean variables x.

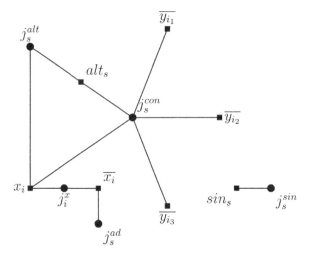

Fig. 2. Representation of $x_i \wedge y_{i_1} \wedge y_{i_2} \wedge y_{i_3}$

The number of follower's facilities $r = l + k$, where l is the number of Boolean variables y.

Now consider how the player's income depends on the choice of the location of the facilities. Let $\kappa_s = x_i \wedge y_{i_1} \wedge y_{i_2} \wedge y_{i_3}$. Let the leader place the facility in x_i or $\overline{x_i}$. If the follower doesn't occupy the alternative location then the leader will receive a income of $16k^2$ from the client j_i^x, otherwise the leader and follower share the income from the client j_i^x and each will receive $8k^2$. In the latter case, the possible additional income of each of the facilities x_i and $\overline{x_i}$ from the clients j_s^{con}, j_s^{alt}, j_s^{ad}, $s = 1, \ldots, k$ will not exceed $k^2 + k$. Thus, if both players place their facilities in x_i and $\overline{x_i}$ their income will not exceed $9k^2 + k$. The maximal income at y_i, $\overline{y_i}$, $i \in \{i_1, i_2, i_3\}$, alt_s, or sin_s doesn't exceed $13k^2 + k$ and the minimal income at y_i, $\overline{y_i}$, alt_s, or sin_s is at least $11k^2 - k$. Therefore, the leader must place own facilities at x_i and $\overline{x_i}$, one facility at each pair $(x_i, \overline{x_i})$, because he knows that in this case the follower set his facilities at y_i, $\overline{y_i}$, alt_s, or sin_s.

Suppose that the leader took all the places near the profitable clients j_i^x. If the follower places the facility in y_i or $\overline{y_i}$, he will receive a income of $12k^2$ from the client j_i^y. The possible income of each of the facilities alt_s and sin_s doesn't exceed $11k^2 + k + 1$. Thus, the follower must place r facilities at y_i and $\overline{y_i}$, one facility at each pair $(y_i, \overline{y_i})$. The remaining facilities of the follower should be placed in locations alt_s and sin_s. It is easy to verify that the follower must select exactly one of locations alt_s or sin_s, for each $s = 1, \ldots, k$ and his choice will depend on the location of the leader's facilities. We note that the income of the follower in sin_s does not depend on the location of the leader's facilities and it is equal to $11k^2 + \varepsilon$. Let the leader open the facility x_i. Then the income of the follower in alt_s is equal to $16k^2 - k - 5k^2 + 10k^2 + k - 10k^2 = 11k^2$ and he prefers to open the facility sin_s. It follows that in this case, the leader receives the client j_s^{alt} and the income of k from it. Let the leader open the facility $\overline{x_i}$.

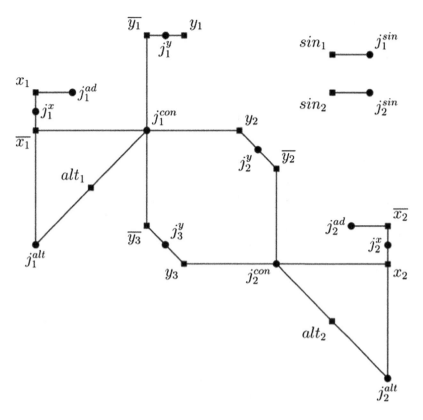

Fig. 3. An example of network for $\varphi(x,y) = (\overline{x_1} \wedge y_1 \wedge \overline{y_2} \wedge y_3) \vee (x_2 \wedge y_2 \wedge \overline{y_3})$

In this case, the leader does not receive the client j_s^{alt} but he receives the client j_s^{ad} the income of k from it. It follows that the total income of the leader from all clients except for conflicting clients is equal to $(16p+1)k^2$ and does not depend on the choice in which of the locations x_i or $\overline{x_i}$ to open the facility. In turn, the follower prefers to open the facility alt_s. Indeed, the income of the follower from client j_s^{alt} in alt_s is equal to $11k^2$ and he get an additional income of at least 1 from client j_s^{con}. It follows that the best location of follower's facilities in vertices alt_s and sin_s, $s = 1, \ldots, k$, is completely determined by the location of leader's facilities and does not depend on the location of profitable follower's facilities. Hence, the income of each player depends on who gets conflicting clients. If the leader opens the facility $\overline{x_i}$ then the client j_s^{con} will be served by the follower. Let the leader open the facility x_i. The follower get the client j_s^{con} if and only if he opens one of the facility $\overline{y_{i_1}}$, $\overline{y_{i_2}}$, or $\overline{y_{i_3}}$. Thus, the leader gets at least one conflicting client if and only if there exists a truth assignment of x such that for all assignments of y the formula $\varphi(x,y)$ is satisfied.

4   Special Cases

In addition to the $\exists\forall 3, 4Sat$ problem, consider $\exists\forall 1, 2Sat$ problem. In this problem, each conjunction contains only one variable from x and at most one variable from y. Obviously, the problem is polynomially solvable. In Theorem 1, we constructed a set of instances of competitive location and pricing problem which corresponds to $\exists\forall 3, 4Sat$ problem. By analogy, we construct a set of instances corresponding to the $\exists\forall 1, 2Sat$ problem. We denote the set as a $CLP2SAT$ problem. Is the $CLP2SAT$ problem polynomially solvable? For the answer, consider a maxmin-1,2-Sat problem. As in $\exists\forall 1, 2Sat$ problem, we are given two vectors $x = (x_1, \ldots, x_p)$ and $y = (y_1, \ldots, y_r)$ of Boolean variables and a formula $\varphi(x, y)$ in the disjunctive normal form. Each conjunction contains only one variable from x and at most one variable from y. We need to find x, at which the total number of satisfied conjunction is maximal for all y. Obviously, the $CLP2SAT$ problem is equivalent to the maxmin-1,2-Sat problem.

Theorem 2. *The maxmin-1,2-Sat problem is NP-hard.*

Proof. Consider the NP-hard Exact Cover by 3-sets problem.

Problem 2 (EC3SET)
Input: We are given a set X, with $|X| = 3q$ (so, the size of X is a multiple of 3), and a collection C of 3-element subsets of X.
Question: Does there exist a subset $\tilde{C}$ of C where every element of X occurs in exactly one member of $\tilde{C}$?

Given an instance of EC3SET and let k be the cardinality of the collection C. We construct the following instance of the maxmin-1,2-Sat problem.

For each subset $C_r = (X_i, X_j, X_l)$, we define boolean variables x_r, y_i, y_j, y_l and introduce conjunctions $(x_r \wedge y_i), (x_r \wedge y_j), (x_r \wedge y_l)$. Additionally, we introduce $3q$ conjunctions $(x_0 \wedge \overline{y_i})$, $i = 1, \ldots, 3q$, k conjunctions $(\overline{x_r} \wedge y_0)$, $r = 1, \ldots, k$ and k identical conjunctions $(x_0 \wedge \overline{y_0})$. We show that an exact cover of X exists if and only if the total number of satisfied conjunctions is equal to $2q + k$.

It is easy to see that we need only consider the case when x_0 and y_0 are equal to 1. Let H be the total number of satisfied conjunctions and h be the number of variables x_r such that $x_r = 1$. If $h < q$ then $H \leq 3h + k - h = 2h + k < 2q + k$. If $h > q$ then $H \leq 3q + k - h = 2q + k - (h - q) < 2q + k$. Let $h = q$. We have

$$H = 3q - s + (k - q) = 2q + k - s,$$

where s is a number of repetitions of variables y in the truth assignment of x. It follows that an exact cover of X exists if and only if the optimum is $2q + k$.

Corollary 1. *The CLP2SAT problem is NP-hard.*

Consider particular cases, when leader's and follower's facilities are opened. In these cases, each client needs to define a facility at which he will be served. It can be done in $O(mn^2)$ times. Therefore, the problem is polynomially solvable.

Note, the competitive location and pricing problem can be solved in $O(C_p^n *$
$C_r^{n-p} * mn^2) = O(mn^{p+r+2})$ by look over through all locations. Consider the
following particular cases: (1) $p = const$; (2) $r = const$; (3) $p, r = const$.

In the first case, the number of leader's location is C_p^n. It is a polynomial of
n. Therefore, the problem belongs to $\Delta_2^P O$.

In the second case, the follower problem is polynomially solvable as $C_r^{n-p} =
Poly(n, p)$. Then, the problem belongs to NPO.

In the third case, the problem can be solved in polynomial time. Then the
problem belongs to PO.

Finally, consider the cases, where $p = 0$ or $r = 0$. Then the problem is
equivalent to the widely known p-median problem, which is strongly NP-hard.
Therefore, these particular cases are strongly NP-hard as well.

5 Conclusion

This paper studies a new optimization model of competitive facility location and
pricing. Results of the computational complexity of the model are presented. A
few numbers of special cases are considered.

There are several interesting areas of research on this problem. The first
of them is connected with the development of exact algorithms for solving the
problem. Here, ideas from [11, 12] can be used. Another area of research is the
development of algorithms for solving on the basis of local search and metaheuristics. Despite the fact that the exponential complexity of local search is
theoretically shown [13], the available experience of using such methods of solution indicates their practical effectiveness [14–18]. It is also important to continue
studying the relationships of this class of problems with the polynomial hierarchy and the approximation hierarchy. Similar results were obtained for a number
of interesting problems of bilevel programming [10, 19–21].

References

1. Hotelling, H.: Stability in competition. Econ. J. **39**(153), 41–57 (1929)
2. Eiselt, H.A., Laporte, G., Thisse, J.-F.: Competitive location models: a framework
   and bibliography. Transp. Sci. **27**, 44–54 (1993)
3. Eiselt, H.A., Laporte, G.: Sequential location problems. Eur. J. Oper. Res. **96**,
   217–242 (1996)
4. Hamacher, H.W., Nickel, S.: Classification of location models. Locat. Sci. **6**, 229–
   242 (1998)
5. Plastria, F.: Sequential location problems. Eur. J. Oper. Res. **129**, 461–470 (2001)
6. Panin, A.A., Pashchenko, M.G., Plyasunov, A.V.: Bilevel competitive facility location and pricing problems. Autom. Remote Control **75**(4), 715–727 (2014)
7. Ausiello, G., Crescenzi, P., Gambosi, G., Kann, V., Marchetti-Spaccamela, A., Protasi, M.: Complexity and Approximation: Combinatorial Optimization Problems
   and Their Approximability Properties. Springer, Heidelberg (1999). https://doi.
   org/10.1007/978-3-642-58412-1

8. Garcia, M.D., Fernandez, P., Pelegrin, B.: On price competition in location-price models with spatially separated markets. TOP **12**, 351–374 (2004)
9. Pelegrin, B., Fernandez, P., Garcia, M.D., Cano, S.: On the location of new facilities for chain epansion under delivered pricing. Omega **40**, 149–158 (2012)
10. Davydov, I., Kochetov, Yu., Plyasunov, A.: On the complexity of the $(r \mid p)$-centroid problem in the plane. TOP. **22**(2), 614–623 (2014)
11. Alekseeva, E., Kochetov, Yu.: Matheuristics and exact methods for the discrete $(r \mid p)$-centroid problem. In: Talbi, El.-G., Brotcorne, L. (eds.) Metaheuristics for Bi-level Optimization. SCI, vol. 482, pp. 189–219. Springer, Heidelberg (2013). https://doi.org/10.1007/978-3-642-37838-6_7
12. Alekseeva, E., Kochetov, Yu., Plyasunov, A.: An exact method for the discrete $(r \mid p)$-centroid problem. J. Global Optim. **63**(3), 445–460 (2015)
13. Alekseeva, E., Kochetov, Yu., Plyasunov, A.: Complexity of local search for the p-median problem. Eur. J. Oper. Res. **191**, 736–752 (2008)
14. Plyasunov, A.V., Panin, A.A.: The pricing problem. I: exact and approximate algorithms. J. Appl. Ind. Math. **7**(2), 241–251 (2013)
15. Davydov, I.A., Kochetov, Y.A., Carrizosa, E.: A local search heuristic for the (rjp)-centroid problem in the plane. Comput. Oper. Res. **52**, 334–340 (2014)
16. Lavlinskii, S.M., Panin, A.A., Plyasunov, A.V.: Comparison of models of planning public-private partnership. J. Appl. Ind. Math. **10**(3), 356–369 (2016)
17. Kochetov, Yu.A., Panin, A.A., Plyasunov, A.V.: Comparison of metaheuristics for the bilevel facility location and mill pricing problem. J. Appl. Ind. Math. **19**(3), 392–401 (2015)
18. Diakova, Z., Kochetov, Yu.: A double VNS heuristic for the facility location and pricing problem. Electron. Notes Discrete Math. **39**, 29–34 (2012)
19. Iellamo, S., Alekseeva, E., Chen, L., Coupechoux, M., Kochetov, Yu.: Competitive location in cognitive radio networks. 4OR **13**(1), 81–110 (2015)
20. Panin, A.A., Plyasunov, A.V.: On complexity of the bilevel location and pricing problems. J. Appl. Ind. Math. **8**(4), 574–581 (2014)
21. Plyasunov, A.V., Panin, A.A.: The pricing problem. II: computational complexity. J. Appl. Ind. Math. **7**(3), 420–430 (2013)

Branch and Bound Method for the Weber Problem with Rectangular Facilities on Lines in the Presence of Forbidden Gaps

Gennady G. Zabudsky$^{(\boxtimes)}$ and Natalia S. Veremchuk$^{(\boxtimes)}$

Sobolev Institute of Mathematics, Omsk, Russia
zabudsky@ofim.oscsbras.ru, n-veremchuk@rambler.ru

Abstract. The problem of location of connected rectangular facilities on parallel lines in the presence of forbidden gaps is studied. The rectangular metric is used. The centers of the placed facilities are connected with the centers of the gaps. The facilities are impossible to place in forbidden gaps. It is necessary to place the facilities on the lines so that the total cost of connections between the facilities and between facilities and gaps was minimized. The problem is an adequate model of many practical situations. It is known that the original continuous problem for one–line variant is reduced to discrete subproblems. In this paper, the review of the properties and the algorithms for solving of the problem on one line are described. The branch and bound method for solving the problem is proposed. Results of computational experiments on comparison of the branch and bound method and a heuristic proposed in [27] are reported. In the experiments, a integer programming model and IBM ILOG CPLEX package are used.

Keywords: Location problem · Connected rectangular facilities
Forbidden gaps

1 Introduction

Models and methods of solution for location problems are intensively developed directions in operations research [5, 10, 11, 15, 27, 28]. The problems have many important applications in different areas: at design of plans of industrial enterprises, in robot motion planning, in service station location, etc. Different statements of such problems are defined by facility sizes (such as points, rectangles), existence of connections between facilities, area in which they are placed (network, plain), various restrictions and the criteria types, metrics and so on.

One of the main subclasses of the location problems in the presence of connections between facilities is the Weber problem. The problem consists in placing new facilities on the plane with respect to existing facilities so as to minimize

G. G. Zabudsky—The work was supported by the program of fundamental scientific research of the SB RAS No. I.5.1., project No. 0314-2016-0019.

A. Eremeev et al. (Eds.): OPTA 2018, CCIS 871, pp. 29–41, 2018.
https://doi.org/10.1007/978-3-319-93800-4_3

the sum of the costs which consists of the costs proportional to the distances between new and existing facilities, and the costs proportional to the distances between new facilities [29].

The Weber problem is a basic location problem in many practical situations. So, for example, the problem is used in automation design of the complex systems and placement of structural elements in the given area. Often at designing of the petrochemical enterprise floorplans, the facilities are the technological equipment [14, 25]. The equipment units can be connected between themselves by different communications, e.g. by a set of pipelines. Often some requirements for a regularity of placement are imposed, e.g. to place the equipment along the so–called *red* lines to create straight roads and facilities for maintenance of equipment [24].

First Weber problem was formulated by Fermat in the beginning 17th century. The problem is: "there are three points on the plane, find the fourth point such that the sum of its Euclidean distances to the given points was minimal." The problem has been solved by Torricelli. In the middle of the 18th century Simpson generalized the problem by including weights for the corresponding distances. In the early 19th century Weber used the generalized model to define an optimum location of a factory which makes one product with two present sources of raw materials and deliveries to one client.

The results of research of the classical Weber problem for point facilities without restrictions are quite well represented in the literature. Weber's rectangular metric problem can be decomposed into two independent subproblems, each of which is a linear programming problem. Dual problem to each of the subproblem is the minimum cost of flow problem in a specific network. Some alternative methods to solving the problems, which are equivalent to the above linear programming problem are also discussed in [3, 19].

The objective function is strictly convex in the Weber problem with Euclidean metric if the fixed facilities are not in the same line. At the same time, the system of equations that are obtained by equating the gradient to zero is nonlinear and it cannot be solved analytically. In [12], a numerical process of obtaining a solution to the problem was described. An approach based on the approximation of the Weber's original continuous problem by a discrete problem was proposed in [7].

The extensions of the Weber problem are defined, e.g. by given facility sizes, the structure of the area in which the facilities are placed, and the type of restrictions on the placement of facilities. If the facility sizes are taken in account (their sizes are commensurate with the size of the placement area), facilities are often replaced by geometric shapes, such as rectangles. Different approaches to the decision of optimal location problem of the rectangles on the plane are developed in [4, 17, 20–22, 24]. In the case of unconnected rectangles, the two–dimensional problem of packing the rectangles in a strip of the minimal length is considered in [20]. The problem can be formulated as a nonlinear mixed–integer program. A tabu search algorithm is proposed for solving the problem. In [17], the algorithm of local optimization for placing rectangles on the plane is described. If it is allowed to rotate the rectangles then for such a problem a heuristic algorithm is proposed in [4]. In [24], the problem of rectangles location on parallel lines is considered. Methods of dynamic programming and integer linear programming are used to search for a set of Pareto–optimal solutions.

The Weber problem with restrictions on facility locations is analyzed in [10, 13,18]. One of the most considerable generalizations of the classical statement of the Weber problem on the plane is related to consideration of the forbidden gaps and barriers [29]. In forbidden gaps the facilities cannot be placed. In the case of renovation of the plant, such gaps can be, for example, existing facilities and technological equipment. Barriers are determined as regions where locating facilities is not allowed. Also, traveling is prohibited in the interior of the barriers. However, traveling is allowed along barrier's border. Here, not only the location problem is solved, but the problem of tracing is solved too [29].

In literature, the location problems in the presence of forbidden gaps for different metrics are considered. Mainly the case of placing one facility with one forbidden gap is considered. The problems with Euclidean metric and forbidden gap in the form of a circle or a convex polygon are studied in [1,8,9]. The rectangular metric for the measurement of distances between facilities is used, for example, in [2]. The location problem in which apart from forbidden gaps account is taken of the fixed facilities on the plane is studied in [16].

Minimax Weber problem on the plane for point facilities with the rectangular metric and the rectangular forbidden gaps is studied in [28]. It is proved that it is enough to consider a subset of admissible solutions to find the optimum of the problem. The branch and bound algorithm for solving the problem is proposed. Computational experiment on comparison of efficiency of the algorithm and application of integer programming model and IBM ILOG CPLEX package is reported. The use of the proved property is effective both in solving the problem by combinatorial methods and by using integer programming [29].

This article is devoted to the Weber problem on parallel lines in the presence of forbidden gaps. The location of lines is fixed. The placed facilities and the forbidden gaps are the rectangles. The centers of the facilities are connected between themselves and with the centers of the gaps. The facilities are impossible to place in forbidden gaps. In addition, a set of rectilinear passages between the lines, which must be preserved when placing facilities is provided. The generalization of mathematical model of nonlinear programming with Boolean variables is proposed. For one–line variant the original continuous problem is reduced to a number of discrete subproblems of smaller dimension. The review of properties and the algorithms for solving the problem is provided. A branch and bound method for solving the subproblems is proposed. Results of computational experiments for the branch and bound method, the heuristic from [27] and for solving the problem using the integer programming model and IBM ILOG CPLEX package are reported.

2 Statement of the Problem and Its Properties

The Weber problem on lines in the presence of forbidden gaps is formulated as follows [29]. Suppose that straight–line segments parallel to OX axis, containing some fixed rectangular areas (forbidden gaps), a set of rectilinear passages between the lines and rectangular facilities are given. Centers of the facilities are connected between themselves and with the centers of the gaps. The goal is to place the facilities on the segments outside of the forbidden gaps so that they do not intersect with each other and with the rectilinear passages and the total cost of connections between the facilities and between facilities and gaps is minimized [29].

Denote the facilities and gaps with the centers at (x_i, y_i) and (b_{1j}, b_{2j}) by X_i and F_j respectively, where $i \in I = \{1, \ldots, n\}$ and $j \in J = \{1, \ldots, m\}$. Denote lengths and heights of facilities and gaps by l_i, d_i and p_j, b_j respectively, where $i \in I$ and $j \in J$. Let $w_{ij} \geq 0$, $u_{ik} \geq 0$ are the specific costs of connections between centers of X_i and F_j, X_i and X_k for $i, k \in I$, $j \in J$, and $i < k$. Let the straight–line segments of length LS and rectilinear passages between the lines be fixed. The left border of each segment is the point $(0, Ly_t)$, where $t \in Q = \{1, \ldots, q\}$ and $Ly_1 < \cdots < Ly_q$. Further we will call these straight–line segments as *lines*. Denote by M_t the maximum height of the facility, which may be placed on the line with number t. These values are determined by the sizes of the passages. Suppose the lines are fixed at such a distance from each other that any facility can be placed on any line. The aim is to place the facilities $X_1, \ldots, X_n$ on the lines outside gaps $F_1, \ldots, F_m$ and so that they do not intersect with each other and with the rectilinear passages and the total cost of the connections between the facilities and between facilities and gaps is minimized. The cost of a connection is determined as the product of the distance and the cost of the connection. Distances are measured in the rectangular metric.

We will introduce the Boolean variables z_{it} for $i \in I$, $t \in Q$, to formulate the conditions of facility's location on the line, so that $z_{it} = 1$ if X_i is placed on the line with the number t, otherwise $z_{it} = 0$ [29].

The nonlinear Boolean programming formulation of the problem is the following:

$$G(x,y) = \sum_{i=1}^{n} \sum_{j=1}^{m} w_{ij}(|x_i - b_{1j}| + |y_i - b_{2j}|) + \sum_{i=1}^{n-1} \sum_{k=i+1}^{n} u_{ik}(|x_i - x_k| + |y_i - y_k|) \to \min,$$
(1)

$$|x_i - b_{1j}| \geq z_{it} \frac{l_i + p_j}{2}, \quad i \in I, \quad j \in JL_t, \quad t \in Q,$$
(2)

$$|x_i - x_k| \geq (z_{it} + z_{kt} - 1)\frac{l_i + l_k}{2}, \quad i, k \in I, \quad i < k, \quad t \in Q,$$
(3)

$$\frac{l_i}{2} \leq x_i \leq LS - \frac{l_i}{2}, \quad i \in I,$$
(4)

$$y_i = \sum_{t=1}^{q} z_{it} Ly_t, \quad i \in I,$$
(5)

$$\sum_{t=1}^{q} z_{it} = 1, \quad i \in I,$$
(6)

$$d_i z_{it} \leq M_t, \quad i \in I, \quad t \in Q, \tag{7}$$

$$z_{it} \in \{0,1\}, \quad i \in I, \quad t \in Q. \tag{8}$$

The first component in (1) indicates the total cost of connections between the facilities and the gaps; the second component indicates the total cost of connections between the facilities themselves. Expressions (2) and (3) are the conditions of disjointness between the facilities and the gaps and between the facilities themselves [29]. Constraints (6) are the requirement that any facility is placed only on one line. Expression (7) are the conditions on height of facility X_i, which is placed on the line with number t.

Denote the range of admissible solutions by B. Range B is disconnected and it consists of r separate blocks B_k of length L_k that contain the facilities X_i, $i \in I$, $B = \bigcup_{k=\overline{1,r}} B_k$. The problem (1)–(8) for $t = 1$ is NP–hard; a feasible solution to the problem can be found by construction of a one–dimensional bin packing [6]. In this case, the facilities with lengths of l_i, $i \in I$, are packed in the containers with sizes L_k, $k = \overline{1,r}$ [29]. It should be noted that fixing or prohibition of the facility location on the lines is possible with the help of assignment of Boolean variables.

The mathematical model for one–line variant is:

$$G(x) = \sum_{i=1}^{n} \sum_{j=1}^{m} w_{ij}|x_i - b_j| + \sum_{i=1}^{n-1} \sum_{k=i+1}^{n} u_{ik}|x_i - x_k| \rightarrow \min, \tag{9}$$

$$|x_i - b_j| \geq \frac{l_i + p_j}{2}, \quad i \in I, j \in J, \tag{10}$$

$$|x_i - x_k| \geq \frac{l_i + l_k}{2}, \quad i, k \in I, i < k, \tag{11}$$

$$\frac{l_i}{2} \leq x_i \leq LS - \frac{l_i}{2}, \quad i \in I. \tag{12}$$

The problem (9)–(12) is studied in [27]. The heuristic algorithm consists of two stages. In the first stage, we find a feasible partition of the facilities into the blocks, and in the second stage, the facilities in the blocks are rearranged so that the total cost of connections is minimized.

Given a feasible location, a *remainder* in the block B_k is a segment of non–zero length between two adjacent elements (facilities, gaps) that do not have a common border or between the border of B_k and an adjacent block. Two elements (facilities, gaps, remainders) are called *glued* if they have a common border [26].

Suppose that $x = (x_1, \ldots, x_n)$ is a feasible solution to the problem (9)–(12); $I_k(x)$ is the set of the facility numbers in the block B_k; $H_k(x)$ is the set of remainders in B_k; n_k is the capacity of the set $I_k(x)$. We will notice that x can be represented as $x = (x^1, \ldots, x^r)$, where x^k are the coordinates of facilities placed in B_k. In [27], it was proved that for a feasible solution x to the problem (9)–(12), we can find another feasible solution x' such that $|H_k(x')| \leq 1$, $k = 1, \ldots, r$ and $G(x') \leq G(x)$. So, it is sufficient to consider no more than one remainder in every block B_k. Thus, the original continuous problem is reduced to discrete problem [29].

Denote by LB_k and RB_k the coordinates of the left and the right borders of B_k (imaginary facilities F_L and F_R) and let $J_L(B_k)$ and $J_R(B_k)$ be the sets of gaps to the left and to the right of the block B_k. Also let $I_L(B_k)$ and $I_R(B_k)$ be the sets of facilities to the left and to the right of the block B_k respectively. Then, for a fixed partition of facilities into the blocks, the objective function $G(x)$ can be represented as [26]

$$G(x) = \sum_{k=1}^{r} G_k(x^k) + Const,$$

where

$$G_k(x^k) = \sum_{s \in I_k(x)} \sum_{t \in I_k(x), t>s} u_{st}|x_s - x_t| + \sum_{s \in I_k(x)} |x_s - LB_k|\left(\sum_{j \in J_L(B_k)} w_{sj} \right.$$
$$\left. + \sum_{i \in I_L(B_k)} u_{si} \right) + \sum_{t \in I_k(x)} |x_t - RB_k|\left(\sum_{j \in J_R(B_k)} w_{tj} + \sum_{i \in I_R(B_k)} u_{ti} \right).$$

The first component of $G_k(x^k)$ is the sum of costs of the connections between facilities in B_k, the second component and the third component are the sums of cost of the connections between facilities from B_k and LB_k, and from B_k and RB_k respectively.

We will call an admissible solution x to the problem (9)–(12) a *local minimum* of the problem if $G(x) \leq G(x')$ for every $x' : I_k(x) = I_k(x')$, $k = 1, \ldots, r$.

Let a partition of facilities into blocks be fixed. Then, in every block B_k it is possible to consider the subproblem of location $n_k + 2$ facilities. In B_k the subproblem contains two imaginary facilities F_L and F_R and n_k placed facilities [26]. Let us denote the total cost of connections between placed facilities in B_k and the facilities F_L and F_R respectively for every $i \in I_k(x)$ by w_{iL} and w_{iR}. Then [26]

$$w_{iL} = \left(\sum_{s \in J_L(B_k)} w_{is} + \sum_{t \in I_L(B_k)} u_{it} \right),$$

$$w_{iR} = \left(\sum_{s \in J_R(B_k)} w_{is} + \sum_{t \in I_R(B_k)} u_{it} \right).$$

The subproblem for B_k can be formulated as

$$G_k(x^k) = \sum_{s \in I_k(x)} \sum_{t \in I_k(x), t>s} u_{st}|x_s - x_t| + \sum_{s \in I_k(x)} w_{sL}|x_s - LB_k|$$
$$+ \sum_{t \in I_k(x)} w_{tR}|x_t - RB_k| \to \min, \tag{13}$$

$$|x_i - x_k| \geq \frac{l_i + l_k}{2}, \quad i, k \in I_k(x), i < k, \tag{14}$$

$$LB_k + \frac{l_i}{2} \le x_i \le RB_k - \frac{l_i}{2}, \quad i \in I_k(x). \tag{15}$$

We will find coordinates x^k of the centers of facilities in B_k, so that the total cost of connections between the facilities and the facilities with facilities F_L and F_R is minimized.

Since $I_k(x) \cap I_l(x) = \emptyset$ for every $k, l = 1, \ldots, r$, so to find a local optimum of the problem (9)–(12) for some fixed partition of facilities into blocks, it is sufficient to find the minimum in r independent subproblems (13)–(15). So, the solution of the original continuous problem is reduced to solving discrete subproblems of smaller dimension [26].

Note that if $u_{st} = 0$, for all $s, t \in I_k(x), s < t$, then for any $k = 1, \ldots, r$, the local optimum of the problem (9)–(12) can be found with the help of the polynomial–time algorithm [27]. If we introduce a partial order on location of facilities in the block, which may be presented in the form of a serial–parallel graph, then in the block the problem may be solved by the polynomial–time algorithm from [23].

Generally, if there exist $s, t \in I_k(x)$, such that $u_{st} > 0$, then for solving the subproblem (13)–(15) for a small value n_k, it is possible to use $n_k!$ permutations of the facilities in the block. For large values of n_k it is possible to apply, e.g. the branch and bound algorithm.

3   Branch and Bound Method and Results of Computational Experiment

Calculation of lower bounds on the objective function and the branching method are the basis for any branch and bound method. Let us describe the branch and bound method on the example of a block B_k.

3.1   Lower Bounds

Let us denote the sets of the facility numbers placed in B_k by NF_l, NF_r. Without loss of generality, we will assume that the facilities in the set NF_l have numbers from 1 to s, and in the set NF_r the numbers of facilities are from $t + 1$ to n_k. Denote by D the set of the admissible locations for the facilities in B_k. Let $\xi(D)$ be the lower bound on function $G_k(x^k)$ for D [26]. Then $\xi(D)$ can be expressed as follows:

$$\xi(D) = \xi_1(D) + \xi_2(D) + \xi_3(D).$$

The value $\xi_1(D)$ is the total cost of connections between the facilities placed in B_k and the facilities with facilities F_L and F_R. Since the coordinates of all these facilities are known, so this value is calculated exactly. The value $\xi_2(D)$ is a lower bound on the total cost of connections between the facilities unplaced in B_k with facilities F_L, F_R and with the facilities placed in B_k. The value $\xi_3(D)$ is a lower bound on the total cost of connections between the facilities unplaced in B_k.

In [26], two methods of calculation of value $\xi_2(D)$ are proposed.

The First Method. For every $i \in I_k(x) \backslash \{NF_l \bigcup NF_r\}$, the total cost of connections between the facilities placed in B_k and the facilities with facilities F_L and F_R is calculated as follows [26]:

$$SL(i) = w_{iL} + \sum_{k \in NF_l} u_{ik}, \quad SR(i) = w_{iR} + \sum_{k \in NF_r} u_{ik}.$$

Later location of the facilities unplaced in B_k is determined by two variants. The facilities are ordered by non–increasing of ratio $SL(i)/l_i$. The facilities consistently are glued together in that order with the most left facility placed in B_k. Without loss of generality, we assume that the glued unplaced facilities have numbers from $s+1$ to t.

Later the facilities are ordered by non–increasing of ratio $SR(i)/l_i$. The facilities consistently are glued together in that order with the most right facility placed in B_k. Without loss of generality, we assume that the glued unplaced facilities have numbers from t to $s+1$ [26]. Then

$$\xi_2(D) = \xi_{2L}(D) + \xi_{2R}(D),$$

where $\xi_{2L}(D)$ and $\xi_{2R}(D)$ are the lower bounds on the total cost of the connections of unplaced facilities with facilities F_L, F_R respectively and with the facilities placed in B_k. The values $\xi_{2L}(D)$ and $\xi_{2R}(D)$ can be calculated the following way:

$$\xi_{2L}(D) = \sum_{q=s+1}^{t} \left(Lw_q \sum_{g=1}^{q-1} l_g + \sum_{i=1}^{s} u_{qi} \sum_{k=i+1}^{q-1} l_k \right),$$

$$\xi_{2R}(D) = \sum_{q=s+1}^{t} \left(Rw_q \sum_{g=q+1}^{n_k} l_g + \sum_{i=t+1}^{n_k} u_{qi} \sum_{k=q+1}^{i-1} l_k \right).$$

The proof that values $\xi_{2L}(D)$ and $\xi_{2R}(D)$ are the lower bounds on the total cost of connections of unplaced facilities with imaginary facilities F_L, F_R and with the facilities placed in B_k is similar to the proof in [21].

The Second Method. The set $I_k(x) \backslash \{NF_l \bigcup NF_r\}$ can be represented as a union of disjoint sets $N_L \bigcup N_C \bigcup N_R$, where by N_L, N_C, N_R are the sets of facility numbers for which we have the inequalities $SL(i) > SR(i)$, $SL(i) = SR(i)$, $SL(i) < SR(i)$ respectively.

Later facilities with numbers from N_L are ordered by non–increasing of ratio $(SL(i) - SR(i))/l_i$. These facilities consistently are glued together in that order with the most left facility placed in B_k. Also facilities with numbers from N_R are ordered by non–increasing of ratio $(SR(i) - SL(i))/l_i$. These facilities consistently are glued together in that order with the most right facility placed in B_k. The facilities with numbers from N_C are placed between sets of the facilities with numbers from N_L and from N_R in any order [26].

Thus, for every $i \in I_k(x) \setminus \{NF_l \bigcup NF_r\}$, the coordinate of the center is determined. Let $I_k(x) \setminus \{NF_l \bigcup NF_r\} = \{s+1, \ldots, t\}$. Define the value Z as

$$Z = \sum_{q=s+1}^{t} \left(Lw_q \sum_{g=1}^{q-1} l_g + \sum_{i=1}^{s} u_{qi} \sum_{k=i+1}^{q-1} l_k + Rw_q \sum_{h=q+1}^{n_k} l_h + \sum_{j=t+1}^{n_k} u_{qj} \sum_{v=q+1}^{j-1} l_v \right).$$

In [26], it was proved that the value Z is a lower bound on the total cost of connections of the unplaced facilities with facilities F_L, F_R and with the facilities placed in B_k.

The calculation of value $\xi_3(D)$ is the most difficult. One of the ways to calculate the value $\xi_3(D)$ is to consider the sets of unplaced facilities that are all connected between themselves. Further, for example, by means of viewing all permutations of any three such facilities, one can find an order of arrangement of facilities in the block with the minimum cost of connections between them.

3.2 Branching

At the first level in the branching tree each of the facilities with numbers from $I_k(x)$ is glued to the left border of the block B_k one by one. At the second level, each of the unplaced facilities is glued to the right border of the block B_k one by one. At the third level, each of the unplaced facilities is glued to the facility placed to the left border of the block B_k one by one. At the subsequent levels, each unplaced facility is glued to the facility placed at the previous level of the branching tree (see Fig. 1).

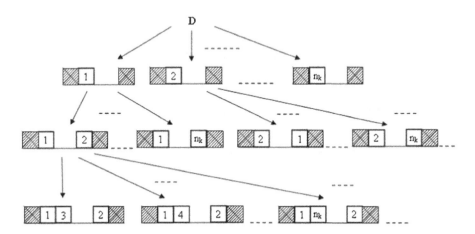

Fig. 1. Branching in the block

In Fig. 1, the borders of the blocks are marked by the shaded squares and the placed facilities are marked by squares with numbers of facilities. The number of vertexes of the branching tree at the first level is equal n_k, at the second level it equals $n_k(n_k - 1)$, at the third level it equals $n_k(n_k - 1)(n_k - 2)$, etc. Note that the height of the branching tree (number of levels) is equal to n_k, and the quantity of its trailing vertexes is $n_k!$, that corresponds to number of possible permutations of facilities in B_k.

3.3   Results of Computational Experiment

A computational experiment on comparison of the solutions obtained by using the branch and bound method (BBA) and the heuristic (A2) from [27] is carried out in this subsection. The experiment was conducted on the computer with the following specifications: Intel CoreTM i5-24502.50 GHz 6.00 GB. The alternative algorithms were implemented in Borland C++ Builder Version 6.0 (Build

Table 1. Comparison of algorithms A2 and BBA

No	n	m	F_{A2}	t_{A2}	F_{BBA}	t_{BBA}	Relative error F, %	$t_{BBA} - t_{A2}$
1	5	3	1148	2	1136	2	1,056	0
2	5	3	968,5	1	968,5	1	0	0
3	5	3	12238	1	11892	1	2,91	0
4	6	3	1844	1	1844	2	0	1
5	10	3	900,25	2	900,25	8	0	6
6	10	3	1080	3	1080	7	0	4
7	10	4	2888	2	2888	4	0	2
8	10	5	1801	2	1801	3	0	1
9	10	6	1253	50	1253	56	0	6
10	15	2	1420	7	1400	901	1,429	894
11	15	3	3502	6	3496	136	0,172	130
12	15	4	4008	5	4008	21	0	16
13	15	5	4468,5	4	4468,5	10	0	6
14	15	6	8464	5	8464	40	0	35
15	15	10	10296	9	10296	20	0	11
16	20	2	20893,5	13	20573,3	4634	1,555	4621
17	20	3	39374,75	12	38810,75	2469	1,453	2457
18	20	10	21533,5	15	21533,5	45	0	30
19	20	15	19443,5	449	19443,5	724	0	275
20	30	5	57934	18	57934	231	0	213
21	30	10	52840	1331	52840	5460	0	4129
22	30	20	128656	26	128656	27	0	1
23	50	10	181909	86	181909	443	0	357
24	50	20	430112	269	430112	453	0	184
25	50	30	756628	708	756628	833	0	125

10.166). The input data were randomly generated. More than 100 instances of the problem were used. The algorithms stops when all local optimal solutions are found or the time allocated for solving the problem ends. Results of comparison of algorithms are presented in Table 1, where F_{A2}, F_{BBA} and t_{A2}, t_{BBA} are the objective function values and the average running time (in seconds) of the heuristic algorithm and the branch and bound algorithm respectively.

The average relative error of algorithm A2 is equal to 3%. The algorithm A2 finds the solutions faster than BBA as follows from Table 1. For example, for instance 11 the relative error of A2 is 0,172% and the running time of A2 is less that of BBA by a factor 20.

Also, a computational experiment on comparison of the solutions obtained by the branch and bound algorithm and IBM ILOG CPLEX 12.2 package using the mixed integer linear programming model is made. Three series of the test problems were randomly generated with a uniform distribution each of which includes 5 problems of the same dimension. For each series, we compared the running times of the branch and bound algorithm and CPLEX package. For the dimensions $|I| = 20$, $|J| = 15$ and $|I| = 50$, $|J| = 20$, we could not obtain the solution within time 1000 s using CPLEX; the average running time for the problems of such dimensions with the proposed branch and bound algorithm is 723 s and 898 s respectively. Note that for the branch and bound algorithm and CPLEX package, the average running time on problems with dimensions $|I| = 5$, $|J| = 3$ is 0,968 s and 0,96 s respectively.

4   Conclusion

In this paper, the NP–hard problem of location of connected rectangular facilities on lines in the presence of rectangular forbidden gaps is considered. A generalization of the one–line variant of the model to several lines is given. A branch and bound algorithm employing lower bounds introduced in [26] is proposed. Results of numerical evaluation of this algorithm on comparison with a heuristic proposed in the paper [27] are reported.

References

1. Bischoff, M., Klamroth, K.: An efficient solution method for Weber problems with barriers based on genetic algorithms. Eur. J. Oper. Res. **177**, 22–41 (2007)
2. Butt, S.E., Cavalier, T.M.: Facility location in the presence of congested regions with the rectilinear distance metric. Socio-Econom. Plan. Sci. **31**, 103–113 (1997)
3. Cabot, A.V., Francis, R.L., Stary, M.A.: A network flow solution to a rectilinear distance facility location problem. AIIE Trans. **II**(2), 132–141 (1970)
4. Erzin, A.I., Cho, J.D.: Concurrent placement and routing in the design of integrated circuits. Autom. Remote Control **64**(12), 1988–1999 (2003)
5. Foulds, L.R., Hamacher, H.W., Wilson, J.M.: Integer programming approaches to facilities layout models with forbidden areas. Ann. Oper. Res. **81**, 405–417 (1998)
6. Garey, M.R., Johnson, D.S.: Computers and Intractability: A Guide to the Theory of NP-Completeness. Freeman, San Francisco (1979). Mir, Moscow (1982)

7. Kacprzyk, J., Stanczak, W.: Discrete approximation of the Weber problem with the Euclidean distance (in Polish). Applicationes Mathematicae **XVIII**, 257–270 (1984)
8. Kafer, B., Nickel, S.: Error bounds for the approximative of restricted planar location problems. Eur. J. Oper. Res. **135**, 67–85 (2001)
9. Katz, N., Cooper, L.: Facility location in the presence of forbidden regions, I: formulation and the case of Euclidean distance with one forbidden circle. Eur. J. Oper. Res. **6**, 166–173 (1981)
10. Klamroth, K.: Single-Facility Location Problems with Barriers. Springer Series in Operations Research. Springer, New York (2002). https://doi.org/10.1007/b98843
11. Kochetov, Yu.A., Panin, A.A., Plyasunov, A.V.: Comparison of metaheuristics for the bilevel facility location and mill pricing problem. Diskret. Anal. Issled. Oper. **22**(3), 36–54 (2015)
12. Kuhn, H.W.: A note on Fermat's problem. Math. Program. **4**, 98–107 (1973)
13. Nickel, S., Puerto, J.: Location Theory: A Unified Approach. Springer, Heidelberg (2005). https://doi.org/10.1007/3-540-27640-8
14. Legkih, S.A., Nagornay, Z.E., Zabudsky, G.G.: Automation of design of plans of production sites and shops of the sewing enterprises (in Russian). Nat. Tech. Sci. **4**, 261–266 (2005)
15. Love, R.F., Wong, J.Y.: On solving a one-dimentional space allocation problem with integer programming. INFORR **14**(2), 139–143 (1976)
16. McGarvey, R.G., Cavalier, T.M.: Constrained location of competitive facilities in the plane. Comput. Oper. Res. **32**, 539–378 (2005)
17. Panyukov, A.V.: The problem of locating rectangular plants with minimal cost for the connecting network. Diskret. Anal. Issled. Oper. Ser. 2 **8**(1), 70–87 (2001)
18. Panyukov, A.V., Shangin, R.E.: Algorithm for the discrete Weber's problem with an accuracy estimate. Autom. Remote Control **77**(7), 1208–1215 (2016)
19. Picard, J.C., Ratliff, D.H.: A cut approach to the rectilinear distance facility location problem. Oper. Res. **26**(3), 422–433 (1978)
20. Rudnev, A.S.: Probabilistic tabu search algorithm for the packing circles and rectangles into the strip. Diskret. Anal. Issled. Oper. **16**(4), 61–86 (2009)
21. Simmons, D.M.: One-dimensional space allocation: an ordering algorithm. Oper. Res. **17**(5), 812–826 (1969)
22. Chen, W.-K.: The VLSI Handbook. CRC Press, Boca Raton (2000)
23. Zabudsky, G.G.: On the problem of the linear ordering of vertices of parallel-sequential graphs. Diskret. Anal. Issled. Oper. **7**(1), 61–64 (2000)
24. Zabudskii, G.G., Amzin, I.V.: Algorithms of compact location for technological equipment on parallel lines (in Russian). Sib. Zh. Ind. Mat. **16**(3), 86–94 (2013)
25. Zabudsky, G.G., Legkih, S.A.: Mathematical model of optimization of flexible modules of processing equipment (in Russian). Appl. Math. Inf. Technol. 20–28 (2005)
26. Zabudsky, G., Veremchuk, N.: About local optimum of the Weber problem on line with forbidden gaps. In: Proceedings of the DOOR 2016, Vladivostok, Russia, 19–23 September 2016, vol. 1623, pp. 115–124. CEUR-WS (2016). CEUR-WS.org. http://ceur-ws.org/Vol-1623/paperco17.pdf
27. Zabudskii, G.G., Veremchuk, N.S.: An algorithm for finding an approximate solution to the Weber problem on a line with forbidden gaps. J. Appl. Ind. Math. **10**(1), 136–144 (2016)

28. Zabudsky, G.G., Veremchuk, N.S.: Solving Weber problem on plane with minimax criterion and forbidden gaps (in Russian). IIGU Ser. Matematika **9**, 10–25 (2014)
29. Zabudsky, G., Veremchuk, N.: Weber problem for rectangles on lines with forbidden gaps. In: IEEE Conference 2016 Dynamics of Systems, Mechanisms and Machines, Omsk, 15–17 November 2016 (2016)

Scheduling and Routing Problems

Inapproximability Lower Bounds for Open Shop Problems with Exact Delays

Alexander Ageev[(⊠)]

Sobolev Institute of Mathematics, pr. Koptyuga 4, Novosibirsk, Russia
ageev@math.nsc.ru

Abstract. We study the two-machine Open Shop problem with exact delays. When all delays are equal to zero this problem converts to the no-wait two-machine Open Shop problem, which is known to be NP-hard. We prove that even the proportionate case of Open Shop problem with exact delays does not admit approximations with ratio $1.5 - \varepsilon$ unless $\mathrm{P} = \mathrm{NP}$. We also consider the very special case when the delays take at most two different values and prove that the existence of a $(1.25 - \varepsilon)$-approximation algorithm for it implies $\mathrm{P} = \mathrm{NP}$.

Keywords: Open Shop · Exact delays · Approximation algorithm
Inapproximability lower bound

1 Introduction

We study the two-machine Open Shop problem with exact delays. An instance of the problem consists of n triples (a_j, l_j, b_j) of nonnegative integers where j is a job in the set of jobs $J = \{1, \ldots, n\}$. Each job j must be processed on each machine in a free order, a_j and b_j are the lengths of operations on machines 1 and 2, respectively. The second operation of job j must start exactly l_j time units after the first operation has been completed. The goal is to minimize makespan. In the standard three-field notation scheme the problem is written as $O2 \mid \text{exact } l_j \mid C_{\max}$. We also investigate the special case when the delays take at most two distinct values which is written as $O2 \mid \text{exact } l_j \in \{L_1, L_2\} \mid C_{\max}$. Scheduling problems with exact delays have evident applications in chemistry manufacturing. In particular, they arise often where there may be an exact technological delay between the completion time of some operation and the starting time of the next operation. The problems with exact delays also arise in command-and-control applications [11,17]. Condotta [6] describes an application related to booking appointments of chemotherapy treatments.

Related Work. When all delays are equal to zero the two-machine Open Shop problem with exact delays is nothing but the no-wait two-machine Open Shop problem written as $O2 \mid \text{no-wait} \mid C_{\max}$. NP-hardness of $O2 \mid \text{no-wait} \mid C_{\max}$ was proved by Giaro [10]. Sidney and Sriskandarajah [17] proposed a heuristic algorithm that solves $O2 \mid \text{no-wait} \mid C_{\max}$ with a tight worst-case ratio bound of $3/2$.

© Springer International Publishing AG, part of Springer Nature 2018
A. Eremeev et al. (Eds.): OPTA 2018, CCIS 871, pp. 45–55, 2018.
https://doi.org/10.1007/978-3-319-93800-4_4

Approximability of the single machine coupled-task and the two-machine Flow Shop problems with exact delays have been investigated in a number of papers ([1–4,7,14–16], see also the survey [5]). In particular, Ageev and Kononov [2], Ageev and Ivanov [3] and Ageev [4] established inapproximability lower bounds for various special cases of these problems (see Table 1).

Our Results. In this paper we present inapproximability lower bounds for two special cases of the two-machine Open Shop problem with exact delays. We prove that even the proportionate case of Open Shop problem with exact delays (where $a_j = b_j$ for all $j \in J$) does not admit approximations with ratio better than $1.5 - \varepsilon$ unless $P = NP$. We also consider the very special case when the delays take at most two different values L_1 and L_2 and prove that the existence of a $(1.25 - \varepsilon)$-approximation algorithm for it implies $P = NP$. Note that the algorithm developed in [2] for solving the two-machine Flow Shop problem with exact delays provides a 3-approximation for our problems. This follows from the proof of Theorem 3 in [2]. Our inapproximability lower bounds are similar to those for the two-machine Flow Shop problem with exact delays established in [2,4]. This seems a bit surprising as the two-machine Open Shop problem with exact delays is NP-hard even in the case of equal delays.

A summary of the approximability results for scheduling problems with exact delays is shown in Table 1.

Table 1. A summary of the approximability results for scheduling problems with exact delays.

Problem	Appr. factor	Inappr. bound
$1 \mid$ exact $l_j \mid C_{\max}$	3.5 [2]	$2 - \varepsilon$ [2]
$1 \mid$ exact $l_j, a_j \leq b_j \mid C_{\max}$	3 [2]	$2 - \varepsilon$ [2]
$1 \mid$ exact $l_j, a_j = b_j \mid C_{\max}$	2.5 [2]	$2 - \varepsilon$ [2]
$1 \mid$ exact $l_j, a_j = b_j = 1 \mid C_{\max}$	1.75 [1]	Strongly NP-hard [18,19]
$1 \mid$ exact $l_j = L \mid C_{\max}$	3 [3]	$1.25 - \varepsilon$ [3]
$1 \mid$ exact $l_j = L, a_j \leq b_j \mid C_{\max}$	2 [3]	$1.25 - \varepsilon$ [3]
$1 \mid$ exact $l_j = L, a_j = b_j \mid C_{\max}$	1.5 [3]	$1.25 - \varepsilon$ [3]
$F2 \mid$ exact $l_j \mid C_{\max}$	3 [2][15]	$1.5 - \varepsilon$ [2]
$F2 \mid$ exact $l_j, a_j \leq b_j \mid C_{\max}$	2 [2][15]	$1.5 - \varepsilon$ [2]
$F2 \mid$ exact $l_j, a_j = b_j = 1 \mid C_{\max}$	1.5 [1]	Strongly NP-hard [18,19]
$F2 \mid$ exact $l_j \in \{L_1, L_2\} \mid C_{\max}$	2 [4]	$1.25 - \varepsilon$ [4]
$O2 \mid$ exact $l_j, a_j = b_j \mid C_{\max}$	3 [2]	$1.5 - \varepsilon$ [this paper]
$O2 \mid$ exact $l_j \in \{L_1, L_2\} \mid C_{\max}$	3 [2]	$1.25 - \varepsilon$ [this paper]

2   Preliminaries

The proofs of both results of the paper are based on reductions from the NP-complete PARTITION problem [9] which we formulate here.

PARTITION

Instance: Nonnegative integers $w_1, \ldots, w_m$ such that $\sum_{k=1}^{m} w_k = 2S$.

Question: Does there exist a subset $X \subseteq \{1, \ldots, m\}$ such that $\sum_{k \in X} w_k = S$?

We assume that there are no missing operations in the sense that a zero processing time implies that the job has to visit the machine for an infinitesimal amount of time $\delta > 0$.

In the proofs we use the fact that the whole construction presenting a feasible schedule can be moved along the time line in both directions. So the length of the schedule is the length of the time interval between the starting time of the first operation (which is not necessarily equal to zero) and the end time of the last one.

Remind that in Open Shop problems operations of each job may be processed in arbitrary order and so we cannot speak of the first and second operations of a job. Instead we speak of operations associated with the corresponding machines.

For the length of a schedule σ we use the standard notation $C_{\max}(\sigma)$; $C_{\max}^*$ will stand for the length of a shortest schedule.

In the proportionate case each job has operations of equal lengths, i.e., $a_j = b_j$ for all $j \in J$. We will refer to it as $O2 \mid$ exact $l_j, a_j = b_j \mid C_{\max}$.

3 Inapproximability Lower Bound for $O2 \mid$ exact $l_j, a_j = b_j \mid C_{\max}$

In this section we establish an inapproximability lower bound for the proportionate case $O2 \mid$ exact $l_j, a_j = b_j \mid C_{\max}$.

To this end we consider the following reduction from PARTITION problem. Let $\mathcal{I}$ be an instance of PARTITION. Construct an instance $\mathcal{I}'$ of $O2 \mid$ exact $l_j \mid C_{\max}$. Set $J = \{1, \ldots, m+2\}$ and

$$a_k = b_k = w_k, \ l_k = 2R + S - a_k \text{ for } k = 1, \ldots m,$$
$$a_{m+1} = b_{m+1} = R, \ l_{m+1} = 0,$$
$$a_{m+2} = b_{m+2} = R, \ l_{m+2} = 0$$

where $R \geq 3S$. We will refer to the jobs in $\{1, \ldots, m\}$ as *small* and to the remaining two jobs as *big*.

Lemma 1. (i) *If $\sum_{k \in X} w_k = S$ for some subset $X \subseteq \{1, \ldots m\}$, then there exists a feasible schedule σ such that $C_{\max}(\sigma) \leq 2R + 2S$.*
(ii) *If there exists a feasible schedule σ such that $C_{\max}(\sigma) \leq 2R + 2S$, then $\sum_{k \in X} w_k = S$ for some subset $X \subseteq \{1, \ldots m\}$.*
(iii) *If $C_{\max}(\sigma) > 2R + 2S$ for some feasible schedule σ, then $C_{\max}(\sigma) \geq 3R$.*

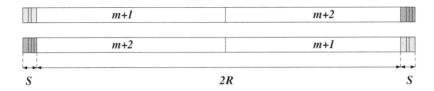

Fig. 1. The shortest schedule when the instance $\mathcal{I}$ has answer "Yes".

Proof. (i) Let $X \subseteq \{1, \ldots, m\}$ such that $\sum_{k \in X} w_k = S$. Then $\sum_{k \in Y} w_k = S$ where $Y = \{1, \ldots, m\} \setminus X$.

To construct the required schedule arrange the big jobs as shown in Fig. 1. To schedule the small jobs we do the following. W.l.o.g. we may assume that the jobs are ordered in such a way that $X = \{1, \ldots, q\}$ and $Y = \{q + 1, \ldots m\}$. Execute all jobs in X first on machine 1 in this order without idles just before the operation of job $m + 1$ on machine 1. Then the next operation of job $i \leq q$ will start at time

$$\sum_{s=1}^{i} w_s + 2R + S - w_i = \sum_{s=1}^{i-1} w_s + 2R + S.$$

It follows that the operations of jobs in X on machine 2 do not overlap and are executed within the interval of length S. Note that the operations of jobs X on machine 1 are also executed within the interval of length S. Schedule the jobs in Y exactly in the same way but process them first on machine 2 and then on machine 1 (see Fig. 1). It is easy to see that the length of the constructed schedule is $2R + 2S$, as required.

(ii) Let σ be a feasible schedule with $C_{\max}(\sigma) \leq 2R + 2S$. Observe that the big jobs in σ are arranged in one of the two ways shown in Fig. 2, since otherwise the length of the schedule is at least $3R$. By the symmetry of the big jobs, we may assume that the case (1) of Fig. 2 holds. Next, we observe that in σ, both operations of the two big jobs are executed within the lag time interval of any small job, since otherwise $C_{\max}(\sigma) \geq 2R + 2R = 4R$. Let X be the subset of small jobs j such that in σ the job j is first processed on machine 1 and then on machine 2. Then $Y = \{1, \ldots, m\} \setminus X$ is the set of small jobs j in σ such that j is first processed on machine 2 and then on machine 1. Assume that $|X| > S$ and let $j^* \in X$ be the first job processed on machine 1 at time 0. Then the operation of j^* on machine 2 must start at time $w_{j^*} + 2R + S - w_{j^*} = 2R + S$. However, as $|X| > S$ the processing of job $m + 1$ on machine 2 finishes at time more than $2R + S$ and we get that the machine 2 operation of job $j*$ overlaps with the machine 2 operation of job $m + 1$, which implies that σ is an infeasible schedule (see Fig. 3). Thus we have that $|X| = |Y| = S$, as required.

(iii) Let σ be a feasible schedule satisfying $C_{\max}(\sigma) > 2R + 2S$. We may assume that the big jobs are arranged as in Fig. 2, since otherwise evidently $C_{\max}(\sigma) \geq 3R$. From (ii) it follows that if for some small job j its lag time interval

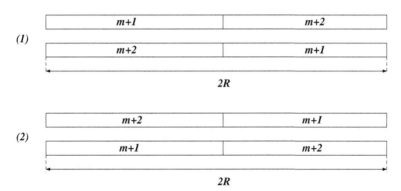

Fig. 2. The two shortest arrangements of the big jobs.

Fig. 3. The schedule σ is infeasible if $|X| \neq |Y|$.

does not contain operations of the big jobs, then $C_{\max}(\sigma) > 2R + 2R + S > 4R$. Finally we have that $C_{\max}(\sigma) \geq 3R$, as required.   □

Set $R = kS$. Then $2R + 2S = 2kS + 2S$. The fraction

$$\frac{3kS}{2kS + 2S} = \frac{3k}{2k + 2}$$

tends to 1.5 as k tends to infinity. Thus Lemma 1 implies

Theorem 1. *The existence of a $(1.5 - \varepsilon)$-approximation algorithm for solving $O2 \,|prop, \; exact \; l_j| \; C_{\max}$ implies $P = NP$.*   □

4   Inapproximability Lower Bound for $O2 \,|\text{exact } l_j \in \{0, L\}| \; C_{\max}$

In this section we establish an inapproximability lower bound for the case when the delay of each job is either 0, or $L > 0$.

To this end consider the following reduction from PARTITION problem.

Consider an instance $\mathcal{I}$ of PARTITION and construct the following instance $\mathcal{I}'$ of $O2 \,|\text{exact } l_j \in \{0, L\}| \; C_{\max}$.

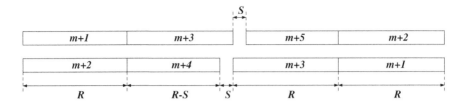

Fig. 4. The shortest partial schedule of the big jobs.

Set $J = \{1, \ldots, m+5\}$ and

$$a_k = b_k = w_k, \ l_k = 2R \text{ for } \quad k = 1, \ldots m,$$
$$a_{m+1} = b_{m+1} = R, \ l_{m+1} = 2R,$$
$$a_{m+2} = R, \ b_{m+2} = R, \ l_{m+2} = 2R,$$
$$a_{m+3} = R, \ b_{m+3} = R, \ l_{m+3} = 0,$$
$$a_{m+4} = 0, \ b_{m+4} = R - S, \ l_{m+4} = 0,$$
$$a_{m+5} = R - S, \ b_{m+5} = 0, \ l_{m+5} = 0,$$

where $R \geq 6S$. We will refer to the jobs in $\{1, \ldots, m\}$ as *small* and to the remaining five jobs as *big*.

Lemma 2. (i) *If* $\sum_{k \in X} w_k = S$ *for some subset* $X \subseteq \{1, \ldots m\}$, *then there exists a feasible schedule* σ *such that* $C_{\max}(\sigma) \leq 4R + 4S$.
(ii) *If there exists a feasible schedule* σ *such that* $C_{\max}(\sigma) \leq 4R + 4S$, *then* $\sum_{k \in X} w_k = S$ *for some subset* $X \subseteq \{1, \ldots m\}$.
(iii) *If* $C_{\max}(\sigma) > 4R + 4S$ *for some feasible schedule* σ, *then* $C_{\max}(\sigma) \geq 5R - S$.

Proof. (i) To construct the required schedule arrange the big jobs in the order shown in Fig. 5. This construction has two idle intervals: A on machine 1 and B on machine 2. The interval A is between the end of the machine 1 operation of job $m + 3$ and the beginning of the machine 1 operation of job $m + 5$. The interval B is between the end of the machine 2 operation of job $m + 4$ and the beginning of the machine 2 operation of job $m+3$. Both intervals have length S.

For scheduling the small jobs we use the following rule. Schedule the small jobs in X in such a way that their first operations are executed within the time interval A in non-increasing order of the lengths. Correspondingly, w.l.o.g. we may assume that $X = \{1, 2, \ldots, q\}$ and $w_1 \leq w_2 \leq \ldots \leq w_q$.

Denote by A' the time interval between the end of the machine 2 operation of job $m + 1$ and the end of the machine 2 operation of job q. It is easy to understand (see Fig. 6) that all the machine 2 operations of jobs $\{1, \ldots, q\}$ fall within A' and the length of A' is equal to

$$\sum_{i=1}^{q} w_i + w_1 + (w_2 - w_1) + (w_3 - w_2) + \ldots + (w_q - w_{q-1}),$$

Fig. 5. Scheduling the small jobs.

which does not exceed $2S$. Now we observe that the construction is symmetric and schedule the jobs in Y quite similarly. Finally, we arrive at the schedule shown in Fig. 6. From the above argument its length does not exceed $4R + 4S$, as required.

Fig. 6. The shortest schedule when the instance $\mathcal{I}$ of Partition has answer "yes".

(ii) Let σ be a feasible schedule with $C_{\max}(\sigma) \leq 4R + 4S$. Observe first that the jobs $m + 1$ and $m + 2$ are identical and we may assume that job $m + 1$ is first processed on machine 1. Then job $m + 2$ is first processed on machine 2, since otherwise clearly $C_{\max}(\sigma) \geq 5R$. Now we observe that job $m+3$ is executed within the time lag interval either of job $m+1$, or of job $m+2$. Due to symmetry we may assume that job $m + 3$ is executed within the time interval of job $m + 1$. Thus we have the configuration shown in Fig. 7. We call this configuration *initial*. Denote by t_0, t_1, t_2, t_3, t_4 the junction times of the operations in the initial configuration (see Fig. 7).

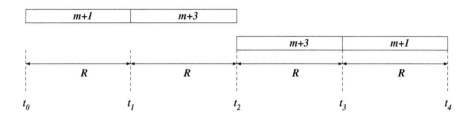

Fig. 7. The initial configuration.

Now our aim is to exclude impossible layouts of job $m + 5$ with respect to the initial configuration.

Observe first that σ has the following property (Q): any small job in σ either completes executing not earlier than $t_1 + S$, or starts executing not later than $t_3 - S$. If property (Q) does not hold then evidently $C_{\max}(\sigma) \geq 5R - S$.

Now observe that job $m + 5$ starts executing later than time t_0 and before time t_4, since otherwise $C_{\max}(\sigma) \geq 4R + R - S = 5R - S$.

Job $m+5$ cannot complete executing at time t_4, since otherwise any possible arrangement of job $m + 2$ gives $C_{\max}(\sigma) \geq 4R + R - S = 5R - S$ (see Fig. 8).

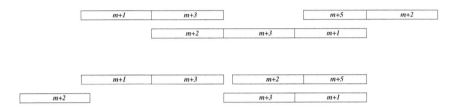

Fig. 8. Job $m + 5$ cannon complete executing at time t_4.

Job $m + 5$ cannot start executing at time t_3, since otherwise any possible arrangement of job $m+2$ also gives $C_{\max}(\sigma) \geq 4R+R-S = 5R-S$ (see Fig. 9).

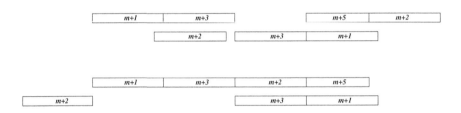

Fig. 9. Job $m + 5$ cannot start executing at time t_3.

Now consider the case when job $m+5$ starts executing at time t_2. Job $m+4$ cannot start executing at time t_0, since otherwise $C_{\max}(\sigma) \geq 4R + R - S = 5R - S$ (see Fig. 10). Job $m + 4$ cannot complete executing at time t_1, since

Fig. 10. The case when job $m + 5$ starts executing at time t_2 and job $m + 4$ starts executing at time t_0.

otherwise $C_{\max}(\sigma) \geq 5R$ (see Fig. 11). Job $m + 4$ cannot start executing at time t_1 since otherwise property (Q) does not hold (see Fig. 12 with the two opposite arrangements of job $m + 2$)). Job $m + 4$ cannot complete executing at time t_2, since otherwise property (Q) does not hold as well (Fig. 13 shows three typical arrangements of job $m + 2$).

Fig. 11. The case when job $m + 5$ starts executing at time t_2 and job $m + 4$ completes executing at time t_1.

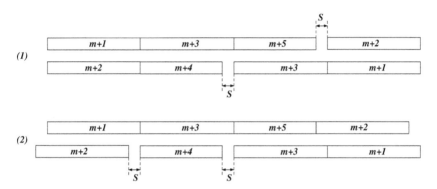

Fig. 12. The two opposite arrangements of job $m + 2$ when job $m + 4$ starts executing at time t_1.

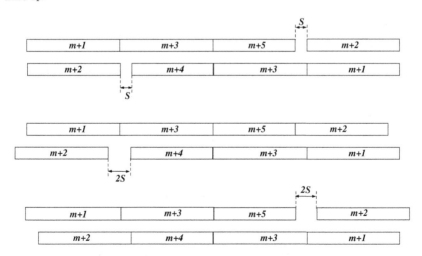

Fig. 13. Three typical arrangements of job $m + 2$ when job $m + 4$ completes executing at time t_2.

Finally we consider the case when job $m + 5$ completes executing at time t_3. Exactly the same argument as in the above case shows that job $m + 4$ starts executing not earlier than time t_1. Otherwise we have $C_{\max}(\sigma) \geq 5R - S$. Job $m + 4$ cannot complete executing at time t_2, since otherwise property (Q) does not hold (see Fig. 14 with the two opposite arrangements of job $m + 2$).

Thus we arrive at the (partial) schedule of big jobs shown in Fig. 4. If the instance $\mathcal{I}$ of PARTITION has answer "No", then at least one of the small jobs is scheduled outside the whole construction of the big jobs, which gives $C_{\max}(\sigma) \geq 6R$. So the only possible schedule that gives $C_{\max}(\sigma) \geq 4R + 4S$ is depicted in Fig. 6 and the instance $\mathcal{I}$ has answer "Yes", as required.

(iii) Follows from the cases that we excluded when proved (ii). If $C_{\max}(\sigma) > 4R + 4S$ for some feasible schedule σ, then $C_{\max}(\sigma)$ is at least $5R - S$. □

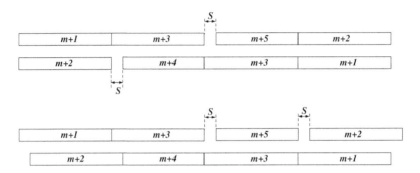

Fig. 14. The two opposite arrangements of job $m + 2$ when job $m + 4$ completes executing at time t_2.

Set $R = kS$. Then $4R + 4S = 4kS + 4S$. The fraction

$$\frac{5kS - S}{4kS + 4S} = \frac{5k - 1}{4k + 4}$$

tends to 1.25 as k tends to infinity. Thus Lemma 2 gives

Theorem 2. *The existence of a* $(1.25 - \varepsilon)$*-approximation algorithm for solving* $O2 \,|exact\; l_j \in \{0, L\}|\; C_{\max}$ *implies* $P = NP$. □

5   Conclusion

In this paper we establish inapproximability lower bounds for two special cases of the two-machine Open Shop problem with exact delays. We omit here the algorithmic issues concerning these problems. Note only that the algorithm developed in [2] for the two-machine Flow Shop with exact delays can be applied to our problems and provides a 3-approximation. Constructing better approximations is a subject of further work.

Acknowledgments. The author would like to thank the anonymous reviewers for their valuable comments and suggestions to improve the quality of the paper.

This research was supported by the Russian Science Foundation, grant 17-11-01021.

References

1. Ageev, A.A., Baburin, A.E.: Approximation algorithms for UET scheduling problems with exact delays. Oper. Res. Lett. **35**(4), 533–540 (2007)
2. Ageev, A.A., Kononov, A.V.: Approximation algorithms for scheduling problems with exact delays. In: Erlebach, T., Kaklamanis, C. (eds.) WAOA 2006. LNCS, vol. 4368, pp. 1–14. Springer, Heidelberg (2007). https://doi.org/10.1007/11970125_1
3. Ageev, A., Ivanov, M.: Approximating coupled-task scheduling problems with equal exact delays. In: Kochetov, Y., Khachay, M., Beresnev, V., Nurminski, E., Pardalos, P. (eds.) DOOR 2016. LNCS, vol. 9869, pp. 259–271. Springer, Cham (2016). https://doi.org/10.1007/978-3-319-44914-2_21
4. Ageev, A.A.: Approximating the 2-machine flow shop problem with exact delays taking two values. J. Global Optim. (To appear). A preliminary version can be found at https://arxiv.org/abs/1711.00081
5. Blazewicz, J., Pawlak, G., Tanas, M., Wojciechowicz, W.: New algorithms for coupled tasks scheduling – a survey. RAIRO Oper. Res. Recherche Operationnelle **46**(4), 335–353 (2012)
6. Condotta, A.: Scheduling with due dates and time lags: new theoretical results and applications. Ph.D. thesis, The University of Leeds, School of Computing, 156 pp. (2011)
7. Condotta, A., Shakhlevich, N.V.: Scheduling coupled-operation jobs with exact time-lags. Discrete Appl. Math. **160**(16–17), 2370–2388 (2012)
8. Farina, A., Neri, P.: Multitarget interleaved tracking for phased array radar. IEEE Proc. Part F Comm. Radar Signal Process. **127**, 312–318 (1980)
9. Garey, M.R., Johnson, D.S.: Computers and Intractability: A Guide to the Theory of NP-Completeness. Freeman, San Francisco (1979)
10. Giaro, K.: NP-hardness of compact scheduling in simplified open and flow shops. Eur. J. Oper. Res. **130**(1), 90–98 (2001)
11. Elshafei, M., Sherali, H.D., Smith, J.C.: Radar pulse interleaving for multi-target tracking. Naval Res. Logist. **51**(1), 79–94 (2004)
12. Izquierdo-Fuente, A., Casar-Corredera, J.R.: Optimal radar pulse scheduling using neural networks. In: IEEE International Conference on Neural Networks, vol. 7, pp. 4588–4591 (1994)
13. Graham, R.L., Lawler, E.L., Lenstra, J.K., Rinnooy Kan, A.H.G.: Optimization and approximation in deterministic sequencing and scheduling: a survey. Ann. Discrete Math. **5**, 287–326 (1979)
14. Hwang, F.J., Lin, B.M.T.: Coupled-task scheduling on a single machine subject to a fixed-job-sequence. Comput. Ind. Eng. **60**(4), 690–698 (2011)
15. Leung, J.Y.-T., Li, H., Zhao, H.: Scheduling two-machine flow shops with exact delays. Int. J. Found. Comput. Sci. **18**(2), 341–359 (2007)
16. Orman, A.J., Potts, C.N.: On the complexity of coupled-task scheduling. Discrete Appl. Math. **72**(1–2), 141–154 (1997)
17. Sherali, H.D., Smith, J.C.: Interleaving two-phased jobs on a single machine. Discrete Optim. **2**(4), 348–361 (2005)
18. Yu, W.: The two-machine shop problem with delays and the one-machine total tardiness problem. Ph.D. thesis, Technische Universiteit Eindhoven, 136 pp. (1996)
19. Yu, W., Hoogeveen, H., Lenstra, J.K.: Minimizing makespan in a two-machine flow shop with delays and unit-time operations is NP-hard. J. Sched. **7**(5), 333–348 (2004)

Exact Solution of One Production Scheduling Problem

Pavel Borisovsky[(✉)]

Sobolev Institute of Mathematics, Omsk, Russia
borisovski@mail.ru

Abstract. In this study, one variant of multi-product scheduling problem is considered. The problem asks to find the optimal selection of a set of tasks to produce a given number of products in required amounts, to allocate the task on units, and to find the order of execution of tasks for each unit. The production rates for each task, the task-unit suitability matrix, and the sequence dependent changeover times for task pairs are given.

For the one-unit problem, two combinatorial algorithms are proposed: a branch-and-bound algorithm and a parallel dynamic programming algorithm. The last one is implemented using the CUDA library for running on a Graphical Processing Unit (GPU). For the multiple-units problem, both approaches are combined in a branch-and-bound algorithm with bounds provided by the dynamic programming procedure.

The algorithms are compared with CPLEX solver applied to the considered problem formulated as a mixed integer linear program. Although, the main limitation of using the proposed algorithms is a requirement of large amount of memory, the experiments showed their superior performance over CPLEX in terms of running time for rather large sized instances. The advantage of parallelization and using the GPU is also demonstrated.

Keywords: Production scheduling · Branch-and-bound
Dynamic programming · GPU computing · CUDA

1   Introduction

A modern chemical production plant is organized as a flexible automated system that contains a number of multipurpose production units and produces a number of products including the final products and intermediate states. The control of the production process involves two main problems: the first one is to choose the most appropriate production plan (the set and the order of reactions to be performed) that can satisfy the market requirements on amount and assortment of the final product, and the second one is to build an optimal production schedule for the chosen plan. In this paper, we consider a simplified multi-product scheduling problem (MPSP) with several production units, which can operate

A. Eremeev et al. (Eds.): OPTA 2018, CCIS 871, pp. 56–67, 2018.
https://doi.org/10.1007/978-3-319-93800-4_5

in different modes. Each mode is referred to as a task and is characterized by the type of produced product and the production rate. The key element of the problem is the presence of sequence dependent changeover times necessary for switching the unit from one mode to another. In case of a real production plant one should consider complex relations between tasks and products, e.g. a partial order on the set of tasks, unit blockage intervals, production recipes where some products are used for production of other products, etc. (see [8,10]). Although the problem considered in this paper does not involve such relations, it can be used as a simplified model to provide a preliminary solution that can be further detailed by other methods. For example, in the model of [10] there are ten main production units (extruders) and many auxiliary units (storage, feed supply, packing). The subproblem for the main units only can be regarded as the MPSP considered here, and its solution helps to build the complete schedule (see [5] for the description of this approach).

The problem is a generalization of the well known parallel machine scheduling problem with sequence dependent setup time (see, e.g. [1]) where all of the given jobs must be processed. The parallel machine scheduling problem with sequence dependent setup time appears when in MPSP we choose exactly one task for each product or if we consider a special case of MPSP, where each product is produced only by one task [6]. Unfortunately, this cannot give a solution approach to MPSP, because the selection of tasks and scheduling are closely related and so for the most effective control they can not be considered separately. Earlier, a genetic algorithm for MPSP was developed in [3] and then used for solving the real life production problem [5].

For the MPSP, several exact algorithms are developed in this paper. For the one-unit problem, two combinatorial algorithms are proposed: a branch-and-bound algorithm and a parallel dynamic programming algorithm. The last one is implemented using the CUDA library for running on a Graphical Processing Unit (GPU). For the multiple-units problem, both approaches are combined in a branch-and-bound algorithm with bounds provided by the dynamic programming procedure. The computational results are given and discussed. The preliminary results of this research for the one-unit problem were presented in [4].

The paper is structured as follows. In Sect. 2, we provide the problem description and its formulation as a mixed integer program (MIP). Section 3 presents the dynamic programming and branch-and-bound algorithms for solving the problem with one production unit. In Sect. 4 a multi-unit problem is considered and the branch-and-bound algorithm is developed. Section 5 presents the experimental results where the algorithms are evaluated and compared with a general purpose MIP solver CPLEX.

2   Problem Formulation

The main objects in mathematical modeling of the production process are products, tasks and units. In this paper, we suppose that all raw materials are available without limit at any time, so we can consider only final products. For each product, its demand is given. Tasks represent the production processes (chemical reactions) and are characterized by the type of produced product and the production rate. We suppose that each task produces only one product. It is possible that one product can be produced by different tasks that differ in technical settings (heat, used catalysts, etc.) and hence may have different production rates and setup times. A task must be performed on a unit suitable for it; here we suppose that there is only one suitable unit for a task. When the unit is switched from one task to another it requires cleaning and setting-up, which is modeled as sequence-dependent changeover times. The problem asks to determine the set of tasks to perform and find the sequence of tasks satisfying all the demands and minimizing the completion time (Makespan).

Introduce the following notation:

$S = \{1, 2, ..., |S|\}$ is a set of products;
$I = \{1, 2, ..., |I|\}$ is a set of tasks;
$U = \{1, 2, ..., |U|\}$ is a set of units;
$D_s > 0$ is the demand for product s, for all $s \in S$;
$s_i \in S$ is the product produced by task i, for all $i \in I$;
$u_i \in U$ is the unit suitable for task i, for all $i \in I$;
$r_i > 0$ is the production rate of task i, for all $i \in I$, i.e. the amount of product s_i produced in one time unit (e.g., one hour);
$M_i = D_s/r_i$ is the duration of task i if it is performed, $i \in I$, $s = s_i$;
$a_{ij} > 0$ is the changeover time necessary to switch the unit from task i to task j, where $u_i = u_j$.

In this paper, we also suppose that when a task is chosen it must produce the whole amount of the corresponding product, i.e. we do not allow to split the production of any product. In case of the one-unit problem when the changeover times satisfy the triangle inequality, this condition holds automatically. Otherwise, this condition is necessary to prevent fake tasks in the solution, i.e. the tasks that produce nothing and are placed only to reduce changeover times. Alternative way to treat such cases would be introducing the minimal duration of each task, but this is not considered in this paper. For the multi-unit problem, if a production of some product can be performed on different units, this involves solving linear program for balancing the load times. In this paper, only the pure combinatorial algorithms are considered, so this case is out of the scope.

MIP Model. For the convenience denote by I_u a set of tasks that can be performed on unit $u \in U$ and by I_s a set of tasks that produce product $s \in S$. The set of event points for each unit u is defined as $K_u = \{1, 2, ..., |I_u|\}$. Introduce the variables:

$x_{ik} \in \{0, 1\}$ such that $x_{ik} = 1$ iff task i is assigned to unit $u = u_i$ at event point $k \in K_u$.
$\delta_i \geq 0$ is the duration of task i;
α_{uk} is the duration of a changeover task on unit u between event points k and $(k+1)$.

The MIP model is as follows.

$$\min C_{\max}, \tag{1}$$

subject to

$$\sum_{i \in I_u} \delta_i + \sum_{k \in K_u} \alpha_{uk} \leq C_{\max}, \quad u \in U, \tag{2}$$

$$\sum_{i \in I_s} \sum_{k \in K_{u_i}} x_{ik} = 1, \quad s \in S, \tag{3}$$

$$\sum_{i \in I_u} x_{ik} \leq 1, \quad k \in K_u, \quad u \in U, \tag{4}$$

$$\sum_{i \in I_s} r_i \delta_i = D_s, \quad s \in S, \tag{5}$$

$$\sum_{k \in K_{u_i}} M_i x_{ik} \geq \delta_i, \quad i \in I, \tag{6}$$

$$\sum_{i \in I_u} x_{i,k-1} \geq \sum_{i \in I_u} x_{i,k}, \quad u \in U, \quad k \in K_u, \quad k > 1, \tag{7}$$

$$\alpha_{uk} \geq \sum_{j \in I_u} a_{ji} x_{j,k-1} - M(1 - x_{ik}), \quad u \in U, \quad i \in I_u, \quad k \in K_u, \quad k > 1, \tag{8}$$

$$\delta_i \geq 0, \quad i \in I, \tag{9}$$

$$\alpha_{uk} \geq 0, \quad u \in U, \quad k \in K_u, \tag{10}$$

$$x_{ik} \in \{0,1\}, \quad i \in I, \quad k \in K_{u_i}. \tag{11}$$

Objective function (1) minimizes total production and changeovers time expressed in (2). Conditions (3) mean that each product is produced by exactly one task placed in one position and (4) mean that each position is occupied by not more than one task. In (5), value $r_i \delta_i$ expresses the amount of product, produced by task i; in the solution it must equal the demand of the product. Conditions (6) guarantee that the duration of task i is zero if the task is not present in a schedule, and if it is included in a schedule the duration does not exceed M_i. Conditions (7) ensure continuous usage of event points, i.e. if some event point is occupied on some unit, then the previous event point is occupied as well (this property is useful for modeling the changeover times). Constraints (8) provide estimations of changeover times. If task i is placed in position k after task j (i.e. $x_{ik} = 1$ and $x_{j,k-1} = 1$) then constraint (8) turns to inequality $\alpha_{uk} \geq a_{ji}$. In case $x_{ik} = 0$ inequality (8) trivially holds for sufficiently large constant M, which can be defined as $M = \max_{i,j} a_{ij}$.

In the special case, where each product corresponds to only one task, and all the production rates equal 1, the problem becomes the well-known NP-hard problem of finding the minimum Hamiltonian path. Note that the described problem can be formulated as a special type of the Vehicle Routing Problem.

Suppose that U is a set of vehicles that must sell the set of goods S in amounts $D_s, s \in S$ in the set of cities I. Each city buys only one type of goods. The distribution process in a city requires certain time that depends on the city and the amount of goods. Unlike the standard Vehicle Routing Problem [9], here it is not required for the vehicles to visit all the cities or to return to the initial point.

3 Solving the One-Unit Problem

3.1 Dynamic Programming Algorithm

The dynamic programming (DP) algorithm is an adaptation of the approach by [7]. Consider the problem (1)–(11) with only one unit. For a set of products $P \subset S$ and operation i that does not produce any product from P let $f(i, P)$ be a makespan of an optimal schedule producing the set of products $\{s_i\} \cup P$ so that operation i is allocated at the first place.

The values $f(i, S)$ can be computed by the dynamic programming algorithm as follows.

- For $|P| = 1$, i.e. $P = \{j\}$ it holds

$$f(i, \{j\}) = M_i + a_{ij} + M_j. \tag{12}$$

- Suppose that all $f(i, P')$ are known for $|P'| < m$, then for all P such that $|P| = m$ they can be found according to Bellman equations:

$$f(i, P) = \min_{j \in P}\{M_i + a_{ij} + f(j, P \setminus \{j\})\}. \tag{13}$$

Below we give the formal description of the dynamic programming algorithm.

Dynamic Programming Algorithm

1 Compute $f(i, S)$:
  1.1 For all tasks i, j such that $s_i \neq s_j$ compute $f(i, \{j\})$ according to (12).
  1.2 For $t = 2$ to $|S|$ do
    1.2.1 Iterate over all $P \subset S$ and $i \in I$ such that $|P| = t$, $s_i \notin P$ and compute $f(i, P)$ according to (13).
2 Build the solution:
  2.1 Choose the task i such that $a_i^0 + f(i, S \setminus \{s_i\})$ is minimal. Let the first task in a sequence $i^{(1)}$ be the chosen task i and let $P = S \setminus \{s_i\}$.
  2.2 For $t := 2$ to $|S|$ do
    2.2.1 Let $j := i^{(t-1)}$. Choose the task i such that $a_{ji} + f(i, P \setminus \{s_i\})$ is minimal. Let the next task in a sequence $i^{(t)}$ be the chosen task i and $P := P \setminus \{s_i\}$.

A serious limitation to application of the DP algorithm is the amount of required memory. Each execution of step 1.2.1 requires to save $|I| \times \binom{|S|-1}{t}$ values. On the other hand, these calculation can be done in parallel, which can be easily implemented for running on a Graphical Processing Unit (GPU). To do so, one must define the so called kernel function (see the CUDA C Programming Guide http://docs.nvidia.com/cuda/cuda-c-programming-guide/index.html), which in our case computes expression (13) for some particular P and i. Then the CUDA framework runs the kernel for all P and i in parallel as much as possible. The iteration over t on Step 1.2 are performed sequentially. Note that it is widely known that GPU computing suits especially well for dynamic programming methods, see, e.g. [2]. The computational experiments, which will be given in the following, showed that the problems of an appropriate size are solved optimally quite fast on the GPU.

3.2   Branch-and-Bound Algorithm

The proposed branch-and-bound algorithm is based on the straightforward enumeration of permutations. Though for the complete search the computation time grows very fast with the problem size, a reasonable pruning strategy can reduce the search space drastically. In this paper, the bounds are constructed using the dynamic programming approach.

Branching Scheme. The branching rule combines branching by products and by tasks. Starting from the empty schedule we repeatedly set each product to be produced at first place. When some product is chosen, we assign a task producing this product. For each fixed product and task, we recursively solve the subproblem on the remaining products.

Minimal Production Time Bound. The first bound simply estimates the total production and changeover time for each product separately. For some product $s \in S$ it is given by:

$$L_M(s) = \min_{j \in I_s} \left\{ M_j + \min_{i \notin I_s} a_{ij} \right\}. \tag{14}$$

For subset $P \subseteq S$ bound $L_M(P)$ is computed as the sum over all products of P.

Dynamic Bound. The second bound is based on the dynamic programming approach of Held and Karp [7]. Suppose the process reached the level k, i.e. the first $k \geq 1$ products are fixed in the partial solution; the remaining products are denoted by subset P. Let the fixed part finishes with task i, and suppose that in the optimal solution of this branch the non-fixed part starts with task j. Define $L_D(i, P)$ as a lower bound on the optimal solution of the subproblem formed by subset P including changeover time a_{ij}.

During the solving process, all the values of $L_D(i, P)$ are stored in memory. The required memory size can be estimated as $O(m2^{n-1})$ which is the most serious limitation on the use of dynamic approach. Initially the values of $L_D(i, P)$ are set to zero and they are updated in the search process according to Bellman equality:

$$L_D(i, P) = \min_{s \in P, j \in I_s} \{a_{ij} + M_j + L_D(j, P \setminus \{s\})\}.$$

For the formal description of the algorithm, introduce task i_0 that will be allocated at the first place. This task does not produce any product and has zero execution and changeover times.

Branch-and-Bound (BB) Algorithm

Procedure $Level(k, i, P, f)$
1. For each state $s \in P$:
    1.1. For each operation $j \in I_s$:
        1.1.1. Let $f := f + a_{ij} + M_j$ be an objective value of the partial solution.
        1.1.2. $L_D(j, P \setminus \{s\}) := \max\{L_D(j, P \setminus \{s\}), L_M(j, P \setminus \{s\})\}$.
        1.1.3. If $f + L_D(j, P \setminus \{s\}) < record$ and the set $P \setminus \{s\} \neq \emptyset$
            then proceed to the next level: call $Level(k + 1, j, P \setminus \{s\}, f)$.
        1.1.4. If $P \setminus \{s\} = \emptyset$, then update $record := \min\{f, record\}$.
        1.1.5. Calculate the bound $L_{sj} = a_{ij} + M_j + L_D(j, P \setminus \{s\})$.
2. Calculate the minimal bound: $L = \min_{s \in P, j \in I_s} L_{sj}$.
3. Update the bound for the current branch: $L_D(i, P) := \max\{L, L_D(i, P)\}$.
End of procedure $Level$.

Start of the algorithm
1. Set $record = +\infty, i = i_0, P = S, f = 0$.
2. Call $Level(1, i, P, f)$.
3. Result is the best found solution.

4   Solving the Multi-unit Problem

In this section, a Branch-and-bound (BB) algorithm for the problem with several units is developed. Introduce some notation. We will say that product s can be produced on unit u if there is some task producing product s that can be executed on unit u. Denote by S_u the set of products that can be produced on unit $u \in U$ and by U_s the set of units that can produce product $s \in S$.

In the proposed algorithm, there is a preliminary procedure that for each unit u enumerates all subsets of S_u and for each subset $Q \subset S_u$ finds an optimal schedule for producing the products from Q on unit u. Let $f^*(u, Q)$ be the objective value of such an optimal schedule. All the possible values $f^*(u, Q)$ are found by the DP algorithm described in Sect. 3.1: the DP is applied for the set S_u and when all values $f(i, P)$ as defined in Sect. 3.1 are available, one can find

$$f^*(u, Q) = \min_{s \in Q, i \in I_s} f(i, Q \setminus \{s\}).$$

In the main part of the algorithm, the branching is performed by products and units. First, the products are sorted by increasing of the number of suitable units, i.e. we suppose that for the set of products $S = \{1, 2, ..., m\}$ it holds $|U_1| \leq |U_2| \leq ... \leq |U_m|$. The partial solutions are represented by assigning units to products: $(w_1, ..., w'_m)$, where $m' \leq m$ and w_s is the unit assigned to product s. For each unit u, such a solution defines the set of products Q_u produced on this unit, so the objective value for this unit is $f^*(u, Q_u)$, and the maximum value over all the units gives the objective of the partial solution.

Branch-and-Bound Algorithm

Procedure $Level(s)$;

1. For each unit $u \in U_s$

   1.1. Assign u for producing s. Let f be an objective value of the current partial solution.

   1.2. If $s = m$, then update $record := \min\{f, record\}$.

   1.3. If $s < m$ and $f < record$ then proceed to the next level: call $Level(s + 1)$.

End of procedure $Level$.

Start of the algorithm

1. Set $record = +\infty$.
2. Call $Level(1)$.
3. Result is the best found solution.

5   Implementation and the Computer Experiments

The algorithms were implemented in C++ and compiled with MS Visual Studio 2008 compiler. The experiments were done on a computer with AMD Phenom 2.8 GHz CPU and GeForce GTS 450 GPU under Windows XP.

5.1   Experiments on One-Unit Problem

The test instances were generated randomly with the following parameters: $D_s \in [10, 200], a_{ij} \in [0, 10], r_i \in [1, 10]$. To eliminate the violations of the triangle inequality, a correction procedure was applied: repeatedly find tasks i, j, k such that $a_{ik} > a_{ij} + a_{jk}$ and set $a_{ik} := a_{ij} + a_{jk}$ until such cases cannot be found.

The correspondence of tasks and products was set in such a way that every product would have at least one task: tasks with number $i < n$ were assigned to products with the same number $s = i$; for the other tasks the product was assigned at random.

In the first test case, the algorithms for the one-unit problem are compared with the general-purpose MILP solver CPLEX, version 12.3 without any parallelization (running on one processor core). The results are given in Table 1. The first column shows the problem size. The next three columns give the results for CPLEX. For the most of the instances, CPLEX was running rather long time and was stopped by the time limit and an approximate solution was returned. Each row corresponds to solution of five random instances. The table shows the predefined time limit, the number of times CPLEX found an optimal solution,

Table 1. Comparison of the algorithms for the one-unit problem on random instances

Size	CPLEX				BB, time		DP, time	
$|S| \times |I|$	Time lim.	# opt.	Ratio	Act. ratio	Simple	Dyn.	CPU	GPU
8×40	100	5	0%	0%	<0.1	<0.1	<0.1	<0.1
10×50	300	0	4.5%	0.1%	<0.1	<0.1	<0.1	<0.1
12×60	600	0	6.9%	0.8%	2.2	<0.1	0.4	<0.1
14×70	600	0	6.1%	0.7%	21	0.2	2.3	0.1
16×80	900	0	6.7%	0.5%	1200	2.8	13.5	0.2
18×90	900	0	7.9%	1%	-	17	76	0.6
20×100	1800	0	7.1%	0.7%	-	95	425	3

the average a posteriori approximation ratio provided by CPLEX, and the actual approximation ratio calculated using optimal solutions found by the other algorithms. For the BB and the DP algorithms, the average solving times in seconds are given. The BB is tested in two cases: using only the simple bound L_M and the combined L_M and L_D bound which is showed in the corresponded columns. One can see that all the proposed algorithms find optimal solutions in rather short time, except for the case of the BB with the simple bound L_M. The simple bound can be used only for small sized problems, for the instances with more that 16 products this version did not stop in a reasonable time. Using of dynamic bound significantly reduces the running time of the BB algorithm so that it suits well for solving rather large problems. The DP algorithm running on the CPU does not show advantages, but on the GPU it gives the best result especially for the largest instances, where it is about ten times faster than the BB algorithm with the dynamic bound. Note that CPLEX provides quite good approximations in the given time, but the approximation ratios provided by CPLEX itself are rather poor.

The second experiment was done on the test instances with the modification making them harder for the simple bound L_M: for each task i find task j producing different product and having maximal production time M_j, and divide a_{ji} by five. Such modification decreases the changeover part in $L_M(s)$ for each s in (14), but as soon as M_j is maximal it is hardly probable for the modified a_{ji} to appear in an optimal solution. Similarly, for each task i a task j producing different product and having minimal production time M_j is found, and a_{ji} is multiplied by five. The results given in Table 2 confirm the arguments. As above, each cell shows the average running time over five independently generated instances of the correspondent size. For the simple bound the slowing down is drastic; the running time for bound LD is the same as for the purely random instances. The performance of the variant with the dynamic bound shows quite good stability. The running time of the DP algorithms does not depend on the particular numerical data, but only on the problem size, so the results for the DP are the same in Tables 1 and 2.

Table 2. Comparison of the algorithms for the one-unit problem on hard instances

Size $|S| \times |I|$	CPLEX				BB, time		DP, time	
	Time lim.	# opt.	Ratio	Act. ratio	Simple	Dyn.	CPU	GPU
8×40	100	5	0%	0%	<0.1	<0.1	<0.1	<0.1
10×50	300	0	6%	0.2%	<0.1	<0.1	<0.1	<0.1
12×60	600	0	7.5%	0.4%	4.5	<0.1	0.4	<0.1
14×70	600	0	6.6%	0.5%	43	0.3	2.3	0.1
16×80	900	0	7.6%	%	>1 h	2.6	13.5	0.2
18×90	900	0	8.1%	%	-	18	76	0.6
20×100	1800	0	7%	1%	-	108	425	3

5.2 Experiments on Multi-unit Problem

The test instances of the multi-unit problem were generated similarly, but with the limit on the number of products that can be produced on one unit. To do this, the task is assigned to a randomly chosen unit only among those units, for which the number of products does not exceed this limit. In our experiments, the limit is set to 22, which was sufficient for the preliminary DP procedure to fit in the available memory of the GPU device.

The results are given in Table 3. As before, each row corresponds to five random instances, the columns for CPLEX have the same meaning as in Table 1. The results show that the BB algorithm can solve rather large instances in a short time provided that the number of products for each unit is limited. Note especially the row with dimension $40 \times 8 \times 150$: the large average running time happened due to one instance, the other four were solved within 15 s. As before, the hard instances were generated and solved, the results given in Table 4 do not show big difference from the ones from Table 3.

Table 3. Comparison of the algorithms for the multi-unit problem on random instances

Size $|S| \times |U| \times |I|$	CPLEX				BB
	Time lim.	# opt.	Ratio	Act. ratio	Time
$30 \times 6 \times 100$	300	2	11.7%	2.7%	<1
$30 \times 6 \times 110$	300	0	22.5%	5.8%	<1
$30 \times 6 \times 120$	300	0	20%	4.6%	1.2
$40 \times 8 \times 140$	900	2	12.9%	1.9%	<1
$40 \times 8 \times 150$	900	0	18.8%	4.1%	75
$40 \times 8 \times 160$	900	0	20.5%	4.3%	7.8

Table 4. Comparison of the algorithms for the multi-unit problem on hard instances

Size	CPLEX				BB
$\lvert S \rvert \times \lvert U \rvert \times \lvert I \rvert$	Time lim.	# opt.	Ratio	Act. ratio	Time
$30 \times 6 \times 100$	300	2	13.7%	4%	<1
$30 \times 6 \times 110$	300	0	22.1%	3.4%	<1
$30 \times 6 \times 120$	300	0	20.2%	4.7%	0.6
$40 \times 8 \times 140$	900	2	14%	4.6%	6
$40 \times 8 \times 150$	900	0	21%	5.3%	80
$40 \times 8 \times 160$	900	0	22%	5.9%	7

6   Conclusion

The algorithms proposed in this paper demonstrated good performance and robustness on the medium-size test instances of the considered multi-product scheduling problem. For the one-unit problem, the branch-and-bound algorithm with dynamic bound has shown better performance compared to CPLEX. The best results in terms of solving time were obtained by the dynamic programming algorithm running on the graphical device. For the one-unit problem, the proposed branch-and-bound algorithm has proven the ability to solve rather large instances in a short time, provided that one unit has a limited number of tasks and products to fit in memory in the preliminary dynamic programming step. Further improvement could be achieved by development of more effective bounds and using special subproblem selection rules; the ideas can be adopted from the broad research on the Traveling Salesman Problem. It would also be useful to apply the proposed algorithm as a local improvement procedure in heuristic algorithms, e.g. the genetic algorithm, tabu search or other metaheuristics.

Acknowledgments. The work was supported by the program of fundamental scientific research of the SB RAS No. I.5.1., project No. 0314-2016-0019.

References

1. Allahverdi, A., Ng, C.T., Cheng, T.C.E., Kovalyov, M.Y.: A survey of scheduling problems with setup times or costs. Eur. J. Oper. Res. **187**, 985–1032 (2008)
2. Berger, K.-E., Galea, F.: An efficient parallelization strategy for dynamic programming on GPU. In: 2013 IEEE 27th International Symposium on Parallel and Distributed Processing Workshops and PhD Forum, Boston, USA, pp. 1797–1806 (2013)
3. Borisovsky, P.A.: A genetic algorithm for one production scheduling problem with setup times. In: Proceedings of XIV Baikal International School-seminar Optimization Methods and Their Applications, vol. 4, pp. 166–172. Melentiev Energy Systems Institute SB RAS, Irkutsk, Russia (2008). (In Russian)

4. Borisovsky, P.A.: A Branch-and-Bound algorithm for one multi-product single-machine scheduling problem. In: Proceeding of 11 International Workshop on CSIT (CSIT 2009), vol 2, pp. 223–227. UFA: USATU Editorial-Publishing Office, Russia (2009)

5. Borisovsky, P.A., Eremeev, A.V., Kallrath, J.: On hybrid method for medium-term multi-product continuous plant scheduling. In: Proceedings of 2017 International Multi-Conference on Engineering, Computer and Information Sciences (SIBIR-CON), Novosibirsk, Russia, pp. 42–47 (2017)

6. Dolgui, A., Eremeev, A.V., Kovalyov, M.Y., Kuznetsov, P.M.: Multi-product lot sizing and scheduling on unrelated parallel machines. IIE Trans. **42**(7), 514–524 (2010)

7. Held, M., Karp, R.M.: A dynamic programming approach to sequencing problems. J. Soc. Ind. Appl. Math. **10**, 196–210 (1962)

8. Ierapetritou, M.G., Floudas, C.A.: Effective continuous-time formulation for short-term scheduling: I. Multipurpose batch processes. Ind. Eng. Chem. Res. **37**, 4341–4359 (1998)

9. Lawler, E.L., Lenstra, J.K., Rinnooy Kan, A.H.G., Shmoys, D.B.: The Traveling Salesman Problem: A Guided Tour of Combinatorial Optimization. Wiley, Chichester (1985)

10. Shaik, M.A., Floudas, C.A., Kallrath, J., Pitz, H.-J.: Production scheduling of a large-scale industrial continuous plant: short-term and medium-term scheduling. Comput. Chem. Eng. **33**, 670–686 (2009)

Towards Tractability of the Euclidean Generalized Traveling Salesman Problem in Grid Clusters Defined by a Grid of Bounded Height

Michael Khachay[1,2,3(✉)] and Katherine Neznakhina[1,3]

[1] Ural Federal University, Ekaterinburg, Russia
mkhachay@imm.uran.ru
[2] Omsk State Technical University, Omsk, Russia
[3] Krasovsky Institute of Mathematics and Mechanics, Ekaterinburg, Russia

Abstract. We consider the Euclidean Generalized Traveling Salesman Problem in Grid Clusters (EGTSP-GC), a special geometric subclass of the famous Generalized TSP, introduced by Bhattacharya et al. They showed that the problem is strongly NP-hard if the number of clusters k belongs to the instance and proposed the first polynomial time algorithm with a fixed approximation ratio. Recently, we proved that EGTSP-GC belongs to PTAS when $k = O(\log n)$ and $k = n - O(\log n)$. Meanwhile, being the special case of GTSP, for any fixed k, EGTSP-GC can be solved to optimality in polynomial time. Therefore, it seems interesting to describe the most general case of the problem sharing this property. Recently, by virtue of generalized pyramidal routes, we provided an optimal algorithm with $O(n^3)$ time complexity bound for the case of EGTSP-GC, whose grid height does not exceed 2. In this paper, we extend this result to the case of EGTSP-GC defined by a grid of any fixed height.

Keywords: Generalized Traveling Salesman Problem
Pseudo-pyramidal tour · Polynomial time solvability

1 Introduction

The motivation of this paper is threefold. Firstly, we are motivated by the famous NP-hardness result [12] obtained by Christos Papadimitriou for the Traveling Salesman Problem (TSP) on the Euclidean plane. Another motivation of this paper stems from recent parametric results both for classic TSP and its well-known modification Generalized Traveling Salesman Problem (GTSP), which are based on Balas precedence constraints [1,2,4] and generalized pyramidal tours [9, 11] and lead to efficient parameterized exact algorithms for these problems. Last

This research was supported by RSF grant 14-01-00109.

(but not the least) motivation comes from recent achievements in computational geometry. In particular, the results concerning a special geometric type of the Euclidean GTSP, where the clusters are induced by cells of a regular planar grid introduced in the recent paper [3].

Theoretical significance of the Papadimitriou's result can hardly be overestimated. Papadimitriou showed that the classic TSP is intractable even in such a specific setting as considered in [12]. Meanwhile, the one-dimensional Euclidean TSP is efficiently solvable. Therefore, the borderline between polynomially solvable instances of the Euclidean TSP and the NP-hard ones lies somewhere near to univariate and two-dimensional settings of this problem, since the known Papadimitriou intractability proof is based on polynomial time reduction of the Exact Cover by 3-Sets (X3C) Problem to the specific, substantially non-flat instances of the Euclidean TSP in the plane. Indeed, for any instance of the X3C, this reduction assigns an appropriate instance of the Euclidean TSP having the following properties:

(i) for any nodes p and q, the distance between them is at least 1;
(ii) the size of the nodeset grows proportionally to $M \times N$, where N is the number of covering sets and M is the size of the groundset of the X3C instance to be reduced;
(iii) the smallest axis-aligned rectangular box enclosing the nodeset (on the plane), whose width and height are proportional to the numbers N and M, respectively, i.e. both the height and the width of this box grow together with the nodeset size and can not be fixed.

Consider a subclass of the Euclidean TSP on the plane consisting of the instances, whose nodeset V satisfies the following additional constraints

separability: for some constants K and $\delta > 0$ and each $V' \subset V$ defining the instance, any time when $|V'| \geq K$, the diameter of V' exceeds δ;
boundedness: V can be enclosed to a bounding-box, one of the sizes of which, e.g. height, is fixed.

In this paper, we give a positive answer to the question: 'Is the aforementioned subclass of the Euclidean TSP tractable?' for a special type of the separability constraint defined by unit regular grid on the plane. Furthermore, we prove polynomial time solvability for a generalization of such a subclass known as Euclidean Generalized Traveling Salesman Problem in Grid Clusters defined by a grid of a fixed height h (EGTSP-GC(h)).

The rest of the paper is structured as follows. In Sect. 2, we present the setting of the EGTSP-GC and remind some known related results. In Section 3 we introduce pseudo-pyramidal tours for the GTSP and show that for any fixed l, an l-pseudo-pyramidal tour of minimal cost can be found efficiently. In Sect. 4, we show that, for any instance of the EGTSP-GC(h), each optimal tour is $l(h)$-pseudo-pyramidal, for some value $l(h)$ independent on the instance and n and come to the final conclusion on tractability both of the EGTSP-GC(h) and the corresponding subclass of the Euclidean TSP in the plane.

2   Euclidean Generalized Traveling Salesman Problem in Grid Clusters

The Generalized Traveling Salesman Problem (GTSP) is a widely known extension of the famous Traveling Salesman Problem (TSP). An instance of the GTSP is defined by a complete edge-weighted graph $G = (V, E, c)$, cost function $c\colon V^2 \to \mathbb{R}_+$, and partition $V_1 \cup \ldots \cup V_k = V$ of the nodeset V onto k non-empty disjoint clusters. A cyclic tour $\tau = v_{i_1}, \ldots, v_{i_k}$ is feasible, if it visits each cluster V_{i_j} exactly once. The goal is to find a feasible tour of the minimum cost

$$C(\tau) = \sum_{j=1}^{k-1} c(v_{i_j}, v_{i_{j+1}}) + c(v_{i_k}, v_{i_1}).$$

We consider a geometric setting of the GTSP known as the Euclidean Generalized Traveling Salesman Problem in Grid Clusters (EGTSP-GC) introduced recently in [3] by Bhattacharya et al. For any instance of the EGTSP-GC, the graph G, cost function c, and clustering $V_1 \ldots, V_k$ have a geometric nature:

(i) the nodeset V of the graph G is a finite subset of the plane
(ii) for any $u, v \in V$, the cost $c(u, v) = \|u - v\|_2$ is defined by the Euclidean distance between these points
(iii) clusters are determined implicitly by non-empty cells of the unit grid[1] of some height h and width w.

As for the general setting of the GTSP, the goal is to find any feasible tour of the minimum cost (length).

Like the general case of the GTSP, the EGTSP-GC NP-hard, if the number of clusters k belongs to the instance. In [3], for this case of the EGTSP-GC and for any $\varepsilon > 0$, a $(1.5 + 8\sqrt{2} + \varepsilon)$-approximation algorithm was proposed. Augmented by some additional constraints, the problem may become approximable much better. For instance, the results of [5] imply that for any instance defined by a grid with a fixed height h, such that the set of non-empty cells is connected, a 2-approximate solution can be found in a polynomial time.

In [7,8], three polynomial time approximation schemes for slow and fast growing dependence of the number of clusters k on the size n of the nodeset were proposed. Actually, first two of them have time complexity bounds of $O(k^2 O(1/\varepsilon)^{2k}) + O(n)$ and $2^{O(k)} k^4 (\log k)^{O(1/\varepsilon)} + O(n)$, respectively, and remain PTAS for $k = O(\log n)$. The last one, for any $\varepsilon > 0$, provides a $(1 + \varepsilon)$-approximate solution in time of $(n/k)^k (\log k)^{O(1/\varepsilon)}$ depending on n polynomially for $k = np - O(\log n)$.

In the sequel, we consider the subclass of the EGTSP-GC defined by grids of height at most h, which is called EGTSP-GC(h). The special case of the EGTSP-GC(h)consisting of the instances, whose clusters has a single node, satisfies the aforementioned separability and boundedness conditions. Indeed, boundedness is valid, obviously. Separability can be represented in terms of the following assertion proven in [3].

---

[1] Any non-empty cell induces a separate cluster, tights are broken arbitrarily.

Assertion 1. *For any subset $V' \subset V$ of size $|V'| \geq 5$, any tree T spanning the subset V' has weight at least 1.*

In [10], we showed that any instance of the EGTSP-GC(2) can be solved to optimality in time of $O(n^3)$. In this paper, we extend this result to the case of any fixed $h \geq 1$.

3   Pseudo-Pyramidal Tours

We proceed with some technical background concerning the general case of the Generalized Traveling Salesman Problem (GTSP).

Any ordering of clusters $V_1, \ldots, V_k$ induces the corresponding partial order on the nodeset V as follows: for any $u \in V_i$ and $v \in V_j$, $u \prec v$ iff $i < j$.

In the sequel, it is convenient to assume that, for any feasible tour $\tau = v_{i_1}, v_{i_2}, \ldots, v_{i_k}$, its vertices are indexed by numbers of the clusters that contain them, i.e. $v_{i_j} \in V_{i_j}$.

We consider a special type of feasible tours that are consistent with the defined order. We call these tours *pseudo-pyramidal* [9].

Definition 1. *A tour $\tau = v_1, v_{i_1}, \ldots, v_{i_r}, v_k, v_{j_{k-r-2}}, \ldots, v_{j_1}$ is called an l-pseudo-pyramidal tour, if $i_p - i_{p+1} \leq l$ and $j_q - j_{q+1} \leq l$ for any $1 \leq p \leq r - 1$ and $1 \leq q \leq k - r - 3$.*

Actually, any l-pseudo-pyramidal tour consists of two chains $v_1, v_{i_1}, \ldots, v_k$ and $v_k, \ldots, v_{j_1}, v_1$ that are 'almost monotonous' with respect to the aforementioned order. We denote them τ^+ and τ^-, respectively. Similarly to the classic pyramidal tours (see, e.g. [6]), pseudo-pyramidal tours of minimum (or maximum) cost can be found efficiently.

Theorem 1. *For any instance of the GTSP with an arbitrary non-negative cost function, a minimum cost l-pseudo-pyramidal tour can be found in time $O(k \cdot l \cdot n^{O(l)})$.*

Proof. Generally, our proof follows to the proof of Theorem 3.7 from [11] for the classic TSP. We start with some necessary notation. For any nodes $u, v \in V$, we introduce ordered pairs $(u, v)^+$ and $(u, v)^-$. Each pair induces a number of subtours connecting in the graph G the nodes u and v. Any such a subtour P is called feasible for the pair $(u, v)^+$ (pair $(u, v)^-$) if P belongs to the chain τ^+ (chain τ^-) of some l-pseudo-pyramidal tour in the graph G.

In the sequel, we consider sets $S = \{p_1, p_2, \ldots, p_m\}$ of node-pairs p_j introduced above such that

(i) $p_1 = (u_1, v_1)^+$, $p_2 = (u_2, u_1)^-$ for some node $u_1 \in V_1$ and nodes v_1, u_2 from other clusters V_{i_1} and V_{i_2};

(ii) all the pairs are mutually disjunctive except the pairs p_1 and p_2.

To any set $S = \{p_1, p_2, \ldots, p_m\}$, we assign a subset $Q = Q(S) \subset V$ comprising all the endpoints of the pairs $p_j \in S$.

Given by an integer $1 \leq i \leq k - 1$ and a set S, we consider collections of feasible subtours $P_1, P_2, \ldots, P_m$ induced by the pairs $p_1, \ldots, p_m$, respectively, visiting all the clusters $V_1, \ldots, V_i$ once (except the cluster V_1, which is visited twice by P_1 and P_2). By $f_l(i, S)$ we denote the total cost of the cheapest collection among them. Evidently,

$$OPT = \min_{u_1 \in V_1, u_k \in V_k} \min_{\{s, t\} \subset V_2 \cup \ldots \cup V_{k-1}} \{f_l(k - 1, \{(u_1, s)^+, (t, u_1)^-\})$$

$$+ w(s, u_k) + w(u_k, t)\} \quad (1)$$

To compute values of f_l we use dynamic programming procedure as follows.

Case 1. Suppose S contains a pair $p = (u, u)^+$ (or $p = (u, u)^-$) for some $u \in V_i$. In this case, $f_l(i, S) = f_l(i - 1, S \setminus \{p\})$.

Case 2. Suppose there exist a pair $p = (u, v)^+ \in S$ such that $u \in V_i$ and $v \in V_j$. Then, in the subtour P induced by the pair p, there is a node t succeeding the node u. Since the resulting tour should be l-pseudo-pyramidal

$$f_l(i, S) = \min_{t \in \cup V_\alpha, \alpha \in [i-l, i) \setminus Q} \{f_l(i - 1, S \cup \{(t, v)^+\} \setminus \{p\}) + w(u, t)\}.$$

Case 3. Suppose $p = (u, v)^- \in S$, where $u \in V_i$ and $v \in V_j$. Then,

$$f_l(i, S) = \min_{t \in \cup V_\alpha, \alpha \in [1, i) \setminus Q} \{f_l(i - 1, S \cup \{(t, v)^-\} \setminus \{p\}) + w(u, t)\}.$$

Cases 4 and 5, where $(v, u)^+ \in S$ and $(v, u)^- \in S$ are similar to Case 3 and Case 4, respectively.

Case 6. For any $p = (u_a, v_a) \in S$, both nodes u_a and va do not belong to V_i. in this case, to compute $f_l(i, S)$, we suppose that some node $u \in V_i$ is an inner vertex of some P_a (defined by elements of S). Denote the predecessor and the successor of u by s and t, respectively. Then,

$$f_l(i, S) = \min \Bigg\{$$

$$\min_{\substack{(u_a, v_a)^+ \in S, \\ s \in \cup V_\alpha, \alpha \in [1, i) \setminus Q' \\ t \in \cup V_\beta, \beta \in [i-l, i) \setminus Q'}} \{f_l(i - 1, S \cup \{(u_a, s)^+, (t, v_a)^+\} \setminus \{p\}) + w(s, u) + w(u, t)\},$$

$$\min_{\substack{(u_a, v_a)^- \in S, \\ s \in \cup V_\alpha, \alpha \in [i-l, i) \setminus Q' \\ t \in \cup V_\beta, \beta \in [1, i) \setminus Q'}} \{f_l(i - 1, S \cup \{(u_a, s)^-, (t, v_a)^-\} \setminus \{p\}) + w(s, u) + w(u, t)\}\Bigg\},$$

for $Q' = Q \setminus \{i_a, j_a\}$, where $u_a \in V_{i_a}$ and $v_a \in V_{j_a}$.

To estimate time complexity of the procedure proposed, we obtain upper bounds for the number of possible states (i, S) and running time for each case, respectively.

The former bound comes from the following observation. By construction, for $i = 1$, there is a unique feasible state $(1, \{(1,1)^+, (1,1)^-\})$. For any $i > 1$, each possible S consists of

(i) two pairs $(u_1, v_1)^+$ and $(u_2, u_1)^-$ exactly for some $u_1 \in V_1$, $v_1 \in V_{i_1}$, and $u_2 \in V_{i_2}$, where $1 \neq i_1 \neq i_2$;

(ii) at most one pair, whose one or both ends belong to V_i;

(iii) at most $l - 1$ pairs featuring the representatives of clusters $V_2, \ldots, V_{i-1}$. Any pair of this kind has a form $(u, v)^+$ or $(v, u)^-$ for some $u \in V_j$, where $j \in [i - 1 - l, i)$.

Therefore, for any $1 \leq i < k$, the number of possible states (i, S) is

$$O\left(n^3 \cdot n^2 \cdot \sum_{z=0}^{l-1} (2n)^z \binom{n}{z}\right).$$

Since, for any z, the value $f_l(i, S)$ can be obtained in time $O(z \cdot n^2)$ and the final computations by formula (2) can be performed in time $O(n^3)$ the overall complexity bound is

$$k \cdot O(n^7) \sum_{z=0}^{l-1} O\left(z \cdot (2n)^z \binom{n}{z}\right) = O(kln^{2l+7}) = O(kln^{O(l)}).$$

Theorem is proved.

4   Optimal Tours of EGTSP-GC(h) are $l(h)$-Pseudo-Pyramidal

In this section, for any fixed h, we show that there exist a number $l = l(h)$, such that all optimal tours of any EGTSP-GC(h) instance are $l(h)$-pseudo-pyramidal. The clusters are numbered left to right and down to up (Fig. 1).

We proceed with some necessary notation. Without loss of generality, we assume that the grid is axis-aligned. For any point p in the plane, by $x(p)$ and $y(p)$ we denote coordinates of the point p. Then, anywhere, we do not distinguish the tour τ and the piece-wise linear curve in the plane induced by this tour. Further, suppose this curve contains some points p and q, by $\tau(p, q)$ we denote a subtour of the τ connecting this points (and starting at the point p).

Theorem 2. *For any instance of the EGTSP-GC(h), an arbitrary minimum cost tour is $(15h^3 + 2h)$-pseudo-pyramidal.*

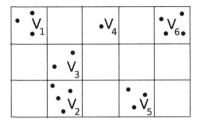

Fig. 1: Cluster numbering.

For the sake of brevity, we provide a short sketch of the proof, its full version will be published in a forthcoming paper.

The main idea of the proof is as follows. Suppose, we are given by an instance of EGTSP-GC(h) defined by a grid of height h and width w. Consider an arbitrary feasible tour τ, which is obviously l-pseudo-pyramidal for some value l. We show that, if $l > 15h^3 + 2h$, the tour τ can be transformed locally to some shorter l'-pseudo-pyramidal tour τ', for $l' \le l$.

Without loss a generality we assume that the edge $\{u, v\}$, for which

$$u \in V_{i_1}, v \in V_{i_2} \text{ and} i_1 - i_2 = l, \tag{2}$$

belongs to the chain τ^+ of the tour τ and the smallest $t \times h$ subgrid T. Furthermore, we assume that in T, the chain τ^+ has the form as presented in Fig. 2.

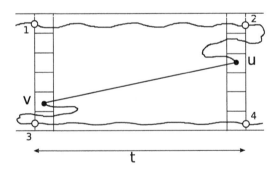

Fig. 2: The subtour $\tau(1, 4)$ and the subgrid T containing the edge $\{u, v\}$.

Namely, for the first time, the chain τ^+ enters T at point 1 and finally leaves it at point 4. We denote this subtour by $\tau(1, 4)$. Of course, the tour τ can leave T before it visits u or after visiting v, once or several times. Nevertheless, we assume that the segments $\tau(1, 2)$ and $\tau(3, 4)$ connecting points 1 and 2 and points 3 and 4, respectively, belong to the subgrid T completely. By virtue of our notation, Eq. (2) and the numbering of clusters $V_1, \ldots, V_k$, we have $t \ge l/h$.

Consider a horizontal projection of the line segment $[u, v]$ connecting the nodes u and v. By construction, its length s satisfies the equation $t - 2 \le s \le t$.

Partition this projection onto 5 equal parts and consider the second and the fourth vertical stripes obtained (of width $s/5$). We call these stripes S_2 and S_4, respectively (see Fig. 3). For any edge $\{p, q\}$ of the subtour $\tau(1, 4)$ (and the corresponding line segment $[p, q]$), denote the length of $[p, q] \cap S_j$ by $C(p, q, S_j)$. Following to Assertion 1 we claim that

(i) in the subtour $\tau(1, 2)$, there exists an edge $\{p_1, q_1\}$, such that $x(p_1) \leq x(q_1)$ and $C(p_1, q_1, S_2) \geq 1/4$;

(ii) in the subtour $\tau(3, 4)$, there exists an edge $\{p_2, q_2\}$, such that $x(p_2) \leq x(q_2)$ and $C(p_2, q_2, S_4) \geq 1/4$;

Further, let $[p_1, q_1] \cap S_2 = [\bar{p}_1, \bar{q}_1]$ and $[p_2, q_2] \cap S_4 = [\bar{p}_2, \bar{q}_2]$. Excluding from the tour τ the edge $\{u, v\}$ and the segments $[\bar{p}_1, \bar{q}_1]$ and $[\bar{p}_2, \bar{q}_2]$ and connecting the points $\bar{p}_1$ with v, $\bar{p}_2$ with q_1, and u with $\bar{q}_2$ directly we obtain a new tour τ', after shortcutting by the triangle inequality.

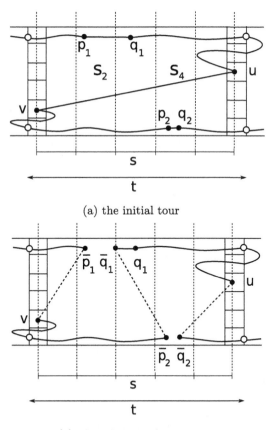

(a) the initial tour

(b) after the transformation

Fig. 3: Shortening the tour τ

Comparing the lengths $C(\tau)$ and $C(\tau')$ of the tours τ and τ', we obtain

$$\Delta C = C(\tau') - C(\tau) \le \sum_{i=1}^{3} \sqrt{(\alpha_i s)^2 + h^2} - s - 1/2. \qquad (3)$$

In Eq. (3), we use notation $\alpha_1 s, \ldots, \alpha_3 s$ for the lengths of horizontal projections of the line segments $[\bar{p}_1, v]$, $[\bar{p}_2, \bar{q}_1]$, and $[u, \bar{q}_2]$, respectively. Since, by construction, $\sum_{i=1}^{3} \alpha_i \le 1$ and any $\alpha_i \ge 1/5$,

$$\Delta C \le \sum_{i=1}^{3} (\sqrt{(\alpha_i s)^2 + h^2} - \alpha_i s) - 1/2 = \sum_{i=1}^{3} \frac{h^2}{\sqrt{(\alpha_i s)^2 + h^2} + \alpha_i s} - 1/2$$

$$\le h^2 \sum_{i=1}^{3} (2\alpha_i s)^{-1} - 1/2 \le (15h^2/s - 1)/2.$$

To obtain $\Delta C < 0$, it is sufficient to ensure

$$s > 15h^2. \qquad (4)$$

Since $s \ge t - 2$, Eq. (4) is valid any time, when $t > 15h^2 + 2$, which follows from the equation

$$l > 15h^3 + 2h. \qquad (5)$$

Thus, we showed that for any l satisfying Eq. (5), l-pseudo-pyramidal tour τ can be shortened. Therefore, for any instance of the EGTSP-GC(h), each optimal tour is $(15h^3 + 2h)$-pseudo-pyramidal. Theorem 2 is proved.

Our main result is a simple consequence of Theorems 1 and 2.

Corollary 1. *For any fixed h, any instance of the EGTSP-GC(h) can be solved to optimality in time $O(k \cdot l(h) \cdot n^{O(l(h))})$, where $l(h) = 15h^3 + 2h$.*

Employing the results of [11] together with Theorem 2, we obtain the similar result for the Euclidean TSP on Grid of height h ETSP-GC(h), which appears to be a special case of the EGTSP-GC(h) with $k = n$.

Corollary 2. *Any instance of the ETSP-GC(h) can be solved to optimality in time $O(2^{l(h)} n^{l(h)+3})$, where $l(h) = 15h^3 + 2h$.*

5   Conclusion

In this paper, we showed that the Euclidean Generalized Traveling Salesman Problem in Grid Clusters (EGTSP-GC(h)) defined by a grid of the bounded height is polynomially solvable. The same result is valid for the special type of the Euclidean TSP on the plane.

The bound $l(h)$ obtained in Theorem 2 seems to be untight and possibly can be improved. In particular, the numerical evaluation carried out on random instances of height 3 and $n \in [100, 750]$ shows that the maximum observed value of $l(3)$ is equal to 4 and does not depend on n.

References

1. Balas, E.: New classes of efficiently solvable generalized traveling salesman problems. Ann. Oper. Res. **86**, 529–558 (1999)
2. Balas, E., Simonetti, N.: Linear time dynamic-programming algorithms for new classes of restricted TSPs: a computational study. INFORMS J. Comput. **13**(1), 56–75 (2001). http://dx.doi.org/10.1287/ijoc.13.1.56.9748
3. Bhattacharya, B., Ćustić, A., Rafiey, A., Rafiey, A., Sokol, V.: Approximation algorithms for generalized MST and TSP in grid clusters. In: Lu, Z., Kim, D., Wu, W., Li, W., Du, D.-Z. (eds.) COCOA 2015. LNCS, vol. 9486, pp. 110–125. Springer, Cham (2015). https://doi.org/10.1007/978-3-319-26626-8_9
4. Chentsov, A.G., Khachai, M.Y., Khachai, D.M.: An exact algorithm with linear complexity for a problem of visiting megalopolises. Proc. Steklov Inst. Math. **295**(1), 38–46 (2016). https://doi.org/10.1134/S0081543816090054
5. Feremans, C., Grigoriev, A., Sitters, R.: The geometric generalized minimum spanning tree problem with grid clustering. 4OR **4**(4), 319–329 (2006). https://doi.org/10.1007/s10288-006-0012-6
6. Gutin, G., Punnen, A.P.: The Traveling Salesman Problem and Its Variations. Springer, Boston (2007). https://doi.org/10.1007/b101971
7. Khachai, M.Y., Neznakhina, E.D.: Approximation schemes for the generalized traveling salesman problem. Proc. Steklov Inst. Math. **299**(1), 97–105 (2017). https://doi.org/10.1134/S0081543817090127
8. Khachay, M., Neznakhina, K.: Towards a PTAS for the generalized TSP in grid clusters. In: AIP Conference Proceedings, vol. 1776(1), p. 050003 (2016). http://dx.doi.org/10.1063/1.4965324
9. Khachay, M., Neznakhina, K.: Generalized pyramidal tours for the generalized traveling salesman problem. In: Gao, X., Du, H., Han, M. (eds.) COCOA 2017. LNCS, vol. 10627, pp. 265–277. Springer, Cham (2017). https://doi.org/10.1007/978-3-319-71150-8_23
10. Khachay, M., Neznakhina, K.: Polynomial time solvable subclass of the generalized traveling salesman problem on grid clusters. In: van der Aalst, W.M.P., Ignatov, D.I., Khachay, M., Kuznetsov, S.O., Lempitsky, V., Lomazova, I.A., Loukachevitch, N., Napoli, A., Panchenko, A., Pardalos, P.M., Savchenko, A.V., Wasserman, S. (eds.) AIST 2017. LNCS, vol. 10716, pp. 346–355. Springer, Cham (2018). https://doi.org/10.1007/978-3-319-73013-4_32
11. Oda, Y., Ota, K.: Algorithmic aspects of pyramidal tours with restricted jump-backs. Interdisc. Inf. Sci. **7**(1), 123–133 (2001)
12. Papadimitriou, C.: Euclidean TSP is NP-complete. Theoret. Comput. Sci. **4**, 237–244 (1977)

Worst-Case Analysis of a Modification of the Brucker-Garey-Johnson Algorithm

Julia Memar[1]([✉]), Yakov Zinder[1], and Aleksandr V. Kononov[2]

[1] University of Technology, Sydney,
PO Box 123, Broadway, NSW 2007, Australia
{julia.memar,yakov.zinder}@uts.edu.au
[2] Sobolev Institute of Mathematics,
Siberian Branch of the Russian Academy of Sciences, Novosibirsk, Russia
alvenko@math.nsc.ru

Abstract. The paper presents the worst-case analysis of a polynomial-time approximation algorithm for the maximum lateness scheduling problem with parallel identical machines, arbitrary processing times and arbitrary precedence constraints. The algorithm is a modification of the Brucker-Garey-Johnson algorithm originally developed as an exact algorithm for the case of the problem with unit execution time tasks and precedence constraints represented by an in-tree. For the case when the largest processing time does not exceed the number of machines, we obtain a worst-case performance guarantee which is tight for arbitrary large instances of the considered maximum lateness problem. It is shown that, if the largest processing time is greater than the number of machines, then the worst-case performance guarantee for the list algorithm, obtained by Hall and Shmoys, is tight.

Keywords: Maximum lateness · Parallel machines
Precedence constraints

1 Introduction

The paper is concerned with the maximum lateness scheduling problem with parallel identical machines. It is assumed that the tasks are partially ordered, have arbitrary processing times, and the preemptions of tasks' processing are not allowed. Even a particular case of the considered problem where all due dates are equal to zero, the number of machines is two, and the processing times are equal to one or two units of time is NP-hard in a strong sense (see [1]).

For the case of the problem when the largest processing time does not exceed the number of machines we obtain a tight worst-case performance guarantee for a polynomial-time approximation algorithm that can be viewed as a modification of the Brucker-Garey-Johnson algorithm [2]. The Brucker-Garey-Johnson algorithm was originally developed as an exact algorithm for the unit execution time tasks and precedence constraints in the form of an in-tree. In order

© Springer International Publishing AG, part of Springer Nature 2018
A. Eremeev et al. (Eds.): OPTA 2018, CCIS 871, pp. 78–92, 2018.
https://doi.org/10.1007/978-3-319-93800-4_7

to stress the origin of the presented approximation algorithm, in what follows, it will be referred to as the Brucker-Garey-Johnson algorithm or simply as the BGJ-algorithm. We also show that when the largest processing time is greater than the number of machines, the worst-case performance guarantee for the list algorithm, obtained in [3], is tight.

The considered scheduling problem can be stated as follows. A set $N = \{1, \ldots, n\}$ of n tasks is to be processed on $m > 1$ identical machines subject to precedence constraints in the form of an anti-reflexive, anti-symmetric and transitive relation on N. If task i precedes task j in this relation, denoted by $i \rightarrow j$, then task i must be completed before task j can be processed. If $i \rightarrow j$, then i is called a predecessor of j and j is called a successor of i. The processing of tasks commences at time $t = 0$. Task $i \in N$ requires p_i units of processing time, where p_i is integer. Each task can be processed on any machine. Each machine can process at most one task at a time. If a machine starts processing task i, then it continues to process this task for p_i units of time, i.e. till the completion. Each task $i \in N$ has an associated due date d_i, where d_i is integer. The goal is to minimise the maximum lateness

$$L_{max}(\sigma) = \max_{j \in N}[C_j(\sigma) - d_j], \tag{1}$$

where $C_j(\sigma)$ is the completion time of task j in schedule σ. In the three-filed notation (see, for example, [4,6]) the considered problem is denoted by $P|\,prec\,|L_{max}$, where P signifies parallel identical machines, $prec$ indicates presence of precedence constraints, and L_{max} specifies the objective function, i.e. the criterion of maximum lateness. If all due dates are zero, the problem is known as the makespan problem and is denoted in the three-field notation by $P|\,prec\,|C_{max}$; if all tasks have unit processing times, then the problem is denoted by $P|\,prec, p_j = 1\,|L_{max}$; if preemptions are allowed, then the problem is denoted by $P|\,prec, prmp\,|L_{max}$.

2   The Existing Literature and Our Contribution

There are two important cases related to the considered problem, which received significant attention in the literature: the case where all tasks have unit execution times and the case where tasks have arbitrary processing times but can be processed with preemptions. Despite its role in the theory and practice (see, for example, [4–6]), much less is known about $P|\,prec\,|L_{max}$, which is studied in this paper. The complexity of this problem is due to the presence of arbitrary precedence constraints, in particular even the $P|\,prec, p_j = 1\,|C_{max}$ problem is NP-hard in the strong sense [1,7]. The $P|\,prec, p_j = 1\,|C_{max}$ problem remains NP-hard in the strong sense when the precedence constraints are in the form of a bipartite graph [8]. Furthermore, $P|\,prec, p_j = 1\,|L_{max}$ is NP-hard in the strong sense even if the precedence constraints are in the form of an out-tree [2]. Other contributors to the complexity are arbitrary processing times and the restriction that a task's processing cannot be preempted, for example $P\,||\,C_{max}$, which does not have precedence constraints, is NP-hard in the strong sense [9].

To the best of authors' knowledge, all known worst-case performance guarantee results, with the only exception of [3], were obtained for $P|prec|C_{max}$ - the particular case of the problem considered in this paper. For example, performance guarantees for the precedence constraints in the form of an in-tree can be found in [10,11]; for the constraints in the form of an out-tree and in the form of chains - in [11]; and for the list algorithm - in [12,13]. Results for $P|prec, p_j = 1|L_{max}$ and for $P|prec, prmp|L_{max}$, related to that established below, were presented in [14] and [15], respectively. A modification of the Brucker-Garey-Johnson algorithm can be also found in [16], but [16] considers scheduling with preemptions, which makes it closer to the original publication [2].

Following [3], it is convenient to replace the problem of the minimization of the maximum lateness by the equivalent problem of the minimization of the criterion G, which will be introduced below. In terms of this new problem our results can be summarised as follows. If the largest processing time $p_{max} \leq m$, then

$$G(\sigma) \leq \left(2 - \frac{1}{m}\right) G(\sigma^*) + \frac{p_{max}}{m} - p_{min}, \tag{2}$$

where σ is the schedule constructed by the BGJ-algorithm, σ^* is an optimal schedule, p_{max} and p_{min} are the largest and the smallest task processing times, correspondingly. The performance guarantee $(2 - \frac{1}{m})$ refines for the BGJ-algorithm the performance guarantee in [3]. Observe that if $p_{max} = p_{min} = 1$, then (2) gives the same worst-case performance guarantee as in [14].

If $p_{max} > m$, we show that (2) does not hold and there exists a sequence of instances of the considered maximum lateness problem such that, for the corresponding sequence of optimal schedules $\sigma_1^*, \sigma_2^*, ..., \sigma_k^*, ...$ and the sequence of schedules $\sigma_1, \sigma_2, ..., \sigma_k, ...$, constructed by the BGJ-algorithm, $G(\sigma_k) \to 2G(\sigma_k^*)$. This complements the result in [17], which established that it is NP-hard to approximate $P|prec|C_{max}$ problem within any factor strictly less than two even in the case of unit processing times.

In what follows we assume that task's largest processing time $p_{max} \leq m$.

3   BGJ-Algorithm

For each task i, the set of all successors of i will be denoted by $K(i)$. That is, $K(i) = \{j \ : i \to j\}$. Let $d = \max_{i \in N} d_i$. Then, the BGJ-algorithm can be described as follows. First, the BGJ-algorithm computes for each $i \in N$ the axillary priority μ_i:

1. For every task i such that $K(i) = \emptyset$, set $\mu_i = d - d_i$.
2. If all tasks $i \in N$ have been assigned μ_i, then stop. Otherwise, select $i \in N$ such that μ_i has not yet been specified and for each $j \in K(i)$, μ_j has been specified.
3. Set

$$\mu_i = \max\left\{d - d_i, \max_{j \in K(i)} (p_j + \mu_j)\right\} \tag{3}$$

and go to Step 2.

After obtaining all μ's, the BGJ-algorithm constructs a schedule, say σ, using $\beta_i = p_i + \mu_i$ as a task's priority. Let t be the earliest time when a machine is available for a task's processing, and t_k signify the earliest time, when machine k, $1 \le k \le m$, is available time for a task's processing.

1. Set $t = t_1 = \ldots = t_m = 0$.
2. If all tasks have been scheduled, then stop.
3. If no unscheduled task can be assigned for processing at time point t, or there is no machine i such that $t_i = t$, then go to Step 6.
4. Among all unscheduled tasks j, which can be assigned for processing at time point t, choose a task with the largest β_j. Let it be task g. Set $C_g(\sigma) = t + p_g$.
5. Choose any machine i with $t_i = t$ and set $t_i = t + p_g$. Go to Step 2.
6. Set $t = \min_{i \in \{k \, : t_k > t\}} t_i$ and then set $t_i = \max\{t, t_i\}$ for all $1 \le i \le m$. Go to Step 3.

Let $S_i(\sigma) = C_i(\sigma) - p_i$ for $i \in N$, i.e. $S_i(\sigma)$ is the starting time of task i in the schedule σ. Define $G(\sigma)$ as

$$G(\sigma) = \max_{i \in N}[S_i(\sigma) + \beta_i] = \max_{i \in N}[C_i(\sigma) + \mu_i]. \tag{4}$$

The following lemma is similar to Lemma 1 in [14] and shows that (1) can be replaced by (4).

Lemma 1. *For any schedule* σ,

$$G(\sigma) = L_{max}(\sigma) + d. \tag{5}$$

Proof. Among all tasks g such that $S_g(\sigma) + \beta_g = G(\sigma)$ select one with the largest $S_g(\sigma)$, say task i. If $\mu_i \ne d - d_i$, then by (3), there exists $j \in K(i)$ such that $\mu_i = p_j + \mu_j = \beta_j$. Observe that $S_j(\sigma) \ge S_i(\sigma) + p_i$. Hence for this j,

$$G(\sigma) = S_i(\sigma) + \beta_i = S_i(\sigma) + p_i + \mu_i \le S_j(\sigma) + \mu_i = S_j(\sigma) + \beta_j$$

which contradicts the choice of i because $S_i(\sigma) < S_j(\sigma)$. Hence, $\mu_i = d - d_i$. On the other hand, (3) implies that, for any g, $\mu_g \ge d - d_g$. Then,

$$L_{max}(\sigma) + d = \max_{g \in N}[C_g(\sigma) + d - d_g] \le \max_{g \in N}[C_g(\sigma) + \mu_g]$$
$$= G(\sigma) = C_i(\sigma) + \mu_i = C_i(\sigma) + d - d_i \le L_{max}(\sigma) + d,$$

which completes the proof. $\qquad\qquad\qquad\qquad\qquad\qquad\qquad\qquad\qquad\square$

4   Schedule's Structure

For any integer t, the slot t is the time interval $[t - 1, t]$. Let σ be a schedule, constructed by the BGJ-algorithm. Consider a task with the smallest $S_j(\sigma)$ among all j such that $S_j(\sigma) + \beta_j = G(\sigma)$. Let it be task g. A task $i \in N$ is *complete*, if $\beta_i \ge \beta_g$. Otherwise, i is *incomplete*. It is easy to see that $S_j(\sigma) \le S_g(\sigma)$ for any complete task j. A slot $t \le S_g(\sigma)$ is *complete* if the number of complete tasks, processed in this slot, equals m. A slot $t \le S_g(\sigma)$, which is not complete, is *incomplete*. An incomplete slot t is Type I if at least one of the following holds:

(t1) in the slot t, at least one machine is idle;

(t2) there exists an incomplete task j such that $S(\sigma) = t - 1$;

(t3) all tasks j, processed in the slot t, have the same starting times $S_j(\sigma)$.

An incomplete slot t, which is not Type I, is Type II.

Lemma 2. *For any Type II slot t, there exists a Type I slot t' such that*

(s1) $t' < t$;

(s2) *any slot τ such that $t' < \tau \leq t$ is Type II;*

(s3) *any incomplete task, processed in slot t, is also processed in slot t'.*

Proof. By the definition of a Type II slot, all machines in slot t are busy and at least one machine processes an incomplete task. Among all such incomplete tasks i choose a task with the largest $S_i(\sigma)$. Let it be task j. Then, any incomplete task, processed in slot t, is also processed in slot $\bar{t} = S_j(\sigma) + 1$. Furthermore, by the definition of a Type I slot, the slot $\bar{t}$ is Type I. Since slot t is Type II, $\bar{t} < t$. The proof is concluded be repeating the procedure for every Type II time slot $\bar{t} < \tau < t$ and choosing t' as the largest integer among all integers $\bar{t}$ such that $\bar{t} < t$ and the slot $\bar{t}$ is Type I . □

For any Type II slot t, the Type I slot t', specified by (s1)-(s2)-(s3), will be referred to as the *supporting* slot for the slot t.

Lemma 3. *For any Type I slot t and any complete task j such that $S_j(\sigma) \geq t$, there exists a task q such that*

$$C_q(\sigma) \geq t \quad and \quad q \to j. \tag{6}$$

Proof. Suppose that either (t1) or (t2) holds or both. Then, the existence of q satisfying (6) follows from the fact that the BGJ-algorithm does not schedule j at $t - 1$. If (t1) and (t2) do not hold, then there are m tasks processed in the slot t and, by virtue of (t3), all these tasks commence their processing at same point in time $t' < t - 1$. Since at least one of these tasks is incomplete and j is not scheduled by the BGJ-algorithm at t', then amongst these m tasks there exists a task q such that $C_q(\sigma) \geq t' + 1$ and $q \to j$. Because the same m tasks are processed in slot t' and in slot t, $C_q(\sigma) \geq t$ which gives (6). □

Corollary 1. *For any Type I slot t and any complete task j such that $S_j(\sigma) \geq t$, there exists a complete task i processed in the slot t and $i \to j$.*

Proof. Among all q, satisfying (6), select a task with the smallest $S_q(\sigma)$, let it be task i. Since $i \to j$, by virtue of (3), $\beta_i \geq p_i + \beta_j > \beta_q$, thus i is a complete task. The proof is concluded by the observation that $S_i(\sigma) < t$, because otherwise by Lemma 3 there exists q such that $C_q(\sigma) \geq t$ and $q \to i \to j$, and therefore $q \to j$, which contradicts the selection of i. □

A sequence of tasks $j_1, ..., j_k$ is a chain if, for each $1 \leq i < k$, $j_i \to j_{i+1}$. The following corollary is a direct consequence of Corollary 1 and the fact that g is complete.

Corollary 2. *There exists a chain of tasks $j_1, ..., j_r$ such that $j_r \to g$ and, for any Type I incomplete slot, the chain has a task that is processed in this time slot.*

Lemma 4. *At least one complete task is processed in each slot t such that $1 \leq t \leq S_g(\sigma)$.*

Proof. If the slot t is complete, then the lemma follows from the definition of a complete slot. If the slot t is a Type I incomplete slot, the lemma follows from Corollary 1 and the fact that g is a complete task. If the slot t is a Type II incomplete slot, then any incomplete task, processed in slot t, is also processed in its supporting slot (see Lemma 2), which by the definition is a Type I slot. As it has been proven above, at least one complete task is processed in it. Hence, the number of incomplete tasks, processed in the supporting slot and therefore in slot t, is less than m, and the lemma follows from the definition of a Type II slot which implies that in this slot all m machines are busy.   □

Let Z be the set of all Type II slots and $l^z = |Z|$. Denote the set of all supporting Type I slots by Y, and let $l^y = |Y|$. Denote by X the set of Type I slots t which are not in Y, and let $l^x = |X|$. Let c^x, c^y and c^z be the total processing time allocated to complete tasks in the slots of sets X, Y and Z, respectively. Similarly, let e^y and e^z be the total processing time allocated to incomplete tasks in the slots of sets Y and Z, respectively.

Lemma 5. *The following statements hold:*

$$c^z + e^z = ml^z; \tag{7}$$
$$e^z \leq e^y(m-1). \tag{8}$$

Proof. By the definition, there is no idle machine in each Type II slot, i.e. the number of tasks processed in the slot is m. Hence the total processing time allocated to complete and incomplete tasks in the l^z Type II slots is ml^z, and (7) holds. Consider a slot t in Y. The slot is associated with a number of consecutive Type II slots, for which t is a supporting slot. By virtue of Lemma 2, each incomplete task processed in each of these Type II slots is also processed in t. Thus the number of these Type II slots does not exceed $p_{max} - 1$. Furthermore, the number of incomplete tasks processed in each of the Type II slots, is not greater than the number of incomplete tasks processed in t. Thus $e^z \leq e^y(p_{max} - 1) \leq e^y(m-1)$.   □

5   Lower Bounds on the Optimal Value of $G(\sigma)$

Denote by σ^* the optimal schedule for $G(\sigma)$. Let a be the task with the maximum completion time among all complete tasks in σ^*. The total time allocated to incomplete tasks in σ in slots $t \leq S_g(\sigma)$ is at least $e^y + e^z$. Denote by e^* the part of this time, which is allocated in slots $t' \leq C_a(\sigma^*)$ in σ^*. It is easy to see that

$$\max_{j \in N} C_j(\sigma^*) \geq C_a(\sigma^*) + \frac{e^y + e^z - e^*}{m}.$$

Denote by $\eta = \frac{e^y + e^z - e^*}{m}$ and by $\delta = \min\{p_a - p_g, C_a(\sigma^*) - C_g(\sigma^*)\}$. Since a is complete, $\beta_a \geq \beta_g$ and $\mu_a \geq \mu_g - (p_a - p_g)$. Thus

$$\begin{aligned}
G(\sigma^*) &\geq \max\{C_a(\sigma^*) + \mu_a, C_g(\sigma^*) + \mu_g, C_a(\sigma^*) + \eta\} \\
&\geq C_a(\sigma^*) + \max\{\mu_g - (p_a - p_g), \mu_g - (C_a(\sigma^*) - C_g), \eta\} \\
&\geq C_a(\sigma^*) + \max\{\mu_g - \delta, \eta\}.
\end{aligned} \tag{9}$$

Let l^c be the number of complete slots in σ. By virtue of (9), Lemmas 4 and 5,

$$\begin{aligned}
G(\sigma^*) &\geq C_a(\sigma^*) + \max\{\mu_g - \delta, \eta\} \\
&\geq l^c + \frac{c^x + c^y + c^z + e^* + p_g}{m} + \eta + \max\{\mu_g - \delta - \eta, 0\} \\
&\geq l^c + \frac{l^x + l^y + c^z + e^* + p_g}{m} + \eta + \max\{\mu_g - \delta - \eta, 0\} \\
&= l^c + \frac{l^x + l^y + p_g}{m} + l^z + \frac{e^y}{m} + \max\{\mu_g - \delta - \eta, 0\}.
\end{aligned} \tag{10}$$

Assume that the first slot in σ is incomplete. Then (10) can be tightened. Let ϱ be the minimum processing time among all j such that $S_j(\sigma) = 0$. All slots t, such that $1 \leq t \leq \varrho$, are Type I slots, since they satisfy the condition (t3). By virtue of Corollary 1, $S_j(\sigma^*) \geq \varrho$ for any complete task j with $S_j(\sigma) \geq \varrho$. Denote by c^ϱ the processing time allocated to complete tasks in Type I slots after point of time ϱ. Then

$$\begin{aligned}
C_a(\sigma^*) &\geq \varrho + l^c + \frac{c^\varrho + c^z + e^* + p_g}{m} \geq \varrho + l^c + \frac{l^x + l^y - \varrho + p_g}{m} + \frac{c^z + e^*}{m} \\
&\geq p_{min}\left(1 - \frac{1}{m}\right) + l^c + \frac{l^x + l^y + p_g}{m} + \frac{c^z + e^*}{m}.
\end{aligned} \tag{11}$$

Thus, (9) and (11) imply that

$$\begin{aligned}
G(\sigma^*) &\geq p_{min}\left(1 - \frac{1}{m}\right) + l^c + \frac{l^x + l^y + p_g}{m} + \frac{c^z + e^*}{m} + \eta + \max\{\mu_g - \delta - \eta, 0\} \\
&= p_{min}\left(1 - \frac{1}{m}\right) + l^c + \frac{l^x + l^y + p_g}{m} + l^z + \frac{e^y}{m} + \max\{\mu_g - \delta - \eta, 0\}.
\end{aligned} \tag{12}$$

Consider a chain $j_1 \to j_2 \to \ldots \to j_r \to g$, satisfying Corollary 2. Then (9) can be presented as:

$$\begin{aligned}
G(\sigma^*) &\geq C_a(\sigma^*) + \max\{\mu_g - \delta, \eta\} = C_g(\sigma^*) + (C_a(\sigma^*) - C_g(\sigma^*)) + \max\{\mu_g - \delta, \eta\} \\
&\geq \sum_{k=1}^{r} p_{j_k} + p_g + \delta + \max\{\mu_g - \delta, \eta\} \geq l^x + l^y + p_g + \max\{\mu_g, \eta + \delta\}.
\end{aligned} \tag{13}$$

If the first slot in σ is complete, then (13) can be tightened. Observe that in this case at least p_{min} first time slots in σ are complete. If $S_{j_1}(\sigma) = 0$, then

$$G(\sigma^*) \geq C_g(\sigma^*) + \beta_g \geq \sum_{k=1}^{r} p_{j_k} + \beta_g \geq p_{min} + l^x + l^y + \beta_g.$$

If $S_{j_1}(\sigma) > 0$, then according to the BGJ-algorithm either there exists $h \to j_1$ or there exist another m tasks q with $\beta_q \geq \beta_{j_1}$ and $S_q(\sigma) < S_{j_1}(\sigma)$. Observe that $\beta_{j_1} \geq \sum_{k=1}^{r} p_{j_k} + \beta_g$, thus in these two cases

$$G(\sigma^*) \geq \max_{j \in \{k:\beta_k \geq \beta_{j_1}\}} S_j(\sigma^*) + \beta_{j_1} \geq p_{min} + \beta_{j_1} \geq p_{min} + l^x + l^y + \beta_g.$$

The last two inequalities and (13) imply that

$$G(\sigma^*) \geq \max\{p_{min} + l^x + l^y + \beta_g, l^x + l^y + p_g + \delta + \eta\}$$
$$= l^x + l^y + p_g + \max\{\mu_g + p_{min}, \delta + \eta\}. \tag{14}$$

6   Worst-Case Performance Guarantee

This lemma helps to simplify the proofs that follow.

Lemma 6. *For nonnegative numbers a, b, h and α, where $\alpha < 1$,*

$$\min\{a, b\} - (1 - \alpha)\max\{a + h, b\} \leq b\alpha - (1 - \alpha)h. \tag{15}$$

Proof. Denote by LHS the left hand side of (15) and consider the following cases:
If $a \leq b - h$, then $LHS = a - (1 - \alpha)b$ and
$a - (1 - \alpha)b \leq b\alpha - h \leq b\alpha - (1 - \alpha)h$.
If $b - h < a \leq b$, then $LHS = a - (1 - \alpha)(a + h)$ and
$a - (1 - \alpha)(a + h) = a\alpha - (1 - \alpha)h \leq b\alpha - (1 - \alpha)h$.
If $a > b$, then $LHS = b - (1 - \alpha)(a + h)$ and
$b - (1 - \alpha)(a + h) < b - (1 - \alpha)(b + h) = b\alpha - (1 - \alpha)h$. $\square$

Theorem 1.

$$G(\sigma) \leq \left(2 - \frac{1}{m}\right)G(\sigma^*) + \frac{p_{max}}{m} - p_{min}, \tag{16}$$

and the bound is tight.

Proof. The value of $G(\sigma)$ can be expressed as:

$$G(\sigma) = C_g(\sigma) + \mu_g = l^c + l^x + l^y + l^z + p_g + \mu_g. \tag{17}$$

If the first slot in σ is complete, then by (10) and (17)

$$G(\sigma) - G(\sigma^*)$$
$$\leq \left(1 - \frac{1}{m}\right)(l^x + l^y + p_g) + \mu_g - \frac{e^y}{m} - \max\{\mu_g - \delta - \eta, 0\}$$
$$= \left(1 - \frac{1}{m}\right)(l^x + l^y + p_g) - \frac{e^y}{m} + \min\{\delta + \eta, \mu_g\}. \tag{18}$$

Furthermore, if the first slot in σ is complete, (14) and (18) imply that

$$G(\sigma)$$
$$\leq \left(2 - \frac{1}{m}\right)G(\sigma^*) - \frac{e^y}{m} + \min\{\delta + \eta, \mu_g\} - \left(1 - \frac{1}{m}\right)\max\{\mu_g + p_{min}, \delta + \eta\}. \tag{19}$$

If the first slot in σ is incomplete, then by (12) and (17)

$$G(\sigma) - G(\sigma^*)$$
$$\leq \left(1 - \frac{1}{m}\right)(l^x + l^y + p_g - p_{min}) + \mu_g - \frac{e^y}{m} - \max\{\mu_g - \delta - \eta, 0\}$$
$$= \left(1 - \frac{1}{m}\right)(l^x + l^y + p_g - p_{min}) - \frac{e^y}{m} + \min\{\delta + \eta, \mu_g\}. \tag{20}$$

It is easy to see that if the first slot in σ is incomplete, (13) and (20) imply the same result as (19):

$$G(\sigma) \leq \left(2 - \frac{1}{m}\right)G(\sigma^*) - \frac{e^y}{m} + \min\{\delta + \eta, \mu_g\} - \left(1 - \frac{1}{m}\right)(p_{min} + \max\{\mu_g, \delta + \eta\})$$
$$\leq \left(2 - \frac{1}{m}\right)G(\sigma^*) - \frac{e^y}{m} + \min\{\delta + \eta, \mu_g\} - \left(1 - \frac{1}{m}\right)\max\{\mu_g + p_{min}, \delta + \eta\}.$$

Let $a = \mu_g$, $b = \delta + \eta$, $h = p_{min}$, $\alpha = \frac{1}{m}$. Then by virtue of (19) and Lemma 6,

$$G(\sigma) \leq \left(2 - \frac{1}{m}\right)G(\sigma^*) - \frac{e^y}{m} + \frac{\delta}{m} + \frac{\eta}{m} - p_{min}\left(1 - \frac{1}{m}\right). \tag{21}$$

Taking into account (8),

$$\frac{\eta}{m} - \frac{e^y}{m} = \frac{e^y + e^z - e^*}{m^2} - \frac{e^y}{m} \leq \frac{e^y + e^y(m-1)}{m^2} - \frac{e^y \times m}{m^2} = 0. \tag{22}$$

Finally, by virtue of (22) and the fact that $\delta \leq p_{max} - p_{min}$, we obtain (16):

$$G(\sigma) \leq \left(2 - \frac{1}{m}\right)G(\sigma^*) + \frac{p_{max} - p_{min}}{m} - p_{min}\left(1 - \frac{1}{m}\right)$$
$$= \left(2 - \frac{1}{m}\right)G(\sigma^*) + \frac{p_{max}}{m} - p_{min}.$$

Observe that by direct substitution of (5) in (16) we obtain the equivalent bound for the criterion of maximum lateness:

$$L_{max}(\sigma) \leq \left(2 - \frac{1}{m}\right)L_{max}(\sigma^*) + \left(1 - \frac{1}{m}\right)d + \frac{p_{max}}{m} - p_{min}.$$

To show that (16) is tight, we consider the partially ordered set of tasks depicted by Fig. 1. The graph constitutes of $km - 1$ identical sections and the last section. The schedule σ constructed by the BGJ-algorithm and the optimal schedule σ^* are depicted by Fig. 2. The values of $G(\sigma)$ and $G(\sigma^*)$ are the following

$$G(\sigma) = C_g(\sigma) + \mu_g = \underbrace{[2p_{min}(m-1) + p_{min}]}_{\text{per section}} \underbrace{(km-1)}_{km-1 \text{ sections}}$$
$$+ \underbrace{2p_{min}(m-1) + p_{max} + p_{min}}_{\text{last section}} + \underbrace{p_{max} - p_{min}}_{\mu_g}$$
$$= 2p_{min}km^2 - p_{min}km + 2p_{max} - p_{min} \tag{23}$$
$$G(\sigma^*) = C_a(\sigma^*) + \mu_a = \underbrace{mp_{min}}_{\text{per section}} \underbrace{(km-1)}_{km-1 \text{ sections}} + \underbrace{mp_{min} + p_{max}}_{\text{last section}} + \underbrace{0}_{\mu_a = 0}$$
$$= kp_{min}m^2 + p_{max} \tag{24}$$

By substituting (23) and (24) in (16), we show that the bound is tight:

$$\left(2 - \frac{1}{m}\right)G(\sigma^*) + \frac{p_{max}}{m} - p_{min} = \left(2 - \frac{1}{m}\right)(kp_{min}m^2 + p_{max}) + \frac{p_{max}}{m} - p_{min}$$
$$= 2kp_{min}m^2 + 2p_{max} - kp_{min}m - \frac{p_{max}}{m} + \frac{p_{max}}{m} - p_{min} = G(\sigma). \qquad \square$$

A p_{min} units task is denoted by p

Each section constitutes of
$m-1$ rows
with $m+1$ p_{min} unit tasks and
for the first $km-1$ sections the
m^{th} row is one p_{min} unit task
and for last section the mth row
is one p_{min} unit task and
m p_{max} units tasks

For p_{min} units tasks in row j
from the top:
$\mu = p_{max} + (km^2 - j - 1)p_{min}$
$\beta = p_{max} + (km^2 - j)p_{min}$

For the p_{min} units task in
the last row:
$\mu = p_{max} - p_{min}$
$\beta = p_{max}$
For the p_{max} units tasks in
the last row:
$\mu = 0$
$\beta = p_{max}$

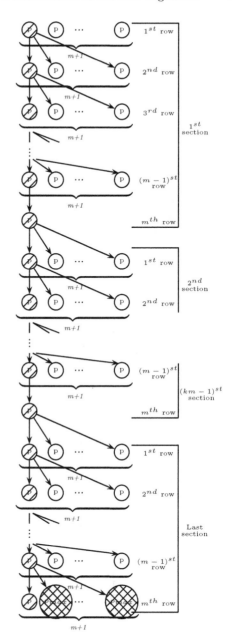

Fig. 1. Set of tasks: $p_{max} \leq m$

7   The Case When $p_{max} > m$

To show that (16) does not hold when $p_{max} > m$ we consider the partially
ordered set of tasks depicted by Fig. 3. The graph constitutes of m identical

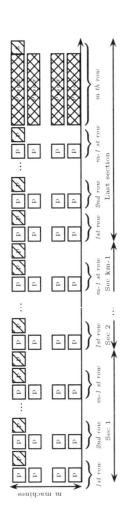

Schedule σ,

constructed by the BGJ-algorithm

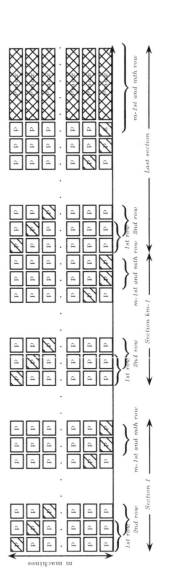

Optimal schedule σ^*

Fig. 2. Schedules: $p_{max} \leq m$

sections and the last section. Let $p_{max} = m + 1$ and $p_{min} = 1$. The schedule σ constructed by the BGJ-algorithm and the optimal schedule σ^* are depicted by Fig. 4.

$$G(\sigma) = C_g(\sigma) + \mu_g = \underbrace{2(m-1)+1}_{1^{st}\ section} + \underbrace{[m+1+2(m-2)+1](m-1)}_{2^{nd}-m^{th}\ sections}$$

$$+ \underbrace{m+1+2(m-2)+m+1+1}_{the\ last\ section} + \underbrace{m^2-1+m}_{\mu_g} = 4m^2 + 2m - 1 \tag{25}$$

$$G(\sigma^*) = C_a(\sigma^*) + \mu_a$$

$$= \underbrace{m}_{per\ section} \times \underbrace{(m+1)}_{m+1\ sections} + \underbrace{m+1}_{long\ tasks\ of\ last\ section} + \underbrace{m^2-1}_{\mu_a = \frac{\Sigma\ of\ incomplete\ tasks}{m}}$$

$$= 2m^2 + 2m \tag{26}$$

When we substitute (25) and (26) in (16), the bound does not hold:

$$\left(2 - \frac{1}{m}\right)G(\sigma^*) + \frac{p_{max}}{m} - p_{min} = \left(2 - \frac{1}{m}\right)(2m^2 + 2m) + \frac{m+1}{m} - 1$$

$$= 4m^2 + 4m - 2m - 2 + 1 + \frac{1}{m} - 1 = G(\sigma) - \left(1 - \frac{1}{m}\right) < G(\sigma).$$

If $p_{max} > m$, we show that there exists a sequence of instances of the considered maximum lateness problem such that, for the corresponding sequence of optimal schedules $\sigma_1^*, \sigma_2^*, ..., \sigma_k^*, ...$ and the sequence of schedules $\sigma_1, \sigma_2, ..., \sigma_k, ...,$ constructed by the BGJ-algorithm, $G(\sigma_k) \to 2G(\sigma_k^*)$. Consider the set of tasks constituting of m p_{max} units tasks and one p_{min} units task. For any task j such that $p_j = p_{max}$ let $\mu_j = 0$, then $\beta_j = p_{max}$. Denote p_{min} units task by g and let $\mu_g = p_{max} - p_{min}$, then $\beta_g = p_{max}$. The BGJ-algorithm could assign $S_j(\sigma) = 0$ for any task j such that $p_j = p_{max}$ and $S_g(\sigma) = p_{max}$. Then

$$G(\sigma) = C_g(\sigma) + \mu_g = \underbrace{p_{max} + p_{min}}_{C_g(\sigma)} + \underbrace{p_{max} - p_{min}}_{\mu_g} = 2p_{max}$$

In the optimal schedule σ^* $S_g(\sigma^*) = 0$, and $S_j(\sigma^*) = 0$ for $m - 1$ tasks j such that $p_j = p_{max}$. For one task i such that $p_i = p_{max}$ $S_i(\sigma^*) = p_{min}$. Then

$$G(\sigma^*) = C_a(\sigma^*) + \mu_a = p_{min} + p_{max};$$

$$G(\sigma) = \frac{2p_{max}}{p_{max} + p_{min}} G(\sigma^*) = (2 - \frac{2p_{min}}{p_{max} + p_{min}})G(\sigma^*) \to 2G(\sigma^*).$$

Acknowledgement. The third author has been partially supported by RFBR grant 17-07-00513A.

Each of the $1^{st} - m^{th}$ sections
constitutes of $m - 1$ rows
with $m + 1$ unit tasks and
the mth row is one unit task
and $m - 1$ $(m + 1)$ units tasks

For unit tasks in row j
from the top:
$\mu = 2m^2 + 2m - j - 1$
$\beta = 2m^2 + 2m - j$
For all $(m + 1)$ units tasks except
　except those in the last row:
$\mu = 0$
$\beta = m + 1$

The last section
constitutes of $m - 1$ rows
with $m + 1$ unit tasks and
the mth row is one unit task
and m $(m + 1)$ units tasks

For the unit task in
the last row:
$\mu = m^2 + m - 1$
$\beta = m^2 + m$
For the $(m + 1)$ units tasks
in the last row:
$\mu = m^2 - 1$
$\beta = m^2 + m$

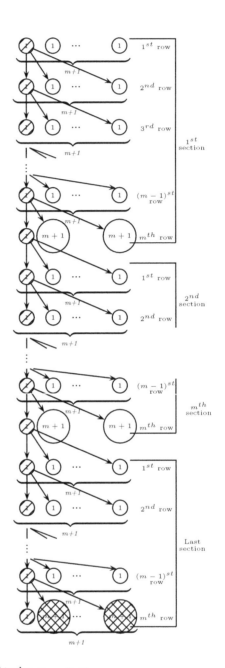

Fig. 3. Set of tasks: $p_{max} > m$

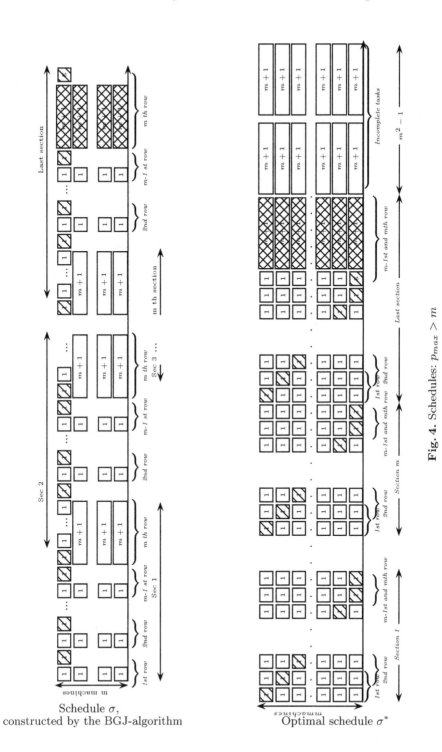

Fig. 4. Schedules: $p_{max} > m$

References

1. Lenstra, J.K., Rinnooy Kan, A.H.G.: Complexity of scheduling under Precedence Constraint. Oper. Res. **26**, 22–35 (1978)
2. Brucker, P., Garey, M.R., Johnson, D.S.: Scheduling equal length tasks under tree-like precedence constraints to minimize maximum lateness. Math. Oper. Res. **2**, 275–284 (1977)
3. Hall, L.A., Shmoys, D.B.: Approximation schemes for constrained scheduling problems. In: Proceedings of 30th IEEE Symposium on Foundations of Computer Science, pp. 134–139 (1989)
4. Brucker, P.: Scheduling Algorithms, 5th edn. Springer, New York (2007)
5. Leung, J.Y.-T.: Handbook of Scheduling: Algorithms, Models and Performance Analysis. Chapman and Hall/CRC, Boca Raton (2004)
6. Pinedo, M.: Scheduling: Theory, Algorithms, and Systems, 4th edn. Springer, New York (2012)
7. Ullman, J.D.: NP-complete scheduling problems. J. Comp. Syst. Sci. **10**, 384–393 (1975)
8. Zinder, Y., Roper, D.: A minimax combinatorial optimisation problem on an acyclic directed graph: polynomial-time algorithms and complexity. In: Aitken Centenary Conference, Dunedin, New Zealand (1995)
9. Garey, M.R., Johnson, D.S.: "Strong" NP-completeness results: motivation, examples, and implication. J. Assoc. Comput. Mach. **25**, 499–508 (1978)
10. Kaufman, M.T.: An almost optimal algorithm for the assembly line scheduling problem. IEEE Trans. Comput. **11**, 1169–1174 (1974)
11. Kunde, M.: Nonpreemtive LP-scheduling on homogeneous multiprocessor systems. SIAM J. Comput. **10**, 151–173 (1981)
12. Graham, R.L.: Bounds on multiprocessing anomalies and related packing algorithms. In: AFIPS Conference Proceedings, vol. 40, pp. 205–217 (1972)
13. Graham, R.L.: Bounds on the performance of scheduling algorithms. In: Coffman, E.G. (ed.) Computer and Job Shop Scheduling Theory, pp. 165–227. Wiley, New York (1976)
14. Singh, G., Zinder, Y.: Worst case performance of two critical path algorithms. Asia-Pacific J. Oper. Res. **17**, 101–122 (2000)
15. Singh, G., Zinder, Y.: Worst case performance of critical path algorithms. Int. Trans. Oper. Res. **7**, 383–399 (2000)
16. Lawler, E.L.: Preemptive scheduling of. precedence-constrained jobs on parallel machines. In: Dempster, M.A.H., Lenstra, J.K., Rinnooy Kan, A.H.G. (eds.) Deterministic and Stochastic Scheduling. NATO Advanced Study Institutes Series, vol. 84, pp. 101–123. Springer, Dordrecht (1982). https://doi.org/10.1007/978-94-009-7801-0_6
17. Svensson, O.: Conditional hardness of precedence constrained scheduling on identical machines. In: Proceedings of the Forty-Second ACM Symposium on Theory of Computing, STOC 2010, pp. 745–754 (2010)

Reduction of the Pareto Set in Bicriteria Asymmetric Traveling Salesman Problem

Aleksey O. Zakharov[1]⬤ and Yulia V. Kovalenko[2,3](✉)⬤

[1] Saint Petersburg State University, St. Petersburg, Russian Federation
a.zakharov@spbu.ru
[2] Department of Mechanics and Mathematics, Novosibirsk State University,
Novosibirsk, Russian Federation
[3] Sobolev Institute of Mathematics, Novosibirsk, Russian Federation
julia.kovalenko.ya@yandex.ru

Abstract. We consider the bicriteria asymmetric traveling salesman problem (bi-ATSP). Optimal solution to a multicriteria problem is usually supposed to be the Pareto set, which is rather wide in real-world problems. We apply to the bi-ATSP the axiomatic approach of the Pareto set reduction proposed by V. Noghin. We identify series of "quanta of information" that guarantee the reduction of the Pareto set for particular cases of the bi-ATSP. An approximation of the Pareto set to the bi-ATSP is constructed by a new multi-objective genetic algorithm. The experimental evaluation carried out in this paper shows the degree of reduction of the Pareto set approximation for various "quanta of information" and various structures of the bi-ATSP instances generated randomly.

Keywords: Reduction of the Pareto set
Multi-objective genetic algorithm · Computational experiment

1 Introduction

The asymmetric traveling salesman problem (ATSP) is one of the most popular problems in combinatorial optimization [2]. Given a complete directed graph where each arc is associated with a positive weight, we search for a circuit visiting every vertex of the graph exactly once and minimizing the total weight. In this paper, we consider the bicriteria ATSP (bi-ATSP) which is a special case of the multicriteria ATSP [5], where an arc is associated to a couple of weights.

The best possible solution to a multicriteria optimization problem (MOP) is usually supposed to be the Pareto set [5,21], which is rather wide in real-world problems, and difficulties arise in choosing the final variant. For that reason numerous methods introduce some mechanism to treat the MOP: utility function, rule, or binary relation, so that methods are aimed at finding an "optimal" solution with respect to this mechanism. However, some approaches do not guarantee that the obtained solution will be from the Pareto set. State-of-the-art methods are the following [8]: multiattribute utility theory, outranking approaches, verbal decision analysis, various iterative procedures with man-machine interface, etc. In this paper, we investigate the axiomatic approach of the Pareto set reduction proposed in [19] which has an alternative idea. Here the author introduced an additional information about the decision maker (DM) preferences in terms of the so-called "quantum of information". The method shows how to construct a new bound of the optimal choice, which is narrower than the Pareto set. Practical applications of the approach could be found in [11, 20].

As far as we know, the axiomatic approach of the Pareto set reduction has not been widely investigated in the case of discrete optimization problems, and an experimental evaluation has not been carried out on real-world instances. We apply this approach to the bi-ATSP in order to estimate its effectiveness, i.e. the degree of the Pareto set reduction and how it depends on the parameters of the information about DM's preferences. We identify series of "quanta of information" that guarantee the reduction of the Pareto set for particular cases of the bi-ATSP.

Originally the reduction is constructed with respect the Pareto set of the considered problem. Due to the strongly NP-hardness of the bi-ATSP we take an approximation of the Pareto set in computational experiments. The ATSP cannot be approximated with any constant or exponential approximation factor already with a single objective function [2]. Moreover, in [1], the non-approximability bounds were obtained for the multicriteria ATSP with weights 1 and 2. The results are based on the non-existence of a small size approximating set. Therefore, meta-heuristics, in particular multi-objective evolutionary algorithms (MOEAs), are appropriate to approximate the Pareto set of the bi-ATSP.

Numerous MOEAs have been proposed to MOPs (see e.g. [3,4,14,28,30,31]). There are three main classes of approaches to develop MOEAs, which are known as Pareto-dominance based (see e.g. SPEA2 [31], NSGA-II [3,4], NSGA-III [28]), decomposition based (see e.g. MOEA/D [14]) and indicator based approaches (see e.g. SIBEA [30]). NSGA-II [4] has one of the best results in the literature on multi-objective genetic algorithms (MOGAs) for the MOPs with two or three objectives. In [3], a fast implementation of a steady-state version of NSGA-II is proposed for two dimensions.

In [9,22], NSGA-II was adopted to the multicriteria symmetric traveling salesman problem, and the experimental evaluation was performed on symmetric instances from TSPLIB library [25]. To the best of our knowledge, there is no adaptation of NSGA-II to the more general problem, where arc weights are non-symmetric. In this paper, we propose a new MOGA based on NSGA-II to solve the bi-ATSP using adjacency-based representation of solutions. A computational experiment is carried out on randomly generated instances. The results of the experiment show the degree of the reduction of the Pareto set approximation for various "quanta of information" and various structures of the problem instances.

2   Problem Statement

An instance of the traveling salesman problem [2] (TSP) is given by a complete graph $G = (V, E)$, where $V = \{v_1, \ldots, v_n\}$ is the set of vertices and set E contains arcs (or edges) between every pair of vertices in V. Each arc (or edge) $e \in E$ is associated with a weight $d(e)$. The aim is to find a Hamiltonian circuit (also called a tour) of minimum weight, where the weight of a tour C is the sum of its arc (or edge) weights $\sum_{e \in C} d(e)$. We denote by $\mathcal{C}$ all possible $(n-1)!$ tours of graph G. If graph G is undirected, we have Symmetric TSP (STSP). If G is a directed graph, then we have Asymmetric TSP (ATSP).

In many situations, however, there is more than one objective function (criterion) to optimize [5,21]. In case of the TSP, we might want to minimize the travel distance, the travel time, the expenses, the number of flight changes, etc. This gives rise to a multicriteria TSP, where Hamiltonian circuits are sought that optimize several objectives simultaneously. For the m-criteria TSP, each arc (or edge) e has a weight $d(e) = (d_1(e), \ldots, d_m(e))$, which is a vector of length m (instead of a scalar). The total weight of a tour C is also a vector $D(C) = (D_1(C), \ldots, D_m(C))$, where $D_j(C) = \sum_{e \in C} d_j(e)$, $j = 1, \ldots, m$. Given this, the goal of the optimization problem could be the following: find a feasible solution which simultaneously minimizes each coordinate. Unfortunately, such an ideal solution rarely exists since objective functions are normally in conflict.

We say that one solution (tour) C^* dominates another solution C if the inequality $D(C^*) \leq D(C)$ holds. The notation $D(C^*) \leq D(C)$ means that $D(C^*) \neq D(C)$ and $D_i(C^*) \leqslant D_i(C)$ for all $i \in I$, where $I = \{1, 2, \ldots, m\}$. This relation $\leq$ is also called *the Pareto relation*. A set of non-dominated solutions is called *the set of pareto-optimal solutions* [5,21] $P_D(\mathcal{C}) = \{C \in \mathcal{C} \mid \nexists C^* \in \mathcal{C} : D(C^*) \leq D(C)\}$. In discrete problems, the set of pareto-optimal solutions is non-empty if the set of feasible solutions is non-empty, which is true for the multicriteria TSP. If we denote $\mathcal{D} = D(\mathcal{C})$, then *the Pareto set* is defined as $P(\mathcal{D}) = \{y \in \mathcal{D} \mid \nexists y^* \in \mathcal{D} : y^* \leq y\}$. We assume that the Pareto set is specified except for a collection of equivalence classes, generated by equivalence relation $C' \sim C''$ iff $D(C') = D(C'')$.

In this paper, we investigate the issue of the Pareto set reduction for the bi-ATSP.

3   Pareto Set Reduction

Axiomatic approach of the Pareto set reduction is applied to both discrete and continuous problems. Due to consideration of the multicriteria ATSP we formulate the basic concepts and results of the approach in terms of notations introduced in Sect. 2. Further, we investigate properties of the bi-ATSP in the scope of the Pareto set reduction.

3.1 Main Approach

According to [19] we consider the extended multicriteria problem $<\mathcal{C}, D, \prec>$:

- a set of all possible $(n-1)!$ tours $\mathcal{C}$;
- a vector criterion $D = (D_1, D_2, \ldots, D_m)$ defined on set $\mathcal{C}$;
- an asymmetric binary preference relation of the DM $\prec$ defined on set $\mathcal{D}$.

The notation $D(C') \prec D(C'')$ means that the DM prefers the solution C' to C''.

Binary relation $\prec$ satisfies some axioms of the so-called "reasonable" choice, according which it is irreflexive, transitive, invariant with respect to a linear positive transformation and compatible with each criteria $D_1, D_2, \ldots, D_m$. The compatibility means that the DM is interested in decreasing value of each criterion when values of other criteria are constant. Also, if for some feasible solutions C', $C'' \in \mathcal{C}$ the relation $D(C') \prec D(C'')$ holds, then tour C'' does not belong to the optimal choice within the whole set $\mathcal{C}$.

In [19], the author established the Edgeworth–Pareto principle: under axioms of "reasonable" choice any set of selected outcomes $Ch(\mathcal{D})$ belongs to the Pareto set $P(\mathcal{D})$. Here the set of selected outcomes is interpreted as some abstract set corresponded to the set of tours, that satisfy all hypothetic preferences of the DM. So, the optimal choice should be done within the Pareto set only if preference relation $\prec$ fulfills the axioms of "reasonable" choice.

In real-life multicriteria problems the Pareto set is rather wide. For this reason V. Noghin proposed a specific information on the DM's preference relation $\prec$ to reduce the Pareto set staying within the set of selected outcomes [18,19]:

Definition 1. *We say that there exists a "quantum of information" about the DM's preference relation $\prec$ if vector $y' \in \mathbb{R}^m$ such that $y'_i = -w_i < 0$, $y'_j = w_j > 0$, $y'_s = 0$ for all $s \in I \setminus \{i, j\}$ satisfies the expression $y' \prec 0_m$. In such case we will say, that the component of criteria i is more important than the component j with given positive parameters w_i, w_j.*

Thus, "quantum of information" shows that the DM is ready to compromise by increasing the criterion D_j by amount w_j for decreasing the criterion D_i by amount w_i. The quantity of relative loss is set by the so-called coefficient of relative importance $\theta = w_j/(w_i + w_j)$, therefore $\theta \in (0, 1)$.

As mentioned before the relation $\prec$ is invariant with respect to a linear positive transformation. Hence Definition 1 is equivalent to the existence of such vector $y'' \in \mathbb{R}^m$ with components $y''_i = \theta - 1$, $y''_j = \theta$, $y''_s = 0$ for all $s \in I \setminus \{i, j\}$ that the relation $y'' \prec 0_m$ holds. Further, in experimental study (Sect. 5) we consider "quantum of information" exactly in terms of coefficient θ.

In [19], the author established the rule of taking into account "quantum of information". This rule consists in constructing a "new" vector criterion using the components of the "old" one and parameters of the information w_i, w_j. Then one should find the Pareto set of "new" multicriteria problem with the same set of feasible solutions and "new" vector criterion. The obtained set will belong to the Pareto set of the initial problem and give a narrower upper bound on the optimal choice, as a result the Pareto set will be reduced.

The following theorem states the rule of applying "quantum of information" and specifies how to evaluate "new" vector criterion upon the "old" one.

Theorem 1 [19]. *Given a "quantum of information" by Definition 1, the inclusions $Ch(\mathcal{D}) \subseteq \hat{P}(\mathcal{D}) \subseteq P(\mathcal{D})$ are valid for any set of selected outcomes $Ch(\mathcal{D})$. Here $\hat{P}(\mathcal{D}) = D(P_{\hat{D}}(\mathcal{C}))$, and $P_{\hat{D}}(\mathcal{C})$ is the set of pareto-optimal solutions with respect to m-dimensional vector criterion $\hat{D} = (\hat{D}_1, \ldots, \hat{D}_m)$, where $\hat{D}_j = \theta D_i + (1 - \theta) D_j$, $\hat{D}_s = D_s$ for all $s \neq j$.*

Thus, "new" vector criterion $\hat{D}$ differs from the "old" one only by less important component j. In [12,17,29] one can find results on applying particular collections of "quanta of information" and scheme to arbitrary collection.

3.2 Pareto Set Reduction in Bi-ATSP

Here we consider the bi-ATSP and its properties with respect to reduction of the Pareto set.

Obviously, the upper bound on the cardinality of the Pareto set $P(\mathcal{D})$ is $(n - 1)!$, and this bound is tight [6]. In [26] authors established the maximum number of elements in the Pareto set for any multicriteria discrete problem, that in the case of the bi-ATSP gives the following upper bound: $|P(\mathcal{D})| \leqslant \min\{l_1, l_2\}$, where l_i is the number of different values in the set $\mathcal{D}_i = D_i(\mathcal{C})$, $i = 1, 2$. In the case of the bi-ATSP with integer weights we get $l_i \leqslant \max\{\mathcal{D}_i\} - \min\{\mathcal{D}_i\} + 1$, where values $\max\{\mathcal{D}_i\}$ and $\min\{\mathcal{D}_i\}$ can be replaced by upper and lower bounds on the objective function D_i, $i = 1, 2$.

Now, we go to establish theoretical results estimating the degree of the Pareto set reduction. Let us consider the case when all elements of the Pareto set lay on principal diagonal of some rectangle in the criterion space.

Theorem 2. *Let $P(\mathcal{D}) = \{(y_1, y_2) : y_2 = a - ky_1, y_1 \in \mathcal{D}_1, y_2 \in \mathcal{D}_2\}$, where a and k are arbitrary positive constants. Suppose the 1st criterion D_1 is more important than the 2nd one D_2 with coefficient of relative importance θ'. If $\theta' \geqslant k/(k+1)$, then the reduction of the Pareto set $\hat{P}(\mathcal{D})$ consists of only one element. In the case of $\theta' < k/(k + 1)$ the reduction does not hold, i.e. $\hat{P}(\mathcal{D}) = P(\mathcal{D})$.*

Theorem 3. *Let in Theorem 2, otherwise, the 2nd criterion D_2 is more important than the 1st one D_1 with coefficient of relative importance θ''. Then the reduction of the Pareto set $\hat{P}(\mathcal{D})$ has only one element if $\theta'' \geqslant 1/(k + 1)$, and $\hat{P}(\mathcal{D}) = P(\mathcal{D})$ if $\theta'' < 1/(k + 1)$.*

Particularly, if the feasible set $\mathcal{D}$ lay on the line $y_2 = a - ky_1$, we have $P(\mathcal{D}) = \mathcal{D}$, and the conditions of Theorems 2 and 3 hold. In such case we say, that *criteria D_1 and D_2 contradict each other with coefficient k*.

Obviously, for any bi-ATSP instance there exists the minimum number of parallel lines with a negative slope, that all elements of the Pareto set belong to them. Thus we have

Corollary 1. Let $P(\mathcal{D}) = \bigcup_{i=1}^{p}\{(y_1, y_2) : y_2 = a_i - ky_1, y_1 \in \mathcal{D}_1, y_2 \in \mathcal{D}_2\}$, where a_i, $i = 1, \ldots, p$, and k are arbitrary positive constants. If criterion D_1 is more important than criterion D_2 with coefficient of relative importance θ' and $\theta' \geqslant k/(k+1)$, or criterion D_2 is more important than criterion D_1 with coefficient of relative importance θ'' and $\theta'' \geqslant 1/(k+1)$, then $|\hat{P}(\mathcal{D})| \leqslant p$.

Further, we identify the condition that guarantees excluding at least one element from the Pareto set.

Proposition 1. Let the criterion D_i is more important than the criterion D_j with coefficient of relative importance θ. Suppose that there exist such tours $C', C'' \in P_D(\mathcal{C})$ that the following inequality holds:

$$\frac{D_i(C') - D_i(C'')}{D_j(C'') - D_j(C')} \geqslant \frac{1-\theta}{\theta}, \tag{1}$$

then $|P(\mathcal{D})| - |\hat{P}(\mathcal{D})| \geqslant 1$. Here $i, j \in \{1, 2\}$, $i \neq j$.

The difficulty in checking inequality (1) is that we should know two elements of the Pareto set. Meanwhile the tours $C_{min_1} = \operatorname{argmin}\{D_1(C),\ C \in \mathcal{C}\}$, $C_{min_2} = \operatorname{argmin}\{D_2(C), C \in \mathcal{C}\}$ are pareto-optimal by definition.

The proofs of Theorems 2, 3 and Proposition 1 are based on geometrical representation of the Pareto set reduction [19]. The results of this subsection are true for any discrete bicriteria problem.

4   Multi-Objective Genetic Algorithm

The genetic algorithm is a random search method that models a process of evolution of a population of *individuals* [24]. Each individual is a sample solution to the optimization problem being solved. Individuals of a new population are built by means of reproduction operators (crossover and/or mutation).

4.1   NSGA-II Scheme

To construct an approximation of the Pareto set to the bi-ATSP we develop a MOGA based on Non-dominated Sorting Genetic Algorithm II (NSGA-II) [4]. The NSGA-II is initiated by generating N random solutions of the initial population. Then the population is sorted based on the non-domination relation (the Pareto relation). All individuals of the population which are not dominated by any other individual compose the first *non-dominated level* and are marked with the *rank* of 1, all individuals which are dominated by at least one individual of the rank $i-1$ compose the i-th non-dominated level and are marked with the rank of i, $i = 2, 3, \ldots$. To get an estimate of the density of solutions surrounding a solution x in a non-dominated level of the population, two nearest solutions on each side of this solution are identified for each of the objectives. The estimation of solution x is called *crowding distance* and it is computed as a normalized perimeter of the cuboid formed in the criterion space by the nearest neighbors.

The NSGA-II is characterized by the population management strategy known as generational model [24]. Here the next population P_t is constructed from the best N solutions of the current population P_{t-1} and an offspring population Q_{t-1} created from P_{t-1} by applying selection, crossover, and mutation. The best solutions are selected using the rank and the crowding distance. Between two solutions with differing non-domination ranks, we prefer the solution with the lower rank. If both solutions belong to the same level, then we prefer the solution with the bigger crowding distance. The formal scheme of the NSGA-II is as follows:

Non-dominated Sorting Genetic Algorithm II

STEP 1. Construct the initial population P_0 of size N and assign $t := 1$. The population P_0 is sorted based on the non-domination relation. The crowding distances of individuals are calculated.

STEP 2. Repeat steps 2.1–2.4 until some stopping criterion is satisfied:

    2.1. Create offspring population Q_{t-1}.

    Steps 2.1.1–2.1.4 are performed N times:

        2.1.1. Choose two parent individuals $\mathbf{p}_1, \mathbf{p}_2$ from the population.

        2.1.2. Apply mutation to $\mathbf{p}_1$ and $\mathbf{p}_2$ and obtain individuals $\mathbf{p}'_1, \mathbf{p}'_2$.

        2.1.3. Create an offspring $\mathbf{p}'$, applying a crossover to $\mathbf{p}'_1$ and $\mathbf{p}'_2$.

        2.1.4. Put individual $\mathbf{p}'$ into population Q_{t-1}.

    2.2. Form a combined population $R_{t-1} := P_{t-1} \cup Q_{t-1}$. The population R_{t-1} is sorted based on the non-domination relation. The crowding distances of individuals are calculated.

    2.3. Construct population P_t from the best individuals of population R_{t-1} using the rank and the crowding distance to select solutions.

    2.4. Set $t := t + 1$.

One iteration of the presented NSGA-II is performed in $O(mN^2)$ time as shown in [4]. In our implementation of the NSGA-II four individuals of the initial population are constructed by a problem-specific heuristic presented in [7] for the ATSP with one criterion. The heuristic first solves the Assignment Problem, and then patches the circuits of the optimum assignment together to form a feasible tour in two ways. So, we create two solutions with each of the objectives. All other individuals of the initial population are generated randomly.

Each parent on Step 2.1.1 is chosen by *s-tournament selection*: sample randomly s individuals from the current population and select the best one by means of the rank and the crowding distance.

4.2 Recombination and Mutation Operators

The experimental results of [7,27] for the TSP indicate that reproduction operators with the adjacency-based representation of solutions have an advantage over operators, which emphasize the order or position of the vertices in parent solutions. We suppose that a feasible solution to the bi-ATSP is encoded as a list of arcs. In the recombination operator on Step 2.1.3 we use a variant of the *Directed Edge Crossover* (DEC), which may be considered as a "direct descendant" of Edge Crossover [27] originally developed for the STSP.

The DEC operator is *respectful* [23], i.e. all arcs shared by both parents are copied into the offspring. The remaining arcs are selected so as the preference is given to those arcs that are contained in at least one of the parents. Arcs are inserted taking into account the non-violation of sub-tour elimination constraints. If the obtained offspring is equal to one of the parents, then the result of the recombination is calculated by applying the well-known *shift mutation* [23] to one of the two parents with equal probability. This approach allows us to avoid creating a clone of parents and to maintain a diverse set of solutions in the population.

The mutation is also applied to each parent on Step 2.1.2 with probability p_{mut}, which is a tunable parameter of the MOGA. We use a mutation operator proposed in [7] for the one-criteria ATSP. It performs a random jump within 3-opt neighborhood, trying to improve a parent solution in terms of one of the criteria. Each time one of two objectives is used in mutation with equal probability.

5   Computational Experiment

This section presents the results of the computational experiment on the bi-ATSP instances. Our MOGA (NSGA-II-biATSP) was programmed in C++ and tested on a computer with Intel Core i5 3470 3.20 GHz processor, 4 Gb RAM.

Various meta-heuristics and heuristics have been developed for the multicriteria STSP, such as Pareto local search algorithms, MOEAs, multi-objective ant colony optimization methods, memetic algorithms and others (see, e.g., [9,10,13,15,22]). However, we have not found in the literature any multi-objective metaheuristic proposed specifically to the multicriteria ATSP and experimentally tested on instances with non-symmetric weights of arcs.

We carried out the preliminary study to evaluate the performance of our GA on bi-ATSP instances generated randomly with $n = 12$. The Pareto sets were found by an exact algorithm [19]. The generational distance [28] and the inverted generational distance [28] were involved as performance metrics. The experimental evaluation showed that the proposed MOGA yields competitive results. The values of metrics decrease not less than 7 times during 5000 iterations, and the final values are approximately 0.6 on average. The number of elements in the final approximation is at least 80% of $|P(\mathcal{D})|$. This indicates the convergence of the approximation obtained by NSGA-II-biATSP to the Pareto set and its diversity. Here the detailed description of the preliminary study is omitted, as the main goal of the paper is to investigate the axiomatic approach of the Pareto set reduction in the case of bi-ATSP.

Note that there exists MOOLIBRARY library [16], which contains instances of some discrete multicriteria problems. However, the multicriteria TSP is not presented in this library, so we generate the bi-ATSP test instances randomly and construct them from the ATSP instances of TSPLIB library, as well.

The reduction of the Pareto set approximation was tested on the following medium-size problem instances of four series with $n = 50$: S50[1,10][1,10], S50[1,20][1,20], S50[1,10][1,20], S50contr[1,2][1,2]. Each series consists of five problems with integer weights $d_1(\cdot)$ and $d_2(\cdot)$ of arcs randomly generated from intervals specified at the ending of the series name. In series S50contr[1,2][1,2] the criteria contradict each other with coefficient 1, i.e. weights are generated so that $d_2(e) = 3 - d_1(e)$ for all $e \in E$. We also took seven ATSP instances of series ftv from TSPLIB library [25]: ftv33, ftv35, ftv38, ftv44, ftv47, ftv55, ftv64. The ftv collection includes instances from vehicle routing applications [25]. These instances compose series denoted by SftvRand, and their arc weights are used for the first criterion. The arc weights for the second criterion are generated randomly from interval $[1, d_1^{\max}]$, where $d_1^{\max}$ is the maximum arc weight on the first criterion. We set the population size $N = 100$, the tournament size $s = 10$, and the mutation probability $p_{mut} = 0.1$. To construct an approximation of the Pareto set A for each instance we run NSGA-II-bi-ATSP once and the run continued for 5000 iterations.

We compare two cases when the 1st criterion is more important than the 2nd criterion (1st-2nd case), and vice versa (2nd-1st case). The degree of the reduction of the Pareto set approximation was investigated with respect to coefficient of relative importance varying from 0.1 to 0.9 by step 0.1. On all instances for each value of θ we re-evaluate the obtained approximation in terms of "new" vector criterion $\hat{D}$ upon the formulae from Theorem 1. Then by the complete enumeration we find the Pareto set approximation in "new" criterion space that gives us the reduction of the Pareto set approximation in the initial criterion space.

The number N^A of elements of the Pareto set approximation A and the percentage of the excluded elements from set A are presented on average over series in Tables 1 and 2. Let Δ_i be the difference between the maximum and minimum values of the Pareto set approximation on the i-th criterion, $i = 1, 2$. The value $\delta_{21} = \Delta_2/\Delta_1$ indicates the ratio between diversities of criteria of set A.

Table 1. Reduction of the Pareto set approximation in the 1st-2nd case

Series	θ									N^A aver	δ_{21} aver
	0.1	0.2	0.3	0.4	**0.5**	0.6	0.7	0.8	0.9		
S50contr[1,2][1,2]	0	0	0	0	**98.04**	98.04	98.04	98.04	98.04	51	1
S50[1,10][1,10]	4.42	17.76	41.64	60.43	**72.32**	78.61	90.29	95.92	97.78	45.8	1.02
S50[1,20][1,20]	5.97	23.09	38.15	59.92	**73.69**	79.91	90.18	94.71	98.05	57.4	1.07
SftvRand	6.63	16.47	27.99	41.31	**58.31**	72.96	86.24	93.7	96.71	61.86	1.56
S50[1,10][1,20]	2.19	9.02	19.18	28.3	**45.91**	61.69	71.79	83.39	95.46	51.6	2.08

For series S50[1,10][1,10], S50[1,20][1,20] when $\theta = 0.5$ approximately 70% of elements of the set A are excluded, and when $\theta = 0.7$ less than 10% of elements are remained. The statement is valid for both 1st–2nd and 2nd–1st cases. Series SftvRand shows different results: in the 1st–2nd case the reduction occurs "almost uniformly", i.e. the value of θ is almost proportional to the degree of the reduction, in the 2nd–1st case the condition $\theta = 0.5$ gives approximately 90% of the excluded elements. On series S50[1,10][1,20] in the 1st–2nd case the

Table 2. Reduction of the Pareto set approximation in the 2nd-1st case

Series	θ									N^A aver	δ_{21} aver
	0.1	0.2	0.3	0.4	**0.5**	0.6	0.7	0.8	0.9		
S50contr[1,2][1,2]	0	0	0	0	**98.04**	98.04	98.04	98.04	98.04	51	1
S50[1,10][1,10]	3.7	19.8	37.92	52.85	**67.38**	80.58	92.92	97.32	97.32	45.8	1.02
S50[1,20][1,20]	7.79	21.1	36.73	54.5	**69.8**	82.93	93.34	97.2	97.91	57.4	1.07
SftvRand	17.42	36.73	59	74.33	**87.37**	92.04	97.67	98.11	98.36	61.86	1.56
S50[1,10][1,20]	19.91	42.97	62.21	77.39	**92.41**	95.54	97.15	98.02	98.02	51.6	2.08

degree of the reduction grows slowly as θ tends to 1 in comparison to other series, and in the 2nd–1st case more than 90% of elements are eliminated at $\theta = 0.5$.

Also, we note that on series S50[1,10][1,20] (SftvRand) for $\theta = 0.5$ the percentage of the excluded elements in the 2nd–1st case is approximately 2 (1.5) times greater than the percentage of the excluded elements in the 1st–2nd case. Note that $\delta_{21} \approx 2$ for series S50[1,10][1,20] and $\delta_{21} \approx 1.5$ for series SftvRand. Therefore, the ratio between diversities of values of the Pareto set approximation on components of criterion influences on the degree of the reduction in the same proportion when $\theta = 0.5$ (each criterion has relatively the same importance).

On series S50contr[1,2][1,2], where the components of criterion contradict each other with coefficient 1, we do not have a reduction when $\theta < 0.5$, and the reduction up to one element takes place when $\theta \geqslant 0.5$. Thus, the results of the experiment confirm the theoretical results of Subsect. 3.2. Moreover, identical character of the reduction for both 1st–2nd and 2nd–1st cases occurs only on series S50[1,10][1,10], S50[1,20][1,20], and S50contr[1,2][1,2], which have the same diversity and distribution with respect to both criteria.

Based on the results of the experiment we suppose that the degree of the reduction of the Pareto set approximation will be similar for the large-size problems with the same structure as the considered instances.

6   Conclusion

We applied to the bicriteria ATSP the axiomatic approach of the Pareto set reduction proposed by V. Noghin. For particular cases the series of "quanta of information" that guarantee the reduction of the Pareto set were identified. An approximation of the Pareto set to the bicriteria ATSP was found by a new generational multi-objective genetic algorithm. The experimental evaluation indicated the degree of reduction of the Pareto set approximation for various "quanta of information" and various problem structures.

Further research may include construction and analysis of new classes of multicriteria ATSP instances with complex structures of the Pareto set. It is also important to consider real-life ATSP instances with real-life decision maker and investigate effectiveness of the axiomatic approach for them. Moreover, developing a faster implementation of the multi-objective genetic algorithm with steady-state replacement and local search procedures has great interest.

Acknowledgement. The research was supported by RFBR grant 17-07-00371 (A. Zakharov) and by the Ministry of Science and Education of the Russian Federation under the 5–100 Excellence Programme (Yu. Kovalenko).

References

1. Angel, E., Bampis, E., Gourvès, L., Monnot, J.: (Non)-approximability for the multi-criteria $TSP(1,2)$. In: Liśkiewicz, M., Reischuk, R. (eds.) FCT 2005. LNCS, vol. 3623, pp. 329–340. Springer, Heidelberg (2005). https://doi.org/10.1007/11537311_29

2. Ausiello, G., Crescenzi, P., Gambosi, G., Kann, V., Marchetti-Spaccamela, A., Protasi, M.: Complexity and Approximation. Springer, Heidelberg (1999). https://doi.org/10.1007/978-3-642-58412-1

3. Buzdalov, M., Yakupov, I., Stankevich, A.: Fast implementation of the steady-state NSGA-II algorithm for two dimensions based on incremental non-dominated sorting. In: Proceedings of the 2015 Annual Conference on Genetic and Evolutionary Computation (GECCO 2015), pp. 647–654 (2015). https://doi.org/10.1145/2739480.2754728

4. Deb, K., Pratap, A., Agarwal, S., Meyarivan, T.: A fast and elitist multi-objective genetic algorithm: NSGA-II. IEEE Trans. Evol. Comput. **6**(2), 182–197 (2002). https://doi.org/10.1109/4235.996017

5. Ehrgott, M.: Multicriteria Optimization. Springer, Heidelberg (2005). https://doi.org/10.1007/3-540-27659-9

6. Emelichev, V.A., Perepeliza, V.A.: Complexity of vector optimization problems on graphs. Optim. J. Math. Program. Oper. Res. **22**(6), 906–918 (1991). https://doi.org/10.1080/02331939108843732

7. Eremeev, A.V., Kovalenko, Y.V.: Genetic algorithm with optimal recombination for the asymmetric Travelling Salesman Problem. In: Lirkov, I., Margenov, S. (eds.) LSSC 2017. LNCS, vol. 10665, pp. 341–349. Springer, Cham (2018). https://doi.org/10.1007/978-3-319-73441-5_36

8. Figueira, J.L., Greco, S., Ehrgott, M.: Multiple Criteria Decision Analysis: State of the Art Surveys. Springer, New York (2005). https://doi.org/10.1007/b100605

9. Garcia-Martinez, C., Cordon, O., Herrera, F.: A taxonomy and an empirical analysis of multiple objective ant colony optimization algorithms for the bi-criteria TSP. Eur. J. Oper. Res. **180**, 116–148 (2007). https://doi.org/10.1016/j.ejor.2006.03.041

10. Jaszkiewicz, A., Zielniewicz, P.: Pareto memetic algorithm with path relinking for bi-objective traveling salesperson problem. Eur. J. Oper. Res. **193**, 885–890 (2009). https://doi.org/10.1016/j.ejor.2007.10.054

11. Klimova, O.N.: The problem of the choice of optimal chemical composition of shipbuilding steel. J. Comput. Syst. Sci. Int. **46**(6), 903–907 (2007). https://doi.org/10.1134/S106423070706007X

12. Klimova, O.N., Noghin, V.D.: Using interdependent information on the relative importance of criteria in decision making. Comput. Math. Math. Phys. **46**(12), 2080–2091 (2006). https://doi.org/10.1134/S0965542506120074

13. Kumar, R., Singh, P.K.: Pareto evolutionary algorithm hybridized with local search for biobjective TSP. In: Abraham, A., Grosan, C., Ishibuchi, H. (eds.) Hybrid Evolutionary Algorithms. SCI, vol. 14, pp. 361–398. Springer, Heidelberg (2007). https://doi.org/10.1007/978-3-540-73297-6_14

14. Li, H., Zhang, Q.: Multiobjective optimization problems with complicated Pareto sets, MOEA/D and NSGA-II. IEEE Trans. Evol. Comput. **13**(2), 284–302 (2009). https://doi.org/10.1109/TEVC.2008.925798

15. Lust, T., Teghem, J.: The Multiobjective Traveling Salesman Problem: a survey and a new approach. In: Coello Coello, C.A., Dhaenens, C., Jourdan, L. (eds.) Advances in Multi-Objective Nature Inspired Computing. SCI, vol. 272, pp. 119–141. Springer, Heidelberg (2010). https://doi.org/10.1007/978-3-642-11218-8_6

16. Multiobjective optimization library: http://home.ku.edu.tr/~moolibrary/. Online accessed 09 Feb 2018

17. Noghin, V.D.: Reducing the Pareto set based on set-point information. Sci. Tech. Inf. Proc. **38**(6), 435–439 (2011). https://doi.org/10.3103/S0147688211050078

18. Noghin, V.D.: Reducing the Pareto set algorithm based on an arbitrary finite set of information "quanta". Sci. Tech. Inf. Proc. **41**(5), 309–313 (2014). https://doi.org/10.3103/S0147688214050086

19. Noghin, V.D.: Reduction of the Pareto Set: An Axiomatic Approach. Springer, Cham (2018). https://doi.org/10.1007/978-3-319-67873-3

20. Noghin, V.D., Prasolov, A.V.: The quantitative analysis of trade policy: a strategy in global competitive conflict. Int. J. Bus. Continuity Risk Manage. **2**(2), 167–182 (2011). https://doi.org/10.1504/IJBCRM.2011.041490

21. Podinovskiy, V.V., Noghin, V.D.: Pareto-optimal'nye resheniya mnogokriterial'nyh zadach (Pareto-Optimal Solutions of Multicriteria Problems). Fizmatlit, Moscow (2007). (in Russian)

22. Psychas, I.D., Delimpasi, E., Marinakis, Y.: Hybrid evolutionary algorithms for the multiobjective traveling salesman problem. Expert Syst. Appl. **42**(22), 8956–8970 (2015). https://doi.org/10.1016/j.eswa.2015.07.051

23. Radcliffe, N.J.: The algebra of genetic algorithms. Ann. Math. Artif. Intell. **10**(4), 339–384 (1994). https://doi.org/10.1007/BF01531276

24. Reeves, C.R.: Genetic algorithms for the operations researcher. INFORMS J. Comput. **9**(3), 231–250 (1997)

25. Reinelt, G.: TSPLIB - a traveling salesman problem library. ORSA J. Comput. **3**(4), 376–384 (1991). https://doi.org/10.1287/ijoc.3.4.376

26. Vinogradskaya, T.M., Gaft, M.G.: Tochnaya verhn'ya otzenka chisla nepodchinennyh reshenii v mnogokriterial'nyh zadachah (The least upper estimate for the number of nondominated solutions in multi-criteria problems). Avtom. Telemekh. **9**, 111–118 (1974). in Russian

27. Whitley, D., Starkweather, T., McDaniel, S., Mathias, K.: A comparison of genetic sequencing operators. In: Proceedings of the Fourth International Conference on Genetic Algorithms, pp. 69–76. Morgan Kaufmann, New York (1991)

28. Yuan, Y., Xu, H., Wang, B.: An improved NSGA-III procedure for evolutionary many-objective optimization. In: Proceedings of the 2014 Annual Conference on Genetic and Evolutionary Computation (GECCO 2014), pp. 661–668 (2014). https://doi.org/10.1145/2576768.2598342

29. Zakharov, A.O.: Pareto-set reducing using compound information of a closed type. Sci. Tech. Inf. Proc. **39**(5), 293–302 (2012). https://doi.org/10.3103/S0147688212050073

30. Zitzler, E., Brockhoff, D., Thiele, L.: The hypervolume indicator revisited: on the design of pareto-compliant indicators via weighted integration. In: Obayashi, S., Deb, K., Poloni, C., Hiroyasu, T., Murata, T. (eds.) EMO 2007. LNCS, vol. 4403, pp. 862–876. Springer, Heidelberg (2007). https://doi.org/10.1007/978-3-540-70928-2_64

31. Zitzler, E., Laumanns, M., Thiele, L.: SPEA2: improving the strength pareto evolutionary algorithm. In: Proceedings of Evolutionary Methods for Design, Optimisation and Control with Application to Industrial Problems, pp. 95–100 (2001)

Optimization Problems in Data Analysis

Randomized Algorithms for Some Clustering Problems

Alexander Kel'manov[1,2], Vladimir Khandeev[1,2(✉)], and Anna Panasenko[1,2]

[1] Sobolev Institute of Mathematics, 4 Koptyug Avenue, 630090 Novosibirsk, Russia
{kelm,khandeev,a.v.panasenko}@math.nsc.ru
[2] Novosibirsk State University, 2 Pirogova Street, 630090 Novosibirsk, Russia

Abstract. We consider two strongly NP-hard problems of clustering a finite set of points in Euclidean Space. Both problems have applications, in particular, in data analysis, data mining, pattern recognition, and machine learning. In the first problem, an input set is given and we need to find a cluster (i.e., a subset) of a given size which minimizes the sum of squared distances between the elements of this cluster and its centroid (the geometric center). Every point outside this cluster is considered as singleton cluster. In the second problem, we need to partition a finite set into two clusters minimizing the sum over both clusters of the weighted intracluster sums of the squared distances between the elements of the clusters and their centers. The center of the first cluster is unknown and determined as the centroid, while the center of the second one is the origin. The weight factors for both intracluster sums are the given clusters sizes. In this paper, we present parameterized randomized algorithms for these problems. For given upper bounds of the relative error and failure probability, the parameter value is defined for which both our algorithms find approximate solutions in a polynomial time. This running time is linear on the space dimension and on the input set size. The conditions are found under which these algorithms are asymptotically exact and have the time complexity that is linear on the space dimension and quadratic on the size of the input set.

Keywords: Euclidean space · Clustering · NP-hardness
Randomized · Approximation algorithm · Asymptotic accuracy

1 Introduction

In the paper, we study two strongly NP-hard problems. The line of the study is the questions of the algorithmic approximability of these problems. Our goal is to construct fast approximation algorithms providing the solution in a linear time and also to find the conditions under which the algorithms guarantee the asymptotically exact solutions.

Despite the intensive research of the concerned problems during the last years and despite the fact that there are effective algorithms with performance guarantees (see the next section) for these problems, there have been no fast algorithms

© Springer International Publishing AG, part of Springer Nature 2018
A. Eremeev et al. (Eds.): OPTA 2018, CCIS 871, pp. 109–119, 2018.
https://doi.org/10.1007/978-3-319-93800-4_9

with the linear time complexity until now. Meanwhile, such algorithms are necessary and demanded (especially in recent years) tools for solving the Big-scaling problems arising, particularly, in Data science, Data mining, Machine learning, Pattern recognition.

The paper has the following structure. Section 2 contains the problems formulation, their interpretation and known algorithmic results. In the same section we announce the obtained results. In Sect. 3, we formulate some statements to justify the algorithms and their properties. Finally, Sect. 4 contains the step-by-step description of the algorithms and justification of their properties (accuracy, time complexity, failure probability). Conditions under which the algorithms are asymptotically exact are established in the same section.

2   Problems Formulation and Related Problems, Known and Obtained Results

Everywhere below $\mathbb{R}$ denotes the set of real numbers, $\| \cdot \|$ denotes the Euclidean norm, and $\langle \cdot, \cdot \rangle$ denotes the scalar product.

The problems under consideration are stated as follows.

Problem 1. Given a set $\mathcal{Y} = \{y_1, \ldots, y_N\}$ in Euclidean space of dimension d and a positive integer M. *Find* a subset $\mathcal{C} \subseteq \mathcal{Y}$ of size M minimizing the value of

$$f(\mathcal{C}) = \sum_{y \in \mathcal{C}} \|y - \overline{y}(\mathcal{C})\|^2 ,$$

where $\overline{y}(\mathcal{C}) = \frac{1}{|\mathcal{C}|} \sum\limits_{y \in \mathcal{C}} y$ is the centroid of $\mathcal{C}$.

Problem 2. Given an N-element set $\mathcal{Y}$ of points in d-dimensional Euclidean space and a positive integer number $M \leq N$. *Find* a partition of $\mathcal{Y}$ into two non-empty clusters $\mathcal{C}$ and $\mathcal{Y} \setminus \mathcal{C}$ such that

$$g(\mathcal{C}) = |\mathcal{C}| \sum_{y \in \mathcal{C}} \|y - \overline{y}(\mathcal{C})\|^2 + |\mathcal{Y} \setminus \mathcal{C}| \sum_{y \in \mathcal{Y} \setminus \mathcal{C}} \|y\|^2 \to \min , \qquad (1)$$

subject to constraint $|\mathcal{C}| = M$.

Problem 1 can be treated as a search in the set $\mathcal{Y}$ for the subset in the form of spherical concentration of M points having the minimum total quadratic variation with respect to their centroid. Since the centroid of a singleton set is equal to the unique element of this set, the problem can be also treated as a partition of $\mathcal{Y}$ into $N - M + 1$ clusters such that the size of one of the clusters is equal to M and the sizes of other clusters are equal to 1.

In Problem 2, it is required to find a 2-partition of $\mathcal{Y}$ so as to minimize the sum of cardinality-weighted intracluster sums of squared distances between the points of the clusters and their centers; the center of cluster $\mathcal{Y} \setminus \mathcal{C}$ is the origin and the center of cluster $\mathcal{C}$ is the unknown centroid $\overline{y}(\mathcal{C})$.

One can treat both problems, in particular, as the problems of Data editing, Data cleaning, Data mining, and Machine learning (see, for example, [1–8] and papers cited therein). Some meaningful interpretations of Problems 1 and 2 can be found in [9–14,16–18].

The interpretation of Problem 1 in the terms of data analysis is the following. There is a table $\mathcal{Y}$ containing the results $\{y_1, \ldots, y_N\}$ of measurements of a tuple y of d significant digital informational characteristics for a family of some objects. Several objects are identical and have the same characteristics; the number M of these objects is known. The remaining objects are various and have different characteristics. There is some error in each measurement result in the table. Moreover, no correspondence between the objects and the table elements is available. It is required, by using a criterium of total quadratic variation, to find a subset $\mathcal{C}$ of the input set $\mathcal{Y}$ corresponding to the set of identical objects and evaluate the tuple $\overline{y}(\mathcal{C})$ of characteristics of these objects based on the measurement results (taking into account that the data have some measurement errors).

Problem 2 can also be treated in a similar way. There is a table $\mathcal{Y}$ containing measurement results for two groups of objects. Each group consists of the homogeneous (in terms of a certain tuple of characteristics) objects. The first group $\mathcal{C}$ contains M objects and the second group $\mathcal{Y}\backslash\mathcal{C}$ contains $(N - M)$ objects. The objects in the first group have unknown characteristics, while the objects in the second group have the given characteristics (in particular, one can consider that all characteristics are equal to zero). It is required, by using the criterium (1), to partition the family $\mathcal{Y}$ into two parts and to evaluate the characteristics of the objects in the first group (taking into account that there is some error in each measurement result).

First, let us recall the known results for each problem. Problem 1 is also known as M-Variance [19]. Strong NP-hardness of this problem is substantiated in [9]. In the same paper, it was shown that there does not exist a fully polynomial time approximation scheme (FPTAS) for this problem unless $P = NP$. The exact algorithms with time complexity $\mathcal{O}(dN^{d+1})$ were proposed in [19,20]. If the space dimension d is fixed, these algorithms are polynomial and their time complexity is $\mathcal{O}(N^{d+1})$.

An exact algorithm for the case of integer inputs was presented in [10]. The time complexity of the algorithm is $\mathcal{O}(dN(2MB+1)^d)$, where B is the maximum absolute coordinate value in the input set. If the space dimension is fixed, the algorithm is pseudopolynomial and its time complexity is $\mathcal{O}(N(MB)^d)$.

In [11], a 2-approximation polynomial algorithm with time complexity $\mathcal{O}(dN^2)$ was presented for the general case of the problem. A polynomial time approximation scheme (PTAS) was proposed in [21]. The time complexity of the scheme is $\mathcal{O}(dN^{2/\varepsilon+1}(9/\varepsilon)^{3/\varepsilon})$, where $\varepsilon > 0$ is a relative error.

In [12], the algorithm was proposed which allows finding a $(1+\varepsilon)$-approximate solution in $\mathcal{O}(dN^2(2\sqrt{d}M/\varepsilon + 2)^d)$ time for given $\varepsilon \in (0,1)$. For fixed space dimension d, the algorithm runs in $\mathcal{O}(N^2(M/\varepsilon)^d)$ time and implements an FPTAS.

An improved approximation scheme that allows finding a $(1+\varepsilon)$-approximate solution in $\mathcal{O}\left(dN^2\left(\sqrt{\frac{2d}{\varepsilon}}+2\right)^d\right)$ time was proposed in [13]. If the space dimension is fixed, the algorithm implements an FPTAS, since its time complexity in this case is $\mathcal{O}(N^2(1/\varepsilon)^{d/2})$. In the same work, an improved approximation scheme were proposed. The time complexity of this scheme is $\mathcal{O}\left(\sqrt{d}N^2\left(\frac{\pi e}{2}\right)^{d/2}\left(\sqrt{\frac{2}{\varepsilon}}+2\right)^d\right)$. In the case of dimension $d = \mathcal{O}(\log N)$, the improved scheme remains polynomial. In this case it implements a PTAS with $\mathcal{O}\left(N^{C\,(1.05+\log(2+\sqrt{\frac{2}{\varepsilon}}))}\right)$ time, where C is a positive constant.

The following results were obtained for Problem 2. First, Problem 2 is close to the known [22–25] *Mini-Sum 2-clustering* problem. In this problem, it is required to find a 2-partition minimizing the value of

$$|\mathcal{C}|\sum_{y\in\mathcal{C}}\|y-\overline{y}(\mathcal{C})\|^2 + |\mathcal{Y}\setminus\mathcal{C}|\sum_{y\in\mathcal{Y}\setminus\mathcal{C}}\|y-\overline{y}(\mathcal{Y}\setminus\mathcal{C})\|^2,$$

where the both centroids are unknown (in Problem 2, only one centroid is unknown). Note that Problem 2 and *Mini-Sum 2-clustering* problem are not equivalent. The strong NP-hardness of the both problems was proved in [14,15]. In addition, in the cited papers it was shown that there are no FPTAS for these problems unless $P = NP$.

In [16], an exact algorithm was constructed for the case of Problem 2. In this case, the input points have integer components. The running time of the algorithm is $\mathcal{O}(dN(2MB + 1)^d)$, where B is the maximum absolute value of coordinates of the input points. If the dimension d of the space is fixed, the algorithm is pseudopolynomial and its running time is $\mathcal{O}(N(MB)^d)$.

An approximation algorithm that allows one to find a 2-approximate solution of the general case of the problem in $\mathcal{O}\left(dN^2\right)$ time was constructed in [17].

In [18], an approximation scheme that allows finding a $(1 + \varepsilon)$-approximate solution in $\mathcal{O}\left(dN^2\left(\sqrt{\frac{2d}{\varepsilon}}+2\right)^d\right)$ time was proposed. It implements an FPTAS in the case of the fixed space dimension, since its time complexity in that case is $\mathcal{O}(N^2(1/\varepsilon)^{d/2})$.

Moreover, in [13], the modification of this algorithm with improved running time $\mathcal{O}\left(\sqrt{d}N^2\left(\frac{\pi e}{2}\right)^{d/2}\left(\sqrt{\frac{2}{\varepsilon}}+2\right)^d\right)$ was proposed. The algorithm implements an FPTAS with $\mathcal{O}(N^2(1/\varepsilon)^{d/2})$ running time in the case of fixed space dimension and remains polynomial for instances of dimension $d = \mathcal{O}(\log N)$. In this case it implements a PTAS with $\mathcal{O}\left(N^{C\,(1.05+\log(2+\sqrt{\frac{2}{\varepsilon}}))}\right)$ time, where C is a positive constant.

In this paper, we present randomized algorithms for Problems 1 and 2. Under assumption $M \geq \beta N$, where $\beta \in (0,1)$ is some constant, and given $\varepsilon > 0$ and $\gamma \in (0,1)$, our algorithms find $(1 + \varepsilon)$-approximate solutions of the problems with probability not less than $1 - \gamma$ in $\mathcal{O}(dN)$ time. The conditions are found under which the algorithms find $(1+\varepsilon_N)$-approximate solutions of the problems in $\mathcal{O}(dN^2)$ time with probability not less than $1-\gamma_N$, where $\varepsilon_N \to 0$ and $\gamma_N \to 0$ as $N \to \infty$, i.e., the conditions under which the algorithms are asymptotically exact.

3   Algorithms Foundations

In order to justify our algorithm we need a few auxiliary assertions.

The probabilistic basis of the algorithms is the following two lemmas [26]. The former is based on Markov inequality; the latter — on Chernoff bound.

Lemma 1. *Let $\mathcal{Z}$ be an N-element set of points in d-dimensional Euclidean space, $\mathcal{C} \subseteq \mathcal{Z}$, $|\mathcal{C}| = M$. Let $\mathcal{T}$ be a multiset obtained by randomly and independently choosing k elements from $\mathcal{Z}$ with replacement. Moreover, let $\overline{z}(\mathcal{C}) = \frac{1}{M} \sum_{z \in \mathcal{C}} z$ and $\overline{z}(\mathcal{T} \cap \mathcal{C}) = \frac{1}{|\mathcal{T} \cap \mathcal{C}|} \sum_{z \in \mathcal{T} \cap \mathcal{C}} z$ be the centroids of set $\mathcal{C}$ and multiset $\mathcal{T} \cap \mathcal{C}$, respectively. Then*

$$\Pr\left(\sum_{z \in \mathcal{C}} \|z - \overline{z}(\mathcal{T} \cap \mathcal{C})\|^2 \geq \left(1 + \frac{1}{\delta t}\right) \sum_{z \in \mathcal{C}} \|z - \overline{z}(\mathcal{C})\|^2 \mid |\mathcal{T} \cap \mathcal{C}| \geq t\right) \leq \delta$$

for any positive integer $t \leq k$ and real $\delta \in (0,1)$.

Lemma 2. *Let the conditions of Lemma 1 hold. Then*

$$\Pr\left(|\mathcal{T} \cap \mathcal{C}| \leq (1 - \nu)\frac{M}{N}k\right) \leq e^{-\frac{\nu^2 Mk}{2N}}$$

for arbitrary $\nu \in (0,1)$.

The proof of the following lemma can be found in [16].

Lemma 3. *Let*

$$S(\mathcal{C}, x) = |\mathcal{C}|\sum_{y \in \mathcal{C}} \|y - x\|^2 + |\mathcal{Y} \setminus \mathcal{C}| \sum_{y \in \mathcal{Y} \setminus \mathcal{C}} \|y\|^2, \ \mathcal{C} \subseteq \mathcal{Y}, \ x \in \mathbb{R}^d \ .$$

Then the next statements are true:

(1) for any nonempty fixed set $\mathcal{C} \subseteq \mathcal{Y}$ the minimum of $S(\mathcal{C}, x)$ over $x \in \mathbb{R}^d$ is reached at the point $x = \overline{y}(\mathcal{C})$;

(2) if $|\mathcal{C}| = M = const$, then for any fixed point $x \in \mathbb{R}^d$ the minimum of $S(\mathcal{C}, x)$ over $\mathcal{C} \subseteq \mathcal{Y}$ is reached at the subset $\mathcal{B}^x$ that consists of M points of $\mathcal{Y}$ at which the function

$$h^x(y) = (2M - N)\|y\|^2 - 2M\langle y, x\rangle, \ y \in \mathcal{Y}, \tag{2}$$

has the smallest values.

4   Randomized Algorithms

Let us formulate the following algorithm for solving Problem 1.

A l g o r i t h m $\mathcal{A}_1$.

Input: a set $\mathcal{Y}$, a positive integer M, a parameter k.

Step 1. Generate a multiset $\mathcal{T}$ by independently and randomly choosing k elements one after another (with replacement) from $\mathcal{Y}$.

Step 2. For each nonempty multisubset $\mathcal{H}$ of $\mathcal{T}$, compute the centroid $\bar{y}(\mathcal{H})$ and form a subset $\mathcal{C}$ that consists of M elements closest (by distance) to $\bar{y}(\mathcal{H})$. Compute $f(\mathcal{C})$.

Step 3. In the family of solutions found at Step 2, choose the subset $\mathcal{C} = \mathcal{C}_{\mathcal{A}_1}$ for which $f(\mathcal{C})$ is minimal. If there are several optimal values, then choose any of them.

Output: the set $\mathcal{C}_{\mathcal{A}_1}$.

The algorithm for solving Problem 2 is similar; the main difference between the algorithms is in constructing a feasible solution of the problem at Step 2.

A l g o r i t h m $\mathcal{A}_2$.

Input: a set $\mathcal{Y}$, a positive integer M, a parameter k.

Step 1. Generate a multiset $\mathcal{T}$ by independently and randomly choosing k elements one after another (with replacement) from $\mathcal{Y}$.

Step 2. For each nonempty multisubset $\mathcal{H}$ of $\mathcal{T}$, compute the centroid $\bar{y}(\mathcal{H})$ and form a subset $\mathcal{C}$ that consists of M elements with the smallest values $h^{\bar{y}(\mathcal{H})}(z)$, $z \in \mathcal{Y}$ (using formula (2)). Compute $g(\mathcal{C})$.

Step 3. In the family of solutions found at Step 2, choose the subset $\mathcal{C} = \mathcal{C}_{\mathcal{A}_2}$ for which $g(\mathcal{C})$ is minimal. If there are several optimal values, then choose any of them.

Output: the set $\mathcal{C}_{\mathcal{A}_2}$.

The following theorem is true.

Theorem 1. *For an arbitrary real $\delta \in (0,1)$ and positive integers $t \le k$, algorithms $\mathcal{A}_1$ and $\mathcal{A}_2$ find $(1 + \frac{1}{\delta t})$-approximate solutions of Problems 1 and 2 in $\mathcal{O}(2^k d(k + N))$ time with a probability of at least $1 - (\delta + \alpha)$, where $\alpha =$*
$$\sum_{i=0}^{t-1} \binom{k}{i} \left(\frac{M}{N}\right)^i \left(1 - \frac{M}{N}\right)^{k-i}.$$

Proof. Let us prove the accuracy bound of algorithm $\mathcal{A}_1$. Let $\mathcal{C}_1^*$ be the optimal solution of Problem 1 and let $\mathcal{C}_{\mathcal{A}_1}$ be the set produced by algorithm $\mathcal{A}_1$.

Assume that the multiset $\mathcal{T}$ is generated so that $|\mathcal{T} \cap \mathcal{C}_1^*| \ge 1$. In this case, the multisubset $\mathcal{H} = \mathcal{T} \cap \mathcal{C}_1^*$ was considered at Step 2 of the algorithm. Let $\mathcal{C}_1$ be the subset of $\mathcal{Y}$ constructed at this step, which consists of M elements closest (by distance) to $\bar{y}(\mathcal{T} \cap \mathcal{C}_1^*)$.

The definition of Step 3 yields

$$f(\mathcal{C}_{\mathcal{A}_1}) \le f(\mathcal{C}_1) . \tag{3}$$

In addition, since for an arbitrary finite set $\mathcal{Z} \subset \mathbb{R}^d$ of points the minimum of $\sum_{y \in \mathcal{Z}} \|y - x\|^2$ over $x \in \mathbb{R}^d$ is reached at the point $x = \frac{1}{|\mathcal{Z}|} \sum_{z \in \mathcal{Z}} z$, for the right side of (3) we get

$$f(\mathcal{C}_1) = \sum_{y \in \mathcal{C}_1} \|y - \overline{y}(\mathcal{C}_1)\|^2 \le \sum_{y \in \mathcal{C}_1} \|y - \overline{y}(\mathcal{H})\|^2 . \tag{4}$$

Since $\mathcal{C}_1$ contains M points closest to $\overline{y}(\mathcal{H})$, we have

$$\sum_{y \in \mathcal{C}_1} \|y - \overline{y}(\mathcal{H})\|^2 \le \sum_{y \in \mathcal{C}_1^*} \|y - \overline{y}(\mathcal{H})\|^2 . \tag{5}$$

Combining (3)–(5), we get that, if $|\mathcal{T} \cap \mathcal{C}_1^*| \ge 1$, the following chain of inequalities is true:

$$f(\mathcal{C}_{\mathcal{A}_1}) \le f(\mathcal{C}_1) \le \sum_{y \in \mathcal{C}_1} \|y - \overline{y}(\mathcal{H})\|^2 \le \sum_{y \in \mathcal{C}_1^*} \|y - \overline{y}(\mathcal{H})\|^2 . \tag{6}$$

Applying Lemma 1 with $\mathcal{Z} = \mathcal{Y}$ and $\mathcal{C} = \mathcal{C}_1^*$, we get that

$$\sum_{y \in \mathcal{C}_1^*} \|y - \overline{y}(\mathcal{H})\|^2 < \left(1 + \frac{1}{\delta t}\right) \sum_{y \in \mathcal{C}_1^*} \|y - \overline{y}(\mathcal{C}_1^*)\|^2 \tag{7}$$

under the condition $|\mathcal{T} \cap \mathcal{C}_1^*| \ge t$ with a probability of at least $1 - \delta$.

Combining (6)–(7) yields that

$$f(\mathcal{C}_{\mathcal{A}_1}) < \left(1 + \frac{1}{\delta t}\right) \sum_{y \in \mathcal{C}_1^*} \|y - \overline{y}(\mathcal{C}_1^*)\|^2 = \left(1 + \frac{1}{\delta t}\right) f(\mathcal{C}_1^*)$$

under the condition $|\mathcal{T} \cap \mathcal{C}_1^*| \ge t$ with a probability of at least $1 - \delta$. In terms of conditional probability it means that

$$\mathrm{Pr}\left(f(\mathcal{C}_{\mathcal{A}_1}) < \left(1 + \frac{1}{\delta t}\right) f(\mathcal{C}_1^*) \,\Big|\, |\mathcal{T} \cap \mathcal{C}_1^*| \ge t\right) \ge 1 - \delta .$$

Therefore, denoting $\alpha = \mathrm{Pr}\left(|\mathcal{T} \cap \mathcal{C}_1^*| < t\right)$, we get

$$\mathrm{Pr}\left(f(\mathcal{C}_{\mathcal{A}_1}) \ge \left(1 + \frac{1}{\delta t}\right) f(\mathcal{C}_1^*)\right)$$

$$= \mathrm{Pr}\left(f(\mathcal{C}_{\mathcal{A}_1}) \ge \left(1 + \frac{1}{\delta t}\right) f(\mathcal{C}_1^*) \text{ and } |\mathcal{T} \cap \mathcal{C}_1^*| \ge t\right) +$$

$$\mathrm{Pr}\left(f(\mathcal{C}_{\mathcal{A}_1}) \ge \left(1 + \frac{1}{\delta t}\right) f(\mathcal{C}_1^*) \text{ and } |\mathcal{T} \cap \mathcal{C}_1^*| < t\right)$$

$$\le \mathrm{Pr}\left(f(\mathcal{C}_{\mathcal{A}_1}) \ge \left(1 + \frac{1}{\delta t}\right) f(\mathcal{C}_1^*) \,\Big|\, |\mathcal{T} \cap \mathcal{C}_1^*| \ge t\right) + \mathrm{Pr}\left(|\mathcal{T} \cap \mathcal{C}_1^*| < t\right)$$

$$\le \delta + \alpha .$$

Finally, equality $\alpha = \sum_{i=0}^{t-1} \binom{k}{i} \left(\frac{M}{N}\right)^i \left(1 - \frac{M}{N}\right)^{k-i}$ follows from the fact that generating the multiset $\mathcal{T}$ can be considered as k independent Bernoulli trials, where each "success" is the result "the element chosen from $\mathcal{Y}$ lies in $\mathcal{C}_1^*$".

The proof of the accuracy bound of algorithm $\mathcal{A}_2$ is partially similar to the proof of the accuracy bound of algorithm $\mathcal{A}_1$. Let $\mathcal{C}_2^*$ be the optimal solution of Problem 2 and let $\mathcal{C}_{\mathcal{A}_2}$ be the set produced by algorithm $\mathcal{A}_2$. Assume that in algorithm $\mathcal{A}_2$ the multiset $\mathcal{T}$ is generated so that $|\mathcal{T} \cap \mathcal{C}_2^*| \geq 1$; in that case, let $\mathcal{C}_2$ be the subset of $\mathcal{Y}$ constructed at Step 2, which consists of M elements with the smallest values $h^{\overline{y}(\mathcal{H})}(z)$, $z \in \mathcal{Y}$, for $\mathcal{H} = \mathcal{T} \cap \mathcal{C}_2^*$.

The definition of Step 3 yields

$$g(\mathcal{C}_{\mathcal{A}_2}) \leq g(\mathcal{C}_2) . \tag{8}$$

Similarly to (4), we obtain

$$g(\mathcal{C}_2) = |\mathcal{C}_2| \sum_{y \in \mathcal{C}_2} \|y - \overline{y}(\mathcal{C}_2)\|^2 + |\mathcal{Y} \setminus \mathcal{C}_2| \sum_{y \in \mathcal{Y} \setminus \mathcal{C}_2} \|y\|^2$$
$$\leq |\mathcal{C}_2| \sum_{y \in \mathcal{C}_2} \|y - \overline{y}(\mathcal{H})\|^2 + |\mathcal{Y} \setminus \mathcal{C}_2| \sum_{y \in \mathcal{Y} \setminus \mathcal{C}_2} \|y\|^2 . \tag{9}$$

Since $\mathcal{C}_2$ contains M points with the smallest values $h^{\overline{y}(\mathcal{H})}(z)$, $z \in \mathcal{Y}$, according to Lemma 3 we have

$$|\mathcal{C}_2| \sum_{y \in \mathcal{C}_2} \|y - \overline{y}(\mathcal{H})\|^2 + |\mathcal{Y} \setminus \mathcal{C}_2| \sum_{y \in \mathcal{Y} \setminus \mathcal{C}_2} \|y\|^2$$
$$\leq |\mathcal{C}_2^*| \sum_{y \in \mathcal{C}_2^*} \|y - \overline{y}(\mathcal{H})\|^2 + |\mathcal{Y} \setminus \mathcal{C}_2^*| \sum_{y \in \mathcal{Y} \setminus \mathcal{C}_2^*} \|y\|^2 . \tag{10}$$

As for algorithm $\mathcal{A}_1$, combining (8)–(10) and applying Lemma 1 yield that

$$g(\mathcal{C}_{\mathcal{A}_2}) < \left(1 + \frac{1}{\delta t}\right) |\mathcal{C}_2^*| \sum_{y \in \mathcal{C}_2^*} \|y - \overline{y}(\mathcal{C}_2^*)\|^2 + |\mathcal{Y} \setminus \mathcal{C}_2^*| \sum_{y \in \mathcal{Y} \setminus \mathcal{C}_2^*} \|y\|^2$$

under the condition $|\mathcal{T} \cap \mathcal{C}_2^*| \geq t$ with a probability of at least $1 - \delta$, where the right side can be bounded from above by $\left(1 + \frac{1}{\delta t}\right) g(\mathcal{C}_2^*)$. The remaining proof is similar to the previous proof.

Let us estimate the time complexity of the algorithms. Step 1 requires $\mathcal{O}(k)$ time. Step 2 is executed 2^k times. The centroid of each multisubset $\mathcal{H}$ in both algorithms is computed in $\mathcal{O}(dk)$ time. In algorithm $\mathcal{A}_1$, the distances between this centroid and points of $\mathcal{Y}$ are computed in $\mathcal{O}(dN)$ time, and M points closest to this centroid are chosen in $\mathcal{O}(N)$ time without sorting (see, e.g., [27]). Similarly, in algorithm $\mathcal{A}_2$, calculation of $h^{\overline{y}(\mathcal{H})}(z)$, $z \in \mathcal{Y}$, requires at most $\mathcal{O}(dN)$ time, and finding M smallest elements in the set of N elements requires $\mathcal{O}(N)$ time without sorting. Step 3 (choosing the least element) requires at most $\mathcal{O}(2^k)$ time. Thus, the time complexity of both algorithms is equal to $\mathcal{O}(2^k d(k + N))$. $\qquad \square$

Corollary 1. *Assume that $M \geq \beta N$, where $\beta \in (0, 1)$ is a constant. Then, given $\varepsilon > 0$ and $\gamma \in (0, 1)$ for the fixed parameter $k = \max\left(\left\lceil \frac{2}{\beta} \left\lceil \frac{2}{\gamma \varepsilon} \right\rceil \right\rceil, \left\lceil \frac{8}{\beta} \ln \frac{2}{\gamma} \right\rceil\right)$, algorithms $\mathcal{A}_1$ and $\mathcal{A}_2$ find $(1+\varepsilon)$-approximate solutions of Problems 1 and 2 in $\mathcal{O}(dN)$ time with probability of at least $1 - \gamma$.*

Proof. Let us prove the corollary for algorithm $\mathcal{A}_1$ (the proof for algorithm $\mathcal{A}_2$ is similar). Let $\delta = \frac{\gamma}{2}$, $t = \lceil \frac{1}{\delta\varepsilon} \rceil = \lceil \frac{2}{\gamma\varepsilon} \rceil$. Note that in that case $k \geq \frac{2t}{\beta}$ and $k \geq \frac{8}{\beta} \ln \frac{2}{\gamma}$. Applying Lemma 2 for $\nu = \frac{1}{2}$ and $\mathcal{C} = \mathcal{C}_1^*$ we get that

$$\Pr\left(|\mathcal{T} \cap \mathcal{C}_1^*| \leq \frac{kM}{2N}\right) \leq e^{-\frac{kM}{8N}}.$$

Then, under the conditions of the corollary the following holds

$$\alpha = \Pr\left(|\mathcal{T} \cap \mathcal{C}_1^*| < t\right) \leq \Pr\left(|\mathcal{T} \cap \mathcal{C}_1^*| < \frac{\beta k}{2}\right) \leq \Pr\left(|\mathcal{T} \cap \mathcal{C}_1^*| \leq \frac{kM}{2N}\right)$$

$$\leq e^{-\frac{kM}{8N}} \leq e^{-\frac{M}{\beta N} \ln \frac{2}{\gamma}} \leq e^{-\ln \frac{2}{\gamma}} = \frac{\gamma}{2}.$$

Therefore, by Theorem 1, for the specified value of k, algorithm $\mathcal{A}_1$ finds the solution of Problem 1 with the relative error $\frac{1}{\delta t} = \left(\frac{\gamma}{2} \left\lceil \frac{2}{\gamma\varepsilon} \right\rceil\right)^{-1} \leq \varepsilon$ in $\mathcal{O}\left(2^k d(k+N)\right)$ time with failure probability of at most $\delta + \alpha \leq \frac{\gamma}{2} + \frac{\gamma}{2} = \gamma$. Since the parameter k is fixed, under the specified conditions the running time of the algorithm is $\mathcal{O}(dN)$. $\square$

Theorem 2. *Let $k = \lceil \log_2 N \rceil$. Assume that $M \geq \beta N$, where $\beta \in (0,1)$ is a constant. Then algorithms $\mathcal{A}_1$ and $\mathcal{A}_2$ find $(1 + \varepsilon_N)$-approximate solutions of Problems 1 and 2 with probability $1 - \gamma_N$ in $\mathcal{O}(dN^2)$ time, where $\varepsilon_N \xrightarrow[N\to\infty]{} 0$, $\gamma_N \xrightarrow[N\to\infty]{} 0$.*

Proof. The time complexity bound of the algorithms under the condition $k = \lceil \log_2 N \rceil$ is obvious.

As in Corollary 1, let us estimate the relative error and the failure probability of algorithm $\mathcal{A}_1$. Under the conditions of Theorem 1, let $\delta = (\log_2 N)^{-1/2}$, $t = \lceil \frac{kM}{2N} \rceil$. With such values of the parameters, the relative error $\varepsilon_N = \frac{1}{\delta t} = (\log_2 N)^{1/2} / \lceil \frac{kM}{2N} \rceil$ can be bounded above by the value $\frac{2}{\beta}(\log_2 N)^{-1/2} \xrightarrow[N\to\infty]{} 0$.

Next, applying Lemma 2 for $\nu = \frac{1}{2}$ and $\mathcal{C} = \mathcal{C}_1^*$ we get that

$$\Pr\left(|\mathcal{T} \cap \mathcal{C}_1^*| \leq \frac{kM}{2N}\right) \leq e^{-\frac{kM}{8N}}.$$

Therefore,

$$\alpha = \Pr\left(|\mathcal{T} \cap \mathcal{C}_1^*| < t\right) \leq \Pr\left(|\mathcal{T} \cap \mathcal{C}_1^*| \leq \frac{kM}{2N}\right)$$

$$\leq e^{-\frac{kM}{8N}} \leq e^{-\frac{\beta \log_2 N}{8}} = N^{-\frac{\beta}{8\ln 2}} \xrightarrow[N\to\infty]{} 0.$$

Thus, for the failure probability γ_N of the algorithm we have $\gamma_N = \delta + \alpha \xrightarrow[N\to\infty]{} 0$. $\square$

5   Conclusion

In this paper we have presented similar randomized algorithms for two different strongly NP-hard quadratic Euclidean 2-clustering problems.

Our algorithms allow finding approximate solutions in a time that is linear on the space dimension and on the input size of the problems for given upper bounds of the relative error, failure probability and for an established parameter value. The conditions are found under which both algorithms are polynomial and asymptotically exact.

In our opinion, these algorithms will be useful, in particular, in Data mining, Pattern recognition, and Machine learning.

Acknowledgments. The study of Problem 1 was supported by the Russian Science Foundation, project 16-11-10041. The study of Problem 2 was supported by the Russian Foundation for Basic Research, projects 16-07-00168 and 18-31-00398, by the Russian Academy of Science (the Program of Basic Research), project 0314-2016-0015, and by the Russian Ministry of Science and Education under the 5-100 Excellence Programme.

References

1. de Waal, T., Pannekoek, J., Scholtus, S.: Handbook of Statistical Data Editing and Imputation. Wiley, Hoboken (2011)
2. Osborne, J.W.: Best Practices in Data Cleaning: A Complete Guide to Everything You Need to Do Before and After Collecting Your Data, 1st edn. SAGE Publication, Inc., Los Angeles (2013)
3. Greco, L.: Robust Methods for Data Reduction Alessio Farcomeni. Chapman and Hall/CRC, Boca Raton (2015)
4. Bishop, C.M.: Pattern Recognition and Machine Learning. Springer, New York (2006)
5. James, G., Witten, D., Hastie, T., Tibshirani, R.: An Introduction to Statistical Learning. Springer, New York (2013). https://doi.org/10.1007/978-1-4614-7138-7
6. Hastie, T., Tibshirani, R., Friedman, J.: The Elements of Statistical Learning, 2nd edn. Springer, New York (2009). https://doi.org/10.1007/978-0-387-84858-7
7. Aggarwal, C.C.: Data Mining: The Textbook. Springer, New York (2015). https://doi.org/10.1007/978-3-319-14142-8
8. Goodfellow, I., Bengio, Y., Courville, A.: Deep Learning (Adaptive Computation and Machine Learning Series). The MIT Press, Cambridge (2017)
9. Kel'manov, A.V., Pyatkin, A.V.: NP-completeness of some problems of choosing a vector subset. J. Appl. Ind. Math. **5**(3), 352–357 (2011)
10. Kel'manov, A.V., Romanchenko, S.M.: Pseudopolynomial algorithms for certain computationally hard vector subset and cluster analysis problems. Autom. Remote Control **73**(2), 349–354 (2012)
11. Kel'manov, A.V., Romanchenko, S.M.: An approximation algorithm for solving a problem of search for a vector subset. J. Appl. Ind. Math. **6**(1), 90–96 (2012)
12. Kel'manov, A.V., Romanchenko, S.M.: An FPTAS for a vector subset search problem. J. Appl. Ind. Math. **8**(3), 329–336 (2014)

13. Kel'manov, A., Motkova, A., Shenmaier, V.: An approximation scheme for a weighted two-cluster partition problem. In: van der Aalst, W.M.P., Ignatov, D.I., Khachay, M., Kuznetsov, S.O., Lempitsky, V., Lomazova, I.A., Loukachevitch, N., Napoli, A., Panchenko, A., Pardalos, P.M., Savchenko, A.V., Wasserman, S. (eds.) AIST 2017. LNCS, vol. 10716, pp. 323–333. Springer, Cham (2018). https://doi.org/10.1007/978-3-319-73013-4_30

14. Kel'manov, A.V., Pyatkin, A.V.: NP-hardness of some quadratic Euclidean 2-clustering problems. Doklady Math. **92**(2), 634–637 (2015)

15. Kel'manov, A.V., Pyatkin, A.V.: On the complexity of some quadratic Euclidean 2-clustering problems. Comput. Math. Math. Phys. **56**(3), 491–497 (2016)

16. Kel'manov, A.V., Motkova, A.V.: Exact pseudopolynomial algorithms for a balanced 2-clustering problem. J. Appl. Ind. Math. **10**(3), 349–355 (2016)

17. Kel'manov, A.V., Motkova, A.V.: Polynomial-time approximation algorithm for the problem of cardinality-weighted variance-based 2-clustering with a given center. Comp. Math. Math. Phys. **58**(1), 130–136 (2018)

18. Kel'manov, A., Motkova, A.: A fully polynomial-time approximation scheme for a special case of a balanced 2-clustering problem. In: Kochetov, Y., Khachay, M., Beresnev, V., Nurminski, E., Pardalos, P. (eds.) DOOR 2016. LNCS, vol. 9869, pp. 182–192. Springer, Cham (2016). https://doi.org/10.1007/978-3-319-44914-2_15

19. Aggarwal, H., Imai, N., Katoh, N., Suri, S.: Finding k points with minimum diameter and related problems. J. Algorithms **12**(1), 38–56 (1991)

20. Shenmaier, V.V.: Solving some vector subset problems by Voronoi diagrams. J. Appl. Ind. Math. **10**(4), 560–566 (2016)

21. Shenmaier, V.V.: An approximation scheme for a problem of search for a vector subset. J. Appl. Ind. Math. **6**(3), 381–386 (2012)

22. Sahni, S., Gonzalez, T.: P-complete approximation problems. J. ACM **23**, 555–566 (1976)

23. Brucker, P.: On the complexity of clustering problems. In: Henn, R., Korte, B., Oettli, W. (eds.) Optimization and Operations Research. LNE, vol. 157, pp. 45–54. Springer, Heidelberg (1978). https://doi.org/10.1007/978-3-642-95322-4_5

24. Indyk, P.: A sublinear time approximation scheme for clustering in metric space. In: Proceedings of the 40th Annual IEEE Symposium on Foundations of Computer Science (FOCS), pp. 154–159 (1999)

25. de la Vega, F., Karpinski, M., Kenyon, C., Rabani, Y.: Polynomial time approximation schemes for metric min-sum clustering. Electronic Colloquium on Computational Complexity (ECCC). Report no. 25 (2002)

26. Kel'manov, A.V., Khandeev, V.I.: A randomized algorithm for two-cluster partition of a set of vectors. Comput. Math. Math. Phys. **55**(2), 330–339 (2015)

27. Wirth, N.: Algorithms + Data Structures = Programs. Prentice Hall, Englewood Cliffs (1976)

An Approximation Polynomial Algorithm for a Problem of Searching for the Longest Subsequence in a Finite Sequence of Points in Euclidean Space

Alexander Kel'manov[1,2]($\boxtimes$), Artem Pyatkin[1,2], Sergey Khamidullin[1],
Vladimir Khandeev[1,2], Yury V. Shamardin[1], and Vladimir Shenmaier[1]

[1] Sobolev Institute of Mathematics, 4 Koptyug Ave., 630090 Novosibirsk, Russia
{kelm,artem,kham,khandeev,orlab}@math.nsc.ru, shenmaier@mail.ru
[2] Novosibirsk State University, 2 Pirogova St., 630090 Novosibirsk, Russia

Abstract. The following problem is considered. Given a finite sequence of Euclidean points, find a subsequence of the longest length (size) such that the sum of squared distances between the elements of this subsequence and its unknown centroid (geometrical center) is at most a given percentage of the sum of squared distances between the elements of the input sequence and its centroid. This problem models, in particular, one of the data analysis problems, namely, search for the maximum subset of elements close to each other in the sense of the bounded from above the total quadratic scatter in the set of time-ordered data. It can be treated as a data editing problem aimed at the removal of extraneous (dissimilar) elements. It is shown that the problem is strongly NP-hard. A polynomial time approximation algorithm is proposed. It either finds out that the problem has no solutions or outputs a 1/2-approximate solution if the length M^* of an optimal subsequence is even, or it outputs a $(M^* - 1)/2M^*$-approximate solution if M^* is odd. Some examples of numerical experiments illustrating the algorithm suitability are presented.

Keywords: Euclidean space · Longest subsequence
Quadratic variation · NP-hard problem
Polynomial-time approximation algorithm

1 Introduction

In the paper we consider a discrete optimization problem that models the search for the maximum subset of objects close to each other in the set of time-ordered measurement results. Our goal is to study computational complexity of the problem and to propose an algorithm solving it.

The study is motivated, on the one hand, by the absence of published results on the problem complexity status and algorithms with guaranteed performance

© Springer International Publishing AG, part of Springer Nature 2018
A. Eremeev et al. (Eds.): OPTA 2018, CCIS 871, pp. 120–130, 2018.
https://doi.org/10.1007/978-3-319-93800-4_10

bounds, and on the other hand, by its importance for the applications related, in particular, to data editing, data cleaning, data mining, machine learning, etc. (see the next section).

The paper is organized as follows. In Sect. 2, we present the mathematical statement of the considered problem and the motivation of our study. The statements of known related problems are also given. In the next section, we show that the problem is strongly NP-hard. In Sect. 4, some preliminary results for the algorithm analysis are proved. The polynomial time approximation algorithm and its analysis can be found in Sect. 5. Finally, Sect. 6 contains some results of numerical experiments illustrating the algorithm suitability.

2   Problem Formulation and Related Problems

Everywhere below denote by $\mathbb{R}$ the set of real numbers and by $\|\cdot\|$ the Euclidean norm.

The problem under consideration is stated as follows.

Problem 1. *Given* a sequence $\mathcal{Y} = (y_1, \ldots, y_N)$ of points in $\mathbb{R}^q$, some positive integer numbers $T_{\min}$, $T_{\max}$, and a number $\alpha \in (0, 1)$. *Find* a subset $\mathcal{M} = \{n_1, \ldots, n_M\} \subseteq \mathcal{N} = \{1, \ldots, N\}$ of largest size such that

$$T_{\min} \leq n_m - n_{m-1} \leq T_{\max} \leq N, \ m = 2, \ldots, M, \tag{1}$$

and

$$F(\mathcal{M}) = \sum_{j \in \mathcal{M}} \|y_j - \overline{y}(\mathcal{M})\|^2 \leq \alpha \sum_{j \in \mathcal{N}} \|y_j - \overline{y}(\mathcal{N})\|^2, \tag{2}$$

where $\overline{y}(\mathcal{M}) = \frac{1}{|\mathcal{M}|} \sum_{i \in \mathcal{M}} y_i$ and $\overline{y}(\mathcal{N}) = \frac{1}{N} \sum_{i \in \mathcal{N}} y_i$ are the centroids (geometrical centers) of the multisets $\{y_i \in \mathcal{Y} \,|\, i \in \mathcal{M}\}$ and $\{y_i \in \mathcal{Y} \,|\, i \in \mathcal{N}\}$ respectively.

The close in statement problem is following.

Problem 2. *Given* a sequence $\mathcal{Y} = (y_1, \ldots, y_N)$ of points in $\mathbb{R}^q$ and positive integer numbers $T_{\min}$, $T_{\max}$ and $M > 1$. *Find* a subset $\mathcal{M} = \{n_1, \ldots, n_M\} \subseteq \mathcal{N} = \{1, \ldots, N\}$ of indices of the sequence elements minimazing $F(\mathcal{M})$ subject to constraints (1).

Problem 2 has the following interpretation (see, for example, [1–4]). There is a time series containing N measurements $y_1, \ldots, y_N$ of q numerical characteristics of some objects. Each measurement result in the time series has an error, and no correspondence between the elements of the time series and the objects is known. Some of these objects have identical characteristics (or one can say that in the time series there are several measurements of one significant object). Other objects are distinguished and have different characteristics (or one can say that in the time series there are some measurements which are treated as "outliers" due to malfunction of the measuring device). The number M of measurements of identical objects is known. In addition, it is known that the time interval between every two consequent results of measuring characteristics of the identical objects is bounded from above and below by some constants $T_{\max}$ and $T_{\min}$. The characteristics of identical objects in contrast to the characteristics of other objects have an important information value. It is required to find the subsequence of measurements which correspond to the identical objects using the criterion of minimum sum of squared distances and to estimate the characteristics of these objects (taking into account the measuring errors in the data).

Problem 1 has a similar interpretation, namely, one needs to find in a sequence $\mathcal{Y}$ a multisubset $\{y_i \in \mathcal{Y} \,|\, i \in \mathcal{M}\}$ of the maximum cardinality whose elements are well concentrated, i. e. total quadratic scatter of points relative to the unknown centroid $\overline{y}(\mathcal{M})$ is at most α times the total quadratic scatter of the input sequence $\mathcal{Y}$ of points relative to its centroid $\overline{y}(\mathcal{N})$. If the points of the input sequence $\mathcal{Y}$ correspond to time-ordered measurement results of characteristics of some objects and these results could contain errors whose dispersion is bounded from above by some threshold (the right part of (2)), then solving Problem 1 would find a multisubset $\{y_i \in \mathcal{Y} \,|\, i \in \mathcal{M}\}$ of the maximum cardinality containing no data with significant (exceeding the threshold) error. The level of the threshold can be regulated by the parameter α.

The difference between Problem 1 and Problem 2 is that in Problem 1, the size M of the sought sequence should be maximized (in Problem 2, this size is given) under the bound on the value of the object function F of Problem 2.

The problems of subsequence search, similar to stated above, are typical for Data editing and Data cleaning problems (see, e.g., [5–7]). In problems of Machine learning and Data mining, cleaning the data from extraneous "outliers" is generally considered as a necessary element [8–15].

For Problem 2, having a direct relation to the above-mentioned problems, several algorithmic results are currently obtained (see [1–4]). For the considered Problem 1, similar by statement to Problem 2, no algorithms with guaranteed performance bounds are known; its complexity status is also open.

In the current paper, we show that Problem 1 is strongly NP-hard and we propose an effective tool for solving it — a polynomial-time approximation algorithm with preciseness bound close to $1/2$.

3   Problem Complexity

Remind the following strongly NP-hard [16] problem.

Problem 3. Given a set $\mathcal{Y} = \{y_1, \ldots, y_N\}$ of points in $\mathbb{R}^q$, and a number $\alpha \in (0,1)$. *Find* a subset $\mathcal{C} \subset \mathcal{Y}$ of largest cardinality such that

$$\sum_{y \in \mathcal{C}} \|y - \overline{y}(\mathcal{C})\|^2 \leq \alpha \sum_{y \in \mathcal{Y}} \|y - \overline{y}(\mathcal{Y})\|^2,$$

where $\overline{y}(\mathcal{C}) = \frac{1}{|\mathcal{C}|} \sum_{y \in \mathcal{C}} y$ and $\overline{y}(\mathcal{Y}) = \frac{1}{|\mathcal{Y}|} \sum_{y \in \mathcal{Y}} y$ are the centroids of the subset $\mathcal{C}$ and the input set $\mathcal{Y}$ respectively.

The following statement is true since Problem 3 is a partial case of Problem 1, when $T_{\min} = 1$ and $T_{\max} = N$.

Proposition 1. *Problem 1 is strongly NP-hard.*

4   Auxiliary Results

In order to justify our algorithm we need several lemmas and an auxiliary polynomially solvable problem.

First, remind the following well-known result (see, for example, [17,18]).

Lemma 1. *Let $\bar{z} = \frac{1}{|\mathcal{Z}|} \sum_{z \in \mathcal{Z}} z$ be the centroid of the finite set $\mathcal{Z} \subset \mathbb{R}^q$ and let a point $x \in \mathbb{R}^q$ satisfy the condition $\|x - \bar{z}\| \le \|z - \bar{z}\|$ for each $z \in \mathcal{Z}$. Then,*

$$\sum_{z \in \mathcal{Z}} \|z - \bar{z}\|^2 \le \sum_{z \in \mathcal{Z}} \|z - x\|^2 \le 2 \sum_{z \in \mathcal{Z}} \|z - \bar{z}\|^2.$$

Note that Lemma 1 holds for the case when $\mathcal{Z}$ is any finite multiset or any sequence of finite size.

Next, we need an exact polynomial-time algorithm for solving the following auxiliary problem.

Problem 4. Given a sequence $\mathcal{Y} = (y_1, \ldots, y_N)$ of points in $\mathbb{R}^q$, a point $x \in \mathbb{R}^q$, positive integer numbers $T_{\min}$, $T_{\max}$ and $M > 1$. Find a subset $\mathcal{M} = \{n_1, \ldots, n_M\} \subseteq \mathcal{N}$ of indexes of the sequence elements such that

$$f^x(\mathcal{M}) = \sum_{i \in \mathcal{M}} \|y_i - x\|^2 \rightarrow \min, \tag{3}$$

while the elements of the tuple $(n_1, \ldots, n_M)$ satisfy the constraints (1).

A dynamic programming scheme is presented in the next lemma and its corollary. This scheme allows to find the optimal solution $\mathcal{M}^x$ of Problem 4. The presented scheme is based on the results of [2,19] and given here for completeness.

Lemma 2. *For any positive integer $M > 1$, such that $(M-1)T_{\min} \le N-1$, and for an arbitrary point $x \in \mathbb{R}^q$ the optimum $f^x_{\min} = \min_{\mathcal{M}} f^x(\mathcal{M})$ of Problem 4 could be found as*

$$f^x_{\min} = \min_{n \in \omega_M} f^x_M(n), \tag{4}$$

where the values of the functions $f^x_M(n)$, $n \in \omega_M$, are calculated using the following recurrent formulas

$$f^x_m(n) = \begin{cases} \|y_n - x\|^2, & \text{if } n \in \omega_1, \ m = 1; \\ \|y_n - x\|^2 + \min_{j \in \gamma^-_{m-1}(n)} f^x_{m-1}(j), & \text{if } n \in \omega_m, \ m = 2, \ldots, M, \end{cases} \tag{5}$$

where

$$\omega_m = \{n \mid 1 + (m-1)T_{\min} \le n \le N - (M-m)T_{\min}\}, m = 1, \ldots, M,$$

$$\gamma_{m-1}^{-}(n) = \{j \mid \max\{1 + (m-2)T_{\min}, n - T_{\max}\} \leq j \leq n - T_{\min}\},$$
$$n \in \omega_m, m = 2, \ldots, M.$$

Corollary 1. *Elements* $n_1^x, \ldots, n_M^x$ *of the optimal tuple* $\mathcal{M}^x$ *can be found by the formulas:*

$$n_M^x = \arg \min_{n \in \omega_M} f_M^x(n), \tag{6}$$

$$n_{m-1}^x = \arg \min_{n \in \gamma_m^{-}(n_m^x)} f_m^x(n), \quad m = M, M-1, \ldots, 2. \tag{7}$$

The algorithm implementing the scheme above is presented below in a step-by-step description.

Algorithm $\mathcal{A}_1$.

Input: a sequence $\mathcal{Y}$, a point x, numbers $T_{\min}$, $T_{\max}$ and M.

Step 1. Compute $\|y_n - x\|^2$ for each $n \in \mathcal{N}$.

Step 2. Using formulas (5), calculate the values $f_m^x(n)$ for each $n \in \omega_m$ while $m = 1, \ldots, M$.

Step 3. Find the minimum $f_{\min}^x$ of the objective function (3) using (4) and the optimal tuple $\mathcal{M}^x = (n_1^x, \ldots, n_M^x)$ by formulas (6), (7).

Output: the tuple $\mathcal{M}^x = (n_1^x, \ldots, n_M^x)$.

Remark 1. In [2,19], it was proved that the algorithm $\mathcal{A}_1$ finds an optimal solution of Problem 4 in $\mathcal{O}(N(M(T_{\max} - T_{\min} + 1) + q))$ time. In this expression, the value $T_{\max} - T_{\min} + 1$ is at most N. Therefore, the running time of the algorithm is estimated as $\mathcal{O}(N(MN + q))$.

Remark 2. In accordance with Lemma 2, $M \in \{2, \ldots, M_{\max}\}$, where

$$M_{\max} = \left\lfloor \frac{N-1}{T_{\min}} \right\rfloor + 1.$$

Further, we need the following easily verifiable property.

Property 1. *If* $\mathcal{M}_1 \subseteq \mathcal{M}_2$, *then* $F(\mathcal{M}_1) \leq F(\mathcal{M}_2)$ *and* $f^x(\mathcal{M}_1) \leq f^x(\mathcal{M}_2)$ *for any fixed* $x \in \mathbb{R}^q$.

Let Ω_2 be the set of all possible subsets $\mathcal{M}$ of size 2, satisfying the condition (1).

In addition, we need the next auxiliary assertion.

Lemma 3. *If in the indices set* $\mathcal{N}$ *of the sequence* $\mathcal{Y}$ *there is a subset* $\mathcal{M} = \{i, k\} \in \Omega_2$ *such that*

$$\|y_i - y_k\|^2 \leq 2\alpha \sum_{j \in \mathcal{N}} \|y_j - \bar{y}(\mathcal{N})\|^2,$$

then Problem 1 has a solution.

Proof. The minimum admissible size M of the set $\mathcal{M}$ in Problem 1 is equal to 2. Due to the well known equality on sum of squared distances between points in a finite set and its centroid and sum of squares of pairwise distances between the points of this set, for every set $\mathcal{M} = \{i, k\}$ the following equality is true:

$$F(\mathcal{M}) = \sum_{j \in \mathcal{M}} \|y_j - \overline{y}(\mathcal{M})\|^2 = \frac{1}{2} \|y_i - y_k\|^2.$$

Now it remains to use Property 1 and note that the inequality (2) is a necessary condition of solvability of Problem 1.   □

5   Approximation Algorithm

The suggested approach for finding an approximate solution of Problem 1 is following. First the algorithm finds out whether an admissible solution exists. Next, for each point y of the input sequence $\mathcal{Y}$ and for each $M \in \{2, \ldots, M_{\max}\}$ an auxiliary Problem 4 (with $x = y$) is solved by the algorithm $\mathcal{A}_1$. A family of admissible (i.e. satisfying the inequality (2)) solutions of Problem 1 is formed from the found solutions (the indices sets). The best (in the sense of the maximum size) of the solutions in this family is outputted. The following algorithm realizes this approach.

Algorithm $\mathcal{A}$.
   Input: a sequence $\mathcal{Y}$ and numbers $T_{\min}, T_{\max}$, and α.
   Step 1. Compute $A = \alpha \sum_{j \in \mathcal{N}} \|y_j - \overline{y}(\mathcal{N})\|^2$.
   Step 2. For all sets $\{i, k\} \in \Omega_2$ calculate $\|y_i - y_k\|^2$. If no element of the set $\{\|y_i - y_k\|^2 \mid \{i, k\} \in \Omega_2\}$ satisfies the condition $\|y_i - y_k\|^2 \leq 2A$, then go to Output 1.
   For each $y \in \mathcal{Y}$ perform Steps 3 and 4.
   Step 3. For every $M = 2, \ldots, M_{\max}$ using the algorithm $\mathcal{A}_1$ find a solution $\mathcal{M}_M^y = \{n_1^y, \ldots, n_M^y\}$ of Problem 4 with $x = y$ and calculate for this solution the value of the objective function $F(\mathcal{M}_M^y)$.
   Step 4. In the family $\{\mathcal{M}_M^y, M = 2, \ldots, M_{\max}\}$ of the sets obtained in Step 3, find a set $\mathcal{M}_{\max}^y$ of maximum cardinality for which $F(\mathcal{M}_M^y) \leq A$.
   Step 5. In the set $\{\mathcal{M}_{\max}^y \mid y \in \mathcal{Y}\}$ of admissible sets constructed in Step 4, choose as an output $\mathcal{M}_A$ the set $\mathcal{M}_{\max}^y$ of maximum cardinality. If there are several such sets choose one with the minimum value of F.
   Output 1: there are no solutions.
   Output 2: the tuple $\mathcal{M}_A$.
   The following theorem is true

Theorem 1. *If there are no solutions of Problem 1, the algorithm $\mathcal{A}$ establishes this fact in $\mathcal{O}(N(T_{\max} - T_{\min} + 1)q)$ time. Otherwise, if the cardinality M^* of the optimal solution of Problem 1 is even, this algorithm finds a 1/2-approximate solution in $\mathcal{O}(N^3(N(T_{\max} - T_{\min} + 1) + q))$ time. If M^* is odd, the algorithm finds a $\frac{1}{2}(1 - \frac{1}{M^*})$-approximate solution in the same time.*

Proof. The correctness of the algorithm and its complexity in the case when Problem 1 has no solutions follows from Lemma 3 and an evident bound $|\Omega_2| \leq N(T_{\max} - T_{\min} + 1)$.

Assume that Problem 1 has an admissible solution. Introduce the following notation for an optimal solution of Problem 1. Let $\mathcal{C}^* = \{y_i \in \mathcal{Y} \mid i \in \mathcal{M}^*\}$ be a multiset containing the elements of the sequence $\mathcal{Y}$ with the indices in $\mathcal{M}^*$ of maximum size M^*. Let $\overline{y}(\mathcal{M}^*)$ be the centroid of the multiset $\mathcal{C}^*$, t be the closest to $\overline{y}(\mathcal{C}^*)$ point in $\mathcal{C}^*$. Then by Lemma 1 we have

$$f^t(\mathcal{M}^*) = \sum_{j \in \mathcal{M}^*} \|y_j - t\|^2 \leq 2 \sum_{j \in \mathcal{M}^*} \|y_j - \overline{y}(\mathcal{M}^*)\|^2 = 2F(\mathcal{M}^*) \leq 2A,$$

where A is defined in Step 1 of the algorithm.

Put $M' = \lfloor M^*/2 \rfloor$. Let $\mathcal{M}_1$ be the first M' indices of the set $\mathcal{M}^*$ and $\mathcal{M}_2$ be the next M' indices of the set $\mathcal{M}^*$. Note that in this case both $\mathcal{M}_1$ and $\mathcal{M}_2$ satisfy the inequalities (1). Besides, by Property 1

$$f^t(\mathcal{M}_1) + f^t(\mathcal{M}_2) = f^t(\mathcal{M}_1 \cup \mathcal{M}_2) \leq f^t(\mathcal{M}^*).$$

Let $\mathcal{M} \in \{\mathcal{M}_i, \ i = 1, 2\}$ satisfy the inequality $f^t(\mathcal{M}_i) \leq f^t(\mathcal{M}^*)/2$. Then

$$f^t(\mathcal{M}) \leq f^t(\mathcal{M}^*)/2 \leq A,$$

i.e. $\mathcal{M}$ is an admissible solution of Problem 1 with the set size M'. Besides, $\mathcal{M}$ is an admissible solution of Problem 4 with $x = t$ and $M = M'$. But then for the set $\mathcal{M}^t_{M'}$, being an optimal solution of this problem, the condition $f^t(\mathcal{M}^t_{M'}) \leq A$ is also true yielding by Lemma 1 the bound

$$F(\mathcal{M}^t_{M'}) \leq f^t(\mathcal{M}^t_{M'}) \leq A.$$

It remains to note that at Step 5 the point t closest to the centroid of the optimal cluster in the family $\{\mathcal{M}^y_{\max} \mid y \in \mathcal{Y}\}$ and the corresponding set $\mathcal{M}^t_{\max}$ of cardinality $|\mathcal{M}^t_{\max}| \geq |\mathcal{M}^t_{M'}| = M'$, clearly, will be considered. So, the solution $\mathcal{M}_A$ found by the algorithm $\mathcal{A}$ also contains at least $M' = \lfloor M^*/2 \rfloor$ elements. Therefore, for even M^* the preciseness of the algorithm is bounded by

$$\frac{\lfloor \frac{M^*}{2} \rfloor}{M^*} = \frac{1}{2},$$

and for odd — by

$$\frac{\lfloor \frac{M^*}{2} \rfloor}{M^*} = \frac{M^* - 1}{2M^*} = \frac{1}{2}\left(1 - \frac{1}{M^*}\right).$$

Evaluate the time complexity of the algorithm in the case when Problem 1 has an admissible solution. Step 1 requires $\mathcal{O}(qN)$ operations, Step 2 can be done in $\mathcal{O}(qN(T_{\max} - T_{\min} + 1))$ operations. The most time consuming Step 3 could require $\mathcal{O}(M_{\max}N^2(M_{\max}(T_{\max} - T_{\min} + 1) + q))$ operations since for each of the N points the algorithm $\mathcal{A}_1$ is fulfilled $M_{\max} - 1$ times for every $M \in \{2, \ldots, M_{\max}\}$ and requires $\mathcal{O}(N(M(T_{\max} - T_{\min} + 1) + q))$ operations for every

M (see Remark 1). The comparison operations in Steps 4 and 5 are fulfilled in $\mathcal{O}(N)$ time, and the calculation time of the values of F is at most $\mathcal{O}(qN)$. Since $M_{\max} \leq N$ (see Remark 2), the total running time of the algorithm is $\mathcal{O}(N^3(N(T_{\max} - T_{\min} + 1) + q))$, that is at most $\mathcal{O}(N^3(N^2 + q))$, since $T_{\max} - T_{\min} + 1 \leq N$. □

6   Examples of Numerical Experiments

The figures presented below show the suitability of the algorithm for the problem of search for a subsequence of similar elements in a data collection. An input sequence of points (out of 500 points) is shown in Fig. 1a (upper tape). Each point of the sequence corresponds to a vertical strip. The subsequence of points found by the algorithm for $\alpha = 0.0006$, $T_{\min} = 2$, $T_{\max} = 20$ is presented in the same Fig. 1a on the lower tape. The found subsequence size is equal to 38.

The points of the same input sequence are presented on a plane in Fig. 1b at the left-hand part. At the right-hand side, one can see a subset (points of darker color) corresponding to the subsequence presented in the Fig. 1a on the lower tape.

(a)

(b)

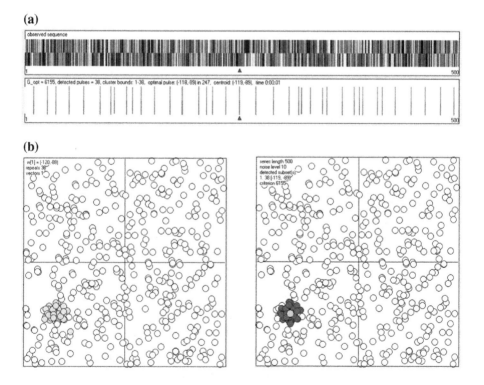

Fig. 1. Example 1 of processing a 2-dimensional sequence

The similar data is presented in Fig. 2a and b for the sequence of length 200 for $\alpha = 0.001$, $T_{\min} = 2$, $T_{\max} = 20$. The found subsequence size is equal to 21.

(a)

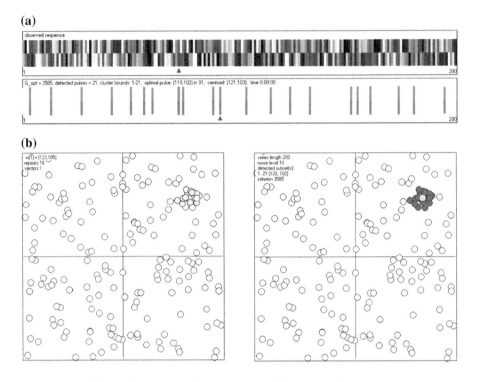

(b)

Fig. 2. Example 2 of processing a 2-dimensional sequence

7   Conclusion

In this paper we have shown the strong NP-hardness of one problem of searching for the longest subsequence (the finite subsequence of the largest size) in a finite sequence of Euclidean points. We have also presented an approximation algorithm for this problem. The proposed algorithm allows to find a 1/2-approximate solution of the problem in a polynomial time if the length M^* of an optimal subsequence is even, or a $(M^* - 1)/2M^*$-approximate solution if M^* is odd.

In our opinion, the presented algorithm would be useful as a tool for solving problems in applications related to Data editing, Data cleaning, Data mining, and Machine learning when the data is time series (signals).

The development of faster approximation algorithms for Problem 1 is of considerable interest. An important direction of study is searching subclasses of Problem 1 for which faster polynomial-time approximation algorithms can be constructed.

Acknowledgments. The study presented in Sects. 2, 3 and 5 was supported by the Russian Science Foundation, project 16-11-10041. The study presented in Sects. 4 and 6 was supported by the Russian Foundation for Basic Research, projects 16-07-00168 and 18-31-00398, by the Russian Academy of Science (the Program of Basic Research), project 0314-2016-0015, and by the Russian Ministry of Science and Education under the 5-100 Excellence Programme.

References

1. Kel'manov, A.V., Pyatkin, A.V.: On the complexity of some problems of choosing a vector subsequence. Zhurnal Vychislitel'noi Matematiki i Matematicheskoi Fiziki **52**(12), 2284–2291 (2012). (in Russian)
2. Kel'manov, A.V., Romanchenko, S.M., Khamidullin, S.A.: Approximation algorithms for some intractable problems of choosing a vector subsequence. J. Appl. Indust. Math. **6**(4), 443–450 (2012)
3. Kel'manov, A.V., Romanchenko, S.M., Khamidullin, S.A.: Exact pseudopolynomial algorithms for some NP-hard problems of searching a vectors subsequence. Zhurnal Vychislitel'noi Matematiki i Matematicheskoi Fiziki **53**(1), 143–153 (2013). (in Russian)
4. Kel'manov, A.V., Romanchenko, S.M., Khamidullin, S.A.: An approximation scheme for the problem of finding a subsequence. Numer. Anal. Appl. **10**(4), 313–323 (2017)
5. de Waal, T., Pannekoek, J., Scholtus, S.: Handbook of Statistical Data Editing and Imputation. Wiley, Hoboken (2011)
6. Osborne, J.W.: Best Practices in Data Cleaning: A Complete Guide to Everything You Need to Do Before and After Collecting Your Data, 1st edn. SAGE Publication, Inc., Los Angeles (2013)
7. Greco, L.: Robust Methods for Data Reduction Alessio Farcomeni. Farcomeni. Chapman and Hall/CRC, Boca Raton (2015)
8. Bishop, C.M.: Pattern Recognition and Machine Learning. Springer, New York (2006)
9. James, G., Witten, D., Hastie, T., Tibshirani, R.: An Introduction to Statistical Learning. Springer, New York (2013). https://doi.org/10.1007/978-1-4614-7138-7
10. Hastie, T., Tibshirani, R., Friedman, J.: The Elements of Statistical Learning. SSS, 2nd edn. Springer, New York (2009). https://doi.org/10.1007/978-0-387-84858-7
11. Aggarwal, C.C.: Data Mining: The Textbook. Springer, Cham (2015). https://doi.org/10.1007/978-3-319-14142-8
12. Goodfellow, I., Bengio, Y., Courville, A.: Deep Learning (Adaptive Computation and Machine Learning Series). The MIT Press, Cambridge (2017)
13. Fu, T.: A review on time series data mining. Eng. Appl. Artif. Intell. **24**(1), 164–181 (2011)
14. Kuenzer, C., Dech, S., Wagner, W. (eds.): Remote Sensing Time Series. RSDIP, vol. 22. Springer, Cham (2015). https://doi.org/10.1007/978-3-319-15967-6
15. Liao, T.W.: Clustering of time series data — a survey. Pattern Recognit. **38**(11), 1857–1874 (2005)
16. Ageev, A.A., Kel'manov, A.V., Pyatkin, A.V., Khamidullin, S.A., Shenmaier, V.V.: Approximation polynomial algorithm for the data editing and data cleaning problem. Pattern Recognit. Image Anal. **27**(3), 365–370 (2017)
17. Kel'manov, A.V., Romanchenko, S.M.: An approximation algorithm for solving a problem of search for a vector subset. J. Appl. Indust. Math. **6**(1), 90–96 (2012)

18. Kel'manov, A.V., Romanchenko, S.M.: An FPTAS for a vector subset search problem. J. Appl. Indust. Math. **8**(3), 329–336 (2014)
19. Kel'manov, A.V., Khamidullin, S.A.: Posterior detection of a given number of identical subsequences in a quasi-periodic sequence. Comput. Math. Math. Phys. **41**(5), 762–774 (2001)

On Vector Summation Problem
in the Euclidean Space

Edward Kh. Gimadi[1,2], Ivan A. Rykov[1,2(✉)], and Yury V. Shamardin[1]

[1] Sobolev Institute of Mathematics,
4 Acad. Koptyug avenue, 630090 Novosibirsk, Russia
orlab@math.nsc.com
[2] Novosibirsk State University,
2 Pirogova Str., 630090 Novosibirsk, Russia
gimadi@math.nsc.ru, rykov-web@yandex.ru

Abstract. We consider a problem of finding a subset of the smallest size in the given set of vectors such that the norm of sum vector is greater or equal to some given value. We show that the problem can be solved optimally with the same complexity as the problem of finding the subset of given cardinality with minimum norm of sum vector.

Keywords: Vector subset · Sum vector · Euclidean space
Exact algorithm

1 Introduction

We consider the following problem: given the set of vectors $V = v_1, \ldots, v_n$ in Euclidean space R^d and a real number B. Find a subset X in the set V of minimum cardinality, provided that the norm of sum vector is greater of equal to B.

$$|X| \to \min_{X \subset V},$$
$$\left\| \sum_{v \in X} v \right\| \geq B. \tag{1}$$

The given Problem 1 is closely related to the Largest m-Vector Sum (m-LVS) Problem considered in papers [2–7]. In this problem for a given cardinality m one needs to find a subset of exactly m vectors so that the Euclidean norm of the sum vector is maximized. The corresponding problem with arbitrary cardinality of the subset is referred to as LVS [7].

m-LVS Problem is NP-hard in general case [2,6]. However, it was proved that for any fixed size of dimension d of the space R^d an optimal solution can be found in polynomial time [4]. In [7] the exact algorithm with the best known complexity $O(n^{d+1})$ was suggested. In [5] an asymptotically exact randomized algorithm with significantly better time complexity $O(nd^{3/2}(8/7 \ln n)^d)$ was introduced.

This research was supported by the Russian Scientific Foundation(project 16-11-10041).

Problem 1 plays a "complementary" role to this problem: a cardinality of the subset is minimized provided that the norm of sum vector remains sufficiently large. We will denote the Problem 1 as B-LVS.

In particular, the decision variant of the Problem 1 can be stated as "given a set V, real value B and integer m, is there a subset of size less or equal to m, s.t. the length sum vector is greater or equal to B?". For the particular case of m equal to n this statement exactly equal to the decision variant of LVS Problem. Thus, B-LVS Problem is also NP-hard.

2   Exact Algorithm for Solving B-LVS Problem

The idea of exact algorithm for solving this problem is based on using the exact algorithm for the problem of finding subset of vectors of the given cardinality with a minimum norm of sum vector, introduced in [4].

Indeed, solving the m-LVS for $m = 1, 2, \ldots n$, we will choose the smallest m, such that the maximal norm of sum vector for the subset of cardinality m is greater or equals B.

In [4] it is shown that m-LVS can be solved optimally with complexity $O(d^2 n^{2d})$. It is easy to show that the B-LVS Problem can be solved with the same complexity.

Theorem 1. *Optimal solution of the Problem 1 can be found in $O(d^2 n^{2d})$ time.*

Proof. Indeed, consider the exact algorithm for solving m-LVS from [4].

Following notation from [1] let us call an *orthogonal hyperplane* for the non-zero vector $v \in R^d$ a hyperplane defined by the equation $(v, x) = 0$. This hyperplane is a $(d-1)$-dimensional linear subspace, which consists of vectors orthogonal to vector v.

We call a *family of solution domains* for the given non-zero vectors $u_1, u_2, \ldots, u_t$ a family of maximal (by inclusion) connected subsets of the space R^d, such that these subsets does not intersect with orthogonal hyperplanes of vectors $u_1, u_2, \ldots, u_t$.

A set of vectors of the space R^d, which contains exactly one vector for each solution domain is called *solution domains representatives* for the vectors $u_1, u_2, \ldots, u_t$.

An estimation of time complexity of the algorithm A_{GPR} is based on the following lemma.

Lemma 1 ([1]). *A family of domain representatives for non-zero vectors $u_1, u_2, \ldots, u_t$ in R^d has cardinality $O(dt^{d-1})$ and can be constructed in $O(d^2 t^d)$ time.*

In context of solving m-LVS we consider $u_{ij} = v_i - v_j$, $1 \leq i < j \leq n$. The full algorithm can be described as follows [4]:

Step 1. For the set of vectors $u_{ij}, 1 \leq i < j \leq n$ construct a family W of solution domains representatives. For each $w \in W$ perform the steps 2–3;

Step 2. Sort the vectors $v_1, v_2, \ldots, v_n$ by non-increasing projection to the direction given by w; denote the corresponding ordering as $v'_1, v'_2, \ldots, v'_n$;

Step 3. Put $U^w = \{v'_1, v'_2, \ldots, v'_m\}$ (i.e., m vectors with maximum projections to w are taken as U^w;

Step 4. Find such $U \in \{U^w \mid w \in W\}$, that

$$F(U) = \left\| \sum_{v \in U} v \right\| = \max\{F(U^w), \mid w \in W\}.$$

U is output as algorithm result.

By applying Lemma 1 to the set of $n(n-1)/2$ vectors $\{u_{ij}\}$ we can see that the time complexity of the step 1 is $O(d^2 n^{2d})$ and that Steps 2–3 are performed $O(dn^{2d-2})$ times. The total complexity of Steps 2–3 is hence defined by the complexity of Step 2 (equal to $O(dn \log n)$) and Step 3 (equal to $O(dn)$) multiplied by $O(dn^{2d-2})$, which does not exceed the complexity of the first step. Step 4 has complexity equal to $O(dn^{2d-2} \cdot dn)$.

Now let us notice that the exact algorithm for solving the Problem 1 (we denote it as $A_{B\text{-LVS}}$) can be obtained from the described algorithm by substituting Steps 3 and 4 by the following steps:

Step 3'. For each $m = 1, 2, \ldots n$ put $U^w_m = \{v'_1, v'_2, \ldots, v'_m\}$;

Step 4'. Considering m = 1, 2, ... n choose $U_m \in \{U^w_m, \mid w \in W,\ m = 1, 2, \ldots n\}$, such

$$F(U_m) = \left\| \sum_{v \in U_m} v \right\| = \max\{F(U^w_m), \mid w \in W,\ m = 1, 2, \ldots n\};$$

for the smallest m, such that $F(U_m) > B$, take U_m as a solution of the Problem 1.

Indeed, the modified algorithm $A_{B\text{-LVS}}$ with new steps 3' and 4' consequentially solves m-LVS for each $m = 1, 2, \ldots n$ and chooses the least m such that the norm of the the longest sum vector among all the subsets of cardinality m exceeds specified threshold B.

At that, the time complexity of Step 3' in the new algorithm equals $O(dn^{d-2} \cdot n \cdot dn)$ which does not exceed the time complexity of Step 1. Similarly, the time complexity of Step 4' equals $O(dn^{2d-1} \cdot dn^2)$, which does not exceed the time complexity of Step 1. Hence, total time complexity of the algorithm is equal to $O(d^2 n^{2d})$.

Thus, we have shown that the algorithm $A_{B\text{-LVS}}$ find optimal solution of the Problem 1 in stated time. The proof of the Theorem 1 is complete.

3   Randomized Algorithm for the B-LVS Problem

In paper [5] a randomized algorithm A_{rand} based on the similar "projecting" idea is introduced. Instead of constructing the set of solution domains representatives this algorithm uniformly generates L random directions and solves

"m projections" subproblem (Steps 2, 3 of the algorithm described in previous section) for each of them.

Recall that a randomized algorithm A is said to have $(\varepsilon_n^A, \delta_n^A)$ estimates (called relative error and failure probability, correspondingly) over the set $\mathfrak{I}_n$ of all the instances with n elements in thee input data, if

$$\mathbb{P}\left\{\frac{f^*(I) - f_A(I)}{f^*(I)} < \varepsilon_n^A \ \middle| \ I \in \mathfrak{I}_n\right\} \geq 1 - \delta_n^A$$

Based on theorem from [5] it is easy to show that with certain value of parameter L algorithm fulfills the condition of the following theorem.

Theorem 2. *Algorithm A_{rand} finds asymptotically optimal solution of m-LVS Problem with the relative error*

$$\varepsilon_n \leq \frac{1}{2\ln^2 n}$$

and the failure probability

$$\delta_n \leq \frac{1}{n^2}$$

in

$$T = O\left(nd^{3/2}(8/7\ln n)^{d+1}\right)$$

time.

It should be noted that this algorithm can't be directly adapted for solving Problem 1, since it is not guaranteed to find the optimal solution of the problem. Hence, when the input boundary value B is set equal to the optimal length of the sum vector (for any cardinality of the subset), an algorithm can fail to find any feasible solution. E.g., consider the problem input given by value B and the following vectors in R^2:

$$v_1 = v_2 = \cdots = v_s = \left(\frac{B}{s}, 0\right)$$

$$v_{s+1} = \cdots = v_{s+s/2} = \left(-\frac{B}{s}, 0\right)$$

for some integer number s.

The optimal solution of the Problem 1 if given by the set of $m^* = s$ vectors with $X^* = (v_1, \ldots v_s)$. In order to obtain solution with the norm greater or equal B for a sampled random direction, one has to sample exactly $(1, 0)$, the probability of which equals 0.

Denote as B_m^* — the optimal value of the m-LVS Problem. We consider conditions when application of A_{rand} to each m will provides a solution of the problem.

Consider the algorithm $A_{B\text{-LVS-rand}}$:

Step 1. For each $m = 1, 2, \ldots n$ find subset $U_m = \{u_1^m, \ldots u_m^m\}$ as result of application of A_{rand} for corresponding m-LVS;

Step 2. Choose the least m for which sum vector for U_m has norm greater or equal to B. Put $U = U_m$ for this m as a solution for Problem 1.

Theorem 3. *If $B \leq \max_m(B_m^*)$ and $B \notin I_m = \left((1 - \frac{1}{2\ln^2 n})B_m^*, B_m^*\right]$ for each $m = 1, 2, \ldots n$, then the algorithm $A_{B\text{-}LVS\text{-}rand}$ finds optimal solution of Problem 1 with failure probability*

$$\delta \leq \frac{1}{n}$$

in

$$T = O(n^2 d^{3/2}(8/7\ln n)^{d+1})$$

time.

Proof. It is easy to see, that the algorithm $A_{B\text{-}LVS\text{-}rand}$ finds an optimal solution under the conditions of the theorem. Indeed, optimal value m^* for the B-LVS Problem is the least m, such that $B \leq B_m^*$. Under conditions of the theorem $B \leq B_m^*$ if and only if $B \leq (1 - \frac{1}{2\ln^2 n})B_m^* \leq C_m$, where $C_m = \sum_i(u_i^m)$ is norm of the sum vector of the solution U_m. The latter inequality holds due to the Theorem 2 in case of no failure. Hence, choosing the least m with $B \leq C_m$ will provide an optimal solution.

The failure probability of solving the Problem m-LVS for each m can be estimated as $1 - (1 - \frac{1}{n^2})^n \leq \frac{1}{n}$ (since the failure is independent for each of n problems). The proof of the Theorem 3 is complete.

4   Conclusion

In this paper we considered the problem of minimizing cardinality of vector subset provided that the norm sum vector exceeds certain boundary value. We showed that the exact algorithm for m-LVS can be adapted for solving this problem without increasing of time complexity. We also showed the conditions on value of B when the randomized algorithm with significantly less complexity can find optimal solution of the problem with high probability (i.e., tending to 1 when n tends to infinity).

Acknowledgments. This research was supported by the Russian Scientific Foundation for Basic Research (project 16-11-10041).

References

1. Baburin, A., Pyatkin, A.: Polynomial algorithms for solving the vector sum problem. J. Appl. Industr. Math. **1**(3), 268–272 (2007)
2. Baburin, A., Gimadi, E., Glebov, N., Pyatkin, A.: The problem of finding a subset of vectors with the maximum total weight. J. Appl. Industr. Math. **2**(1), 32–38 (2008)
3. Gimadi, E., Kel'manov, A., Kel'manova, M., Khamidullin, S.: A posteriori detecting a quasiperiodic fragment in a numerical sequence. Pattern Recogn. Image Anal. **18**(1), 30–42 (2008)

4. Gimadi, E., Pyatkin, A., Rykov, I.: On polynomial solvability of some problems of a vector subset choice in a euclidean space of fixed dimension. J. Appl. Industr. Math. **4**(48), 48–53 (2010)
5. Gimadi, E., Rykov, I.: A randomized algorithm for finding a subset of vectors with the maximum euclidean norm of their sum. J. Appl. Industr. Math. **9**(3), 351–357 (2015)
6. Pyatkin, A.: On the complexity of the maximum sum length vectors subset choice problem. J. Appl. Industr. Math. **4**(4), 549–552 (2010)
7. Shenmaier, V.: Solving some vector subset problems by voronoi diagrams. J. Appl. Industr. Math. **10**(4), 550–566 (2016)

Fast Numerical Evaluation of Periodic Solutions for a Class of Nonlinear Systems and Its Applications for Parameter Estimation Problems

Ivan Y. Tyukin[1,3,4(✉)], Jehan Mohammed Al-Ameri[1,2],
Alexander N. Gorban[1,4], Jeremy Levesley[1], and Valery A. Terekhov[3]

[1] Department of Mathematics, University of Leicester, Leicester, UK
I.Tyukin@le.ac.uk
[2] College of Science, Department of Mathematics,
University of Basrah, Basrah, Iraq
[3] Department of Automation and Control Processes,
Saint-Petersburg State Electrotechnical University, Saint-Petersburg, Russia
[4] Lobachevsky State University of Nizhny, Novgorod, Russia

Abstract. Fast numerical evaluation of forward models is central for a broad range of inverse problems. Here we propose a method for deriving computationally efficient representations of periodic solutions of parameterized systems of nonlinear ordinary differential equations. These representations depend on parameters of the system explicitly, as quadratures of parameterized computable functions. The method applies to systems featuring both linear and nonlinear parametrization, and time-varying right-hand side. In addition, it opens possibilities to invoke scalable parallel computations and suitable function approximation schemes for numerical evaluation of solutions for various parameter values. Application of the method to the problem of parameter estimation of nonlinear ordinary differential equations is illustrated with a numerical example for the Morris–Lecar system.

Keywords: Parameter estimation · Nonlinear parametrization
Adaptive observers · Time-varying systems

1 Introduction

The problem of state and parameter estimation of systems of ordinary differential equations (ODEs) has been in the focus of attention for many decades. Many frameworks for addressing this problem have been developed to date, including but not limited to shooting methods [6], sensitivity functions [1], splines [29] and adaptive observers [3,4,10,18,26,28] (see also [16,24] for system-identification take on the problem).

© Springer International Publishing AG, part of Springer Nature 2018
A. Eremeev et al. (Eds.): OPTA 2018, CCIS 871, pp. 137–151, 2018.
https://doi.org/10.1007/978-3-319-93800-4_12

Notwithstanding significant progress in this area in both theoretical and applied directions, there is a fundamental yet practical issue with this problem affecting further progress. The issue is that, in general, expressing state variables of systems of ordinary differential equations as explicit *known functions* of parameters and initial conditions or their quadratures is an challenging mathematical problem. Thus sequential numerical approximation of solutions over time is typically involved in the estimation process. The problem, however, is that this process is slow and does not scale well with computational resources available. At the same time there are problems such as e.g. real-time estimation of kinetic parameters of neural membranes [23] that do require fast estimation of model parameters. Hence new approaches are needed.

Here we provide a method enabling us to address the above fundamental challenges for a class of systems with nonlinear parameterziation. The main idea of the method is to present an observed quantity as an integral that is explicitly (a) computable and (b) explicitly dependent on the parameters entering original ODE model nonlinearly. Doing so enables to benefit from computational advantages of prefix sum algorithms [5] and thus alleviate the issues of scalability and real-time. Our preliminary work in this direction [19,27] showed that employing the tools of adaptive observer design [11,18] provides a feasible solution for a relevant class of systems. We demonstrate that further improvement might be achieved by replacing certain integrals with their approximations by e.g. Radial Basis Functions [22].

The contribution is organized as follows. Section 2 specifies main technical assumptions and details mathematical statement of the problem. Section 3 presents main ingredients of the method. In Sect. 4 we discuss these results in relation to the possibility of replacing some integrals in the representation with their approximations. Section 5 presents a numerical example, and Sect. 6 provides a brief conclusion.

2   Problem Formulation

Throughout the paper the following notational agreements are used. Symbol $\mathbb{R}$ denotes the field of real numbers, and $\mathbb{R}^n$ stands for the n-dimensional real space. Let $x \in \mathbb{R}^n$, then $\|x\|$ is the Euclidean norm of x: $\|x\| = \sqrt{x_1^2 + \cdots + x_n^2}$. $\mathcal{C}^r$ denotes the space of continuous functions which are differentiable at least r times. By $L_\infty^n[t_0, T]$ or, when n is clear from the context, $L_\infty[t_0, T]$ we denote the space of all functions $f : [t_0, T] \to \mathbb{R}^n$ such that $\|f\|_{\infty,[t_0,T]} = \sup_{t \in [t_0,T]} \|f(t)\| < \infty$, and $\|f\|_{\infty,[t_0,T]}$ stands for the $L_\infty^n[t_0, T]$ norm of $f(\cdot)$.

2.1   System Definition

Consider the following class of nonlinear systems

$$\dot{x} = F(y,t)x + \Psi(y,t)\theta + g(y,\lambda,t)$$
$$y(t) = C_1^T x; \ x(t_0) = x_0, \tag{1}$$

where $x \in \mathbb{R}^n$ and $y \in \mathbb{R}$ are the state and the output of the system, respectively, $F(y,t) \in \mathbb{R}^{n \times n}$ is a known matrix dependent on y and t; $\lambda \in \Omega_\lambda$, $\Omega_\lambda \subset \mathbb{R}^p$, $\theta \in \Omega_\theta$, $\Omega_\theta \subset \mathbb{R}^m$ are parameters, and $C_1 \in \mathbb{R}^n$: $C_1 = \text{col}(1, 0, \cdots, 0)$. With regards to the sets Ω_θ, Ω_λ, they are allowed to be arbitrary subsets of $\mathbb{R}^m$ and $\mathbb{R}^p$, respectively. Unless stated otherwise, no other prior knowledge about the sets Ω_θ, Ω_λ is assumed.

Other technical assumptions are detailed in Assumption 1 below.

Assumption 1. *The following properties hold for* (1):

1. *the solution of* (1) *is defined for all* $t \geq t_0$, *and it is T-periodic, $T > 0$;*
2. *the function F is continuous, bounded, and $F(y(\cdot), \cdot)$ is T-periodic;*
3. *exact values of parameters λ and θ are unknown;*
4. *the values of $y(t)$ for $t \in [t_0, t_0 + T]$ are available and known;*
5. *the function $\Psi : \mathbb{R} \times \mathbb{R} \to \mathbb{R}^{n \times m}$ is such that $\Psi(y(\cdot), \cdot)$ is T-periodic and is in $L_\infty[t_0, \infty) \cap \mathcal{C}^0$;*
6. *the function $g : \mathbb{R} \times \mathbb{R}^p \times \mathbb{R} \to \mathbb{R}^n$ is such that $g(y(\cdot), \lambda, \cdot)$ is T-periodic and is in $L_\infty[t_0, \infty) \cap \mathcal{C}^1$ for all $\lambda \in \Omega_\lambda$;*
7. *the observability Gramian matrix*

$$G(T, t_0) = \int_{t_0}^{t_0 + T} \Phi_A(s, t_0) C C^T \Phi_A^T(s, t_0) ds, \ C = \text{col}(1, 0, \ldots, 0),$$

where $\Phi_A(t, t_0)$, is the normalized (i.e. $\Phi_A(t_0, t_0) = I_{n+m}$) fundamental solution matrix of

$$\dot{x} = A(y(t), t)x,$$
$$A(y(t), t) = \begin{pmatrix} F(y(t), t) & \Psi(y(t), t) \\ 0 & 0 \end{pmatrix}, \tag{2}$$

is of full-rank, i.e. $\text{rank}(G(T, t_0)) = n + m.$

The class of Eq. (1) accommodates a broad set of technical and natural systems ranging from models of [2], dynamics of populations [14], and neural membranes [20]. In case the solutions are periodic it also may, after suitable modifications [27], include systems

$$\dot{x} = F(y, t)x + \Psi(y, t)\theta + g(y, q, \lambda, t)$$
$$\dot{q} = v(y, \lambda, t)q + \omega(y, \lambda, t) \tag{3}$$
$$y = C_1^T x; \ x(t_0) = x_0, \ q(t_0) = q_0,$$

in which the functions $v(y(\cdot), \lambda, \cdot)$, $\omega(y(\cdot), \lambda, \cdot)$ are continuous.

For notational convenience (cf. [25]), in what follows, we will combine the state variable x and parameters θ entering the right-hand side of (1) linearly into a single variable χ and rewrite the system accordingly:

$$\dot{\chi} = A(y, t)\chi + \begin{pmatrix} g(y, \lambda, t) \\ 0 \end{pmatrix}, \ y(t) = C^T \chi, \ \chi(t_0) = \chi_0. \tag{4}$$

In (4) $\chi = (x, \theta)$ is the combined state vector, matrix $A(y, t)$ is defined as in (2), and $C \in \mathbb{R}^{n+m}$ is $C = \text{col}(1, 0, \cdots, 0)$. Let us now proceed with the formal definition of the problem considered in this contribution.

2.2 Problem Statement

Consider system (4) and suppose that the values of $y(t)$ for $t \in [t_0, t_0 + T]$ are known and available *a-priori*. These values will depend on the parameters λ and initial condition χ_0 which themselves are assumed to be *unknown*. The question is if there exists an operator $\mathcal{F}$ mapping $y(\cdot)$ over $[t_0, t_0 + T]$ into an efficiently computable quantity that does depend on the parameters λ explicitly?

In particular, we are seeking for an $\mathcal{F}(\lambda, [y], t)$ such that

$$
\begin{aligned}
C^T \chi(t; t_0, \chi_0, \lambda) &= \mathcal{F}(\lambda, [y], t), \ \forall \ t \in [t_0, t_0 + T], \ \lambda \in \Omega_\lambda, \\
\mathcal{F}(t, \lambda, [y]) &= \pi(t, \lambda, [y]) + \int_{t_0}^{t} p(\tau, \lambda, y(\tau), [y]) d\tau,
\end{aligned}
\tag{5}
$$

in which the functionals π and p are known and computable, e.g. in quadratures. The functionals π, p must not depend on χ_0 as a parameter, but nevertheless have to ensure the required representation (5).

In what follows, (Theorem 2 in Sect. 3) we demonstrate that finding the required representations $\mathcal{F}(\lambda, [y], t)$ is possible, subject to some mild technical conditions largely contained in Assumptions 1, 2. When such a representation is found one can employ numerous off-line numerical optimisation techniques to infer the values of λ, θ, and initial conditions from the values of y in the interval $[t_0, t_0 + T]$. We will illustrate this step with an example in Sect. 5 in which the Nelder–Mead algorithm [21] will be used for this purpose.

3 Observer-Based Explicit Parametrized Representations of Periodic Solutions

The problem of existence of representations (5) in the context of parameter estimation is hardly viable without assessing parameter identifiability [9] of (4). The corresponding sufficient conditions are derived below.

3.1 Indistinguishable Parametrizations of (4)

We begin with the following technical lemma [19] (cf. [28]).

Lemma 1. *Consider the following class of system*

$$
\dot{\chi} = A_0(t)\chi + u(t) + d(t), \ y = C\chi, \ \chi(t_0) = \chi_0, \ \chi_0 \in \mathbb{R}^\ell,
\tag{6}
$$

where

$$A_0(t) = \begin{pmatrix} \alpha_1(t) & \beta_2(t) & \beta_3(t) & \cdots & \beta_\ell(t) \\ \alpha_2(t) & & & & \\ \vdots & & A_0^*(t) & & \\ \alpha_\ell(t) & & & & \end{pmatrix} \quad and\ u, d, \alpha : \mathbb{R} \to \mathbb{R}^\ell,\ \beta : \mathbb{R} \to \mathbb{R}^{\ell-1},$$

$u \in \mathcal{C}^1,\ d, \alpha, \beta \in \mathcal{C}, \alpha = \mathrm{col}(\alpha_1(t), \alpha_2(t), \dots, \alpha_\ell(t)),\ \beta = (\beta_2(t), \beta_3(t), \dots, \beta_\ell(t)),$ *and assume that solutions of* (6) *are globally bounded in forward time.*

Let, *in addition:*

1. $u, \dot{u}, d, \alpha, \beta$ *be bounded:* $\max\{\|u(t)\|, \|\dot{u}(t)\|\} \le B,\ \|d(t)\| \le \Delta_\xi, \|\alpha(t)\| \le M_1, \|\beta(t)\| \le M_2$ *for all* $t \ge t_0.$

2. *there exist a* $b : \mathbb{R} \to \mathbb{R}^{\ell-1}, b \in \mathcal{C}, \|b(t)\| \le M_3$ *such that the zero solution of the system*

$$\dot{z} = \Lambda(t)z, \qquad \Lambda(t) = A_0^*(t) - b(t)\beta(t),$$

is uniformly exponentially stable, and let $\Phi_\Lambda(t, t_0)$ *be the corresponding fundamental solution:* $\Phi_\Lambda(t_0, t_0) = I_\ell.$

Then the following statements hold:

1. *If the solution of* (6) *is globally bounded for all* $t \ge t_0$ *then, for* T *sufficiently large, there are* $k_1, k_2 \in \mathcal{K}:$
   $\|y(t)\|_{\infty, [t_0, t_0+T]} \le \epsilon \Rightarrow \exists\ t'(\epsilon, x_0) \ge t_0 : \|h(\tau) + u_1(\tau)\|_{\infty, [t', t_0+T]} \le k_1(\epsilon) + k_2(\Delta_\xi),$ *where* $h(t) = \beta(t)z,$

$$\dot{z} = \Lambda(t)z + Gu, \\ G = \begin{pmatrix} -b(t) & I_{\ell-1} \end{pmatrix}, \quad z(t_0) = 0, \tag{7}$$

2. *If* $d(t) \equiv 0,$ *then* $y(t) = 0$ *for all* $t \in [t_0, t_0+T]$ *implies existence of* $P \in \mathbb{R}^{\ell-1}:$

$$\beta(t)\Phi_\Lambda(t, t_0)P + h(t) + u_1(t) = 0 \tag{8}$$

for all $t \in [t_0, t_0 + T].$

According to Lemma 1 the set of parameters:

$$\mathcal{E}(\lambda) = \{\lambda' \in \mathbb{R}^p \mid \exists\ p \in \mathbb{R}^{\ell-1} : \eta(t, p, \lambda', \lambda) = 0,\ \forall t \in [t_0, t_0 + T]\} \tag{9}$$

where

$$\eta(t, p, \lambda', \lambda) = \beta(t)\Phi_\Lambda(t, t_0)p + g_1(y(t), \lambda', t) - g_1(y(t), \lambda, t) \\ + \beta(t) \int_{t_0}^t \Phi_\Lambda(t, \tau)G(\tau) \begin{pmatrix} g(y(\tau), \lambda', \tau) - g(y(\tau), \lambda, \tau) \\ 0 \end{pmatrix} d\tau,$$

and Λ is defined as in (7), contains parameters λ' producing measurements $y(t) = C^T \chi(t; t_0, \chi_0, \lambda')$ that are indistinguishable from $C^T \chi(t; t_0, \chi_0, \lambda)$ on the interval $[t_0, t_0 + T]$. If the set $\mathcal{E}(\lambda)$ contains more than one element then the system (4) may not be uniquely identifiable on $[t_0, t_0 + T]$. Notwithstanding existence and possible utility of systems that are not uniquely identifiable, we will nevertheless focus on systems (4) that are uniquely identifiable on $[t_0, t_0+T]$. Thus we assume that the following holds:

Assumption 2. *For every $\lambda \in \Omega_\lambda$, the set $\mathcal{E}(\lambda)$ consists of just one element.*

3.2   Auxiliary Observer in the Differential Form

In addition to (4) consider the following *auxiliary* system:

$$
\begin{aligned}
\dot{\hat{\chi}} &= A(y(t),t)\hat{\chi} + \begin{pmatrix} g(y(t),\lambda',t) \\ 0 \end{pmatrix} - R^{-1}C(C^T\hat{\chi} - y), \\
\dot{R} &= -\delta R - A(y(t),t)^T R - RA(y(t),t) + CC^T \\
\hat{\chi}(t_0) &= \hat{\chi}_0 \in \mathbb{R}^{n+m}, \;\; R(t_0) \in \mathbb{R}^{(n+m)\times(n+m)},
\end{aligned}
\tag{10}
$$

where $\hat{\chi} \in \mathbb{R}^{n+m}$ is the observer's state, $R(t_0)$ is a positive-definite symmetric matrix, and $\delta \in \mathbb{R}_{>0}$ is a positive parameter. Solutions of (10) are defined for all $t \geq t_0$ (see items (1), (2) in Assumption 1), and hence, [11], $R(t)$ is given by

$$
\begin{aligned}
R(t) = {}& e^{-\delta(t-t_0)}\Phi_A(t_0,t)^T R(t_0)\Phi_A(t_0,t) + \\
& \int_{t_0}^t e^{-\delta(t-s)}\Phi_A(s,t)^T CC^T \Phi_A(s,t)ds.
\end{aligned}
\tag{11}
$$

It is clear that $R(t)$ is non-singular for all $t \geq t_0$, symmetric, and positive-definite. Furthermore, if the value of the parameter $\delta > 0$ is chosen so that

$$
\|e^{-\delta(t-t_0)/2}\Phi_A(t_0,t)\| \leq De^{-a(t-t_0)}, \;\; a > 0,
\tag{12}
$$

then $R(t)$ is bounded. In what follows the following additional assumption is instrumental:

Assumption 3. *There exist $t_1 \geq t_0$ and $\alpha(\delta) > 0$ such that*

$$
\phi(t,\delta) = \int_{t_0}^t e^{-\delta(t-s)}\Phi_A(s,t)^T CC^T \Phi_A(s,t)ds \geq \alpha(\delta)I_{n+m}
$$

for all $t \geq t_1$.

The next theorem specifies asymptotic behaviour of the observer system (10) (adapted from [11]).

Theorem 1. *Consider (10) and suppose that $\delta > 0$ be chosen so that both (12) and Assumption 3 hold, and $\lambda' = \lambda$. Then there exists a $t_2 \geq t_0$, such that:*

$$
\|\hat{\chi}(t;\hat{\chi}_0) - \chi(t;\chi_0)\| \leq ke^{-\delta(t-t_0)}
$$

for all $t \geq t_2$, where k is a constant dependent on δ, t_0, χ_0 and the initial state $\hat{\chi}_0$ of the observer system (10).

Theorem 1 states the variable $\hat{\chi}(t)$ asymptotically tracks $\chi(t)$, and that the difference between the two converges to zero exponentially. Here, however, we are interested in establishing finite-time relationships (5). To do so we need another technical result establishing sufficient conditions for the existence of unique periodic solutions of R. The result is provided in Lemma 2 [19].

Lemma 2. *Consider (10) with $A(y(t), t)$ being T-periodic. Then, for sufficiently large $\delta > 0$, there exists a unique symmetric $R(t_0)$ ensuring that the function $R(t)$ defined by (11) is T-periodic. If, in addition, (12) and Assumption 3 hold then $R(t_0)$ is positive-definite.*

3.3   Integral Parametrization of Periodic Solutions of (4)

For notational convenience, let us rewrite auxiliary observer Eq. (10) as:

$$\dot{\hat{\chi}} = (A(t) - R^{-1}CC^T)\hat{\chi} + \begin{pmatrix} g(y(t), \lambda', t) \\ 0 \end{pmatrix} + R^{-1}Cy(t)$$

$$\dot{R} = -\delta R - A(y(t), t)^T R - RA(y(t), t) + CC^T$$

$$\hat{\chi}(t_0) = \hat{\chi}_0 \in \mathbb{R}^{n+m}, \ R(t_0) \in \mathbb{R}^{(n+m) \times (n+m)}, \quad (13)$$

and additionally consider dynamics of the linear part of the first equation:

$$\xi = \left(A(y(t), t) - R^{-1}(t)CC^T \right)\xi. \quad (14)$$

Let $\Phi(t, s)$ be the normalized fundamental solution matrix of (14), i.e. $\Phi(t, t) = I_{n+m}$ and $\Phi(s, t) = \Phi(t, s)^{-1}$.

Theorem 2. *Consider system (13) and suppose that Assumptions 1 and 2 hold. In addition, suppose that condition (12) hold and the values of δ and the initial condition $R(t_0)$ in (13) are chosen such that $R(t) > 0$ is T-periodic.*
Consider the function $\hat{y} : \mathbb{R}^p \times \mathbb{R} \to \mathbb{R}$:

$$\hat{y}(\lambda', t) = C^T \Big(\Phi(t, t_0)\hat{\chi}_0 +$$
$$\int_{t_0}^t \Phi(t, \tau) \left(R^{-1}(\tau)Cy(\tau) + \begin{pmatrix} g(y(\tau), \lambda', \tau) \\ 0 \end{pmatrix} \right) d\tau \Big) \quad (15)$$

where

$$\hat{\chi}_0 = (I_{n+m} - \Phi(t_0 + T, t_0))^{-1} \int_{t_0}^{t_0+T} \Phi(t_0 + T, \tau)$$
$$\times \left(R^{-1}(\tau)Cy(\tau) + \begin{pmatrix} g(y(\tau), \lambda', \tau) \\ 0 \end{pmatrix} \right) d\tau. \quad (16)$$

Then

$$\hat{y}(\lambda', t) = C\chi(t; t_0, \chi_0, \lambda) \ \forall \ t \in [t_0, t_0 + T] \Leftrightarrow \lambda = \lambda'.$$

The proof can be found in [19].

4   Discussion

One of the immediate computational advantages of the method is that the proposed integral representations offer a possibility to employ parallel calculations. In addition, the method offers reduction of dimensionality of the problem due to incorporating linearly parameterized part of the model into internal variables of the proposed representations. These internal variables are uniquely determined by parameters entering the model nonlinearly and are computed as a part of the representation.

In what follows we will show that further computational improvements might be possible and are practically viable (as illustrated with an example) if certain variables in the representations are replaced by their reasonable sparse Radial Basis Function approximations.

One of the key steps justifying incorporation of relevant class of equations specified by (3) into the setting focusing on (1) was an assumption that the variable $q(t; q_0, \lambda, y)$ is expressible as a known function of parameters, initial conditions, and t. For example, if q relates to a single first-order equation then such function can be computed as follows:

$$q(t; q_0, \lambda, y) = e^{\int_{t_0}^t \upsilon(y(\tau), \lambda, \tau)d\tau} q_0 + e^{\int_{t_0}^t \upsilon(y(\tau), \lambda, \tau)d\tau}$$
$$\times \int_{t_0}^t e^{-\int_{t_0}^\tau \upsilon(y(s), \lambda, s)ds} \omega(y(\tau), \lambda, \tau)d\tau \quad (17)$$
$$q_0 = (1 - e^{-\int_{t_0}^{t_0+T}(\upsilon(y(s), \lambda, s)ds)})^{-1} \int_{t_0}^{t_0+T} e^{-\int_z^t (\upsilon(y(s), \lambda, s))ds} \omega(y(z), \lambda, z)dz.$$

If the original problem is governed by (3) then availability of $q(t; q_0, \lambda, y)$ is required in our explicit parameter-dependent representation. One way to resolve the problem is to numerically evaluate all integrals involved. This, however, may not always be optimal. An alternative could be to use computationally efficient approximations of $q(t; q_0, \lambda, y)$ instead.

A possible class of approximations is the class of Radial Basis Functions (RBF) which are known to be efficient for approximating scattered datasets [7]. Recall that Radial Basis Functions are those functions that exhibit radial symmetry, that is, may be seen to depend only (apart from some known parameters) on the distance $r = \|X - X_c\|$ between the centre of the function, X_c, and a generic point X. These functions may be generically represented in the form $\phi(r)$, where the function ϕ is a real-valued function of a real non-negative argument. The functions ϕ may be both globally or compactly supported, and Table 1 presents some relevant examples. The Gaussian and the inverse multi-quadric are positive definite, so that the matrices which arise in interpolation problems are invertible. The other functions are conditionally positive definite, and a polynomial needs to be appended in general so that the interpolation problem is well-posed [7].

Let $X \in \mathbb{R}^d$ be a vector accommodating relevant measurement parameters, i.e. t and λ. In other words, $X = (t, \lambda)$. Consider $X_c = \{X_{c_1}, X_{c_2}, \cdots, X_{c_M}\}$. The centres X_c could be selected from the given data samples or derived via

Table 1. Some commonly used radial basis functions. Parameter α, called "local shape parameter", controls the shape of the radial basis function.

Infinitely smooth RBFs	Functional form, $\phi(r)$	Parameters
Polyharmonic Spline	r^k	$k > 0, k \notin 2\mathbb{N}$
Gaussian	$e^{-(\alpha r)^2}$	$\alpha > 0$
Multiquadric(MQ)	$(1 + \alpha^2 r^2)^{k/2}$	$k > 0, k \notin 2\mathbb{N}, \alpha > 0$
Inverse multiquadric	$(1 + \alpha^2 r^2)^{k/2}$	$k < 0, k \notin 2\mathbb{N}, \alpha > 0$

clustering algorithms. Let

$$S(X) = \sum_{j=1}^{M} \omega_j \phi(\|X - X_{c_j}\|) + p(X), \ X \in \mathbb{R}^d, \tag{18}$$

where p is a polynomial, be an RBF approximation of $q(t; q_0, \lambda, y)$ or simply $q(t, \lambda)$, where ω_j are unknown coefficients that need to be determined. The polynomial p is appended when ϕ is not positive definite. It is well-known that, for a broad range of $\phi(\cdot)$, any continuous function on a bounded domain can be approximated by sums (18) with arbitrary accuracy in L_p-norm, $p > 1$, subject to the choice of parameters X_{c_j}, ω_j, and M [22].

The following heuristics is proposed to replace repeat evaluations (17) of $q(t, \lambda)$ with their RBF approximations in a generic optimisation routine for inferring the values of θ and λ.

Algorithm 1 *[Parameter inference with approximated variables].*

1. *Initialisation: set $\hat{\lambda}$ as an initial guess of λ.*
2. *A set of M samples $X_i = (t_{n_i}, \lambda_{m_i})$ is randomly drawn from a relevant domain or chosen in accordance with some pre-defined process. The domain, in general, may depend on $\hat{\lambda}$.*
3. *Group spatially close points using a suitable clustering algorithm (e.g. [8, 12, 13, 15]), and set the centres X_{cj} as the centres of these clusters.*
4. *Determine parameters ω_j in (18) as the minimizer of $\sum_{i=1}^{N}(S(X_i) - q(t_i, \lambda_i))^2$, $N > 0$. Note that adjustments of the shape parameter, α, might be needed to ensure good approximation.*
5. *Using representation (15) and approximant (18) define:*

$$\begin{aligned} \tilde{y}(\hat{\lambda}, t) &= F(t, t_0, \theta, \hat{\lambda}, \hat{q}(\hat{\lambda}, t)) \\ \hat{q}(\hat{\lambda}, t) &= \sum_{k=1}^{M} \omega_k \phi(\|(t, \hat{\lambda}) - (t_{c_k}, \lambda_{c_k})\|). \end{aligned} \tag{19}$$

The function $\tilde{y}(\hat{\lambda}, t)$ is an approximation of $\hat{y}(\hat{\lambda}, t)$.
6. *Use $\tilde{y}(\hat{\lambda}, t)$, to produce a refined guess of $\hat{\lambda}$ and return to Step 1 if required.*

In the next section we illustrate an application the method (with and without Algorithm 1) to the problem of parameter estimation for the Morris–Lecar system.

5   Example

5.1   Direct Application of the Method

Consider the following simple point model of neural membrane activity [20]:

$$\dot{x} = g_{Ca}m_\infty(x)(x - E_{Ca}) + g_K q(x + E_K) + g_L(x + E_L) + I$$
$$\dot{q} = -\tau(x)^{-1}q + \tau(x)^{-1}w_\infty(x), \ y = x, \tag{20}$$

where

$$m_\infty(x) = 0.5\left(1 + \tanh\left(\frac{x - V_1}{V_2}\right)\right), \ w_\infty(x) = 0.5\left(1 + \tanh\left(\frac{x + V_3}{V_4}\right)\right)$$
$$\tau(x) = T_0/\left(\cosh\left(\frac{x + V_3}{2V_4}\right)\right).$$

Here x is the measured voltage, q is the recovery variable. Parameters E_{Ca}, E_K, E_L are the Nernst potentials of which the nominal values are assumed to be *known*: $E_{Ca} = 55.17, E_K = -110.14, E_L = 49.49$; other parameters may vary from one cell to another and thus are considered *unknown*.

Assume that the model operates in the oscillatory regime which corresponds to periodic solutions of (20). For practically relevant values of T_0, V_3, V_4 and measurements $x(\cdot)$ the integral $\int_{t_0}^{t_0+T} \tau(x(s))^{-1}ds > 0$, where T is the period of oscillations. Given that $x(\cdot)$ is T-periodic, the variable q can be expressed as:

$$q(t) = e^{-\int_{t_0}^t \tau(x(s))^{-1}ds}q_0 + \int_{t_0}^t e^{-\int_z^t \tau(x(s))^{-1}ds}\tau(x(z))^{-1}w_\infty(x(z))dz$$

$$q_0 = \left(1 - e^{-\int_{t_0}^t \tau(x(s))^{-1}ds}\right)^{-1}\int_{t_0}^{t_0+T} e^{-\int_z^{t_0+T} \tau(x(s))^{-1}ds}\tau(x(z))^{-1}w_\infty(x(z))dz.$$

Denoting $g(t, \lambda, [y]) = g_{Ca}m_\infty(x)(x - E_{Ca}) + g_K q(x + E_K)$, $\Psi(t, y) = (y(t), 1)$, and combining parameters as $\theta = (g_L, I)$, $\lambda = (V_1, V_2, V_3, V_4, T_0, g_{Ca}, g_K)$ we can rewrite (20) in the form of Eq. (4) with

$$A(y(t), t) = \begin{pmatrix} 0 & y(t) & 1 \\ 0 & 0 & 0 \\ 0 & 0 & 0 \end{pmatrix}.$$

For this system and chosen nominal parameter values, the period of oscillations is $T = 15.1692$. For convenience, the integration interval was set to $[0, 15.1692]$. Numerical evaluations of integrals and solutions of all auxiliary differential equations have been performed on equi-spaced grids with the step size of 0.0002.

According to Theorem 2, explicit parameter-dependent representation of the observed quantity, $\hat{y}(\lambda, t)$, is defined by (15), where $C = (1, 0, 0)$, $\chi = \text{col}(x, \theta)$, and the fundamental solution (3×3)-matrices $\Phi(t, t_0)$ and $\Phi_A(t, t_0)$ are computed for the linear systems $\dot{\chi} = (A(y(t), t) - R^{-1}(t)CC^T)\chi$, $\dot{R} = -\delta R - A(y(t), t)^T R - RA(y(t), t) + CC^T$, and $\dot{\chi} = A(y(t), t)\chi$, respectively, by the Improved Euler method for $t \in [0, 15.1692]$. The value of δ was set as $\delta = 2$, and numerical approximations of matrices $\Phi_A(t, t_0)$ were used to compute the matrices $R(t)$ in accordance with Eq. (11). The value of $R(t_0)$ in (11) was so that $R(t)$ is periodic (see Lemma 2).

Figure 1 shows the relative error, $e(t) = (\hat{y}(\lambda, t) - y(t))/\|y\|_{\infty, [t_0, t_0+\infty]}$, between the proposed numerical representation (15) and simulated $y(t)$ (Runge–Kutta, step size 0.0002) for nominal parameter values.

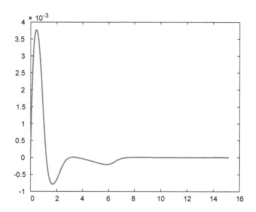

Fig. 1. Relative error $e(t) = (\hat{y}(\lambda, t) - y(t))/\|y\|_{\infty, [t_0, t_0+\infty]}$ as a function of t.

The parameterized representations were later used, in combination with the Nelder–Mead algorithm [21] to recover the values of parameters λ and θ. Results of the estimation process after 3000 steps are shown in Table 2. The process took less than 10 minutes on a standard PC in Matlab R2015a.

Table 2. True (first row) and Estimated (second row) of λ and θ, and the value of x_0

Vector $\lambda = (V_1, V_2, V_3, V_4, T_0, g_{Ca}, g_K)$

V_1	V_2	V_3	V_4	T_0	g_{Ca}	g_K
-1	15	-10	14.5	3	-1.1	-2
-0.999	14.999	-10.000	14.500	3.000	-1.100	-2.000

Vector $\theta = (g_L, I)$ and x_0

g_L	I	x_0
-0.5	10	21.96388
-0.49982	9.99345	21.96166

To assess potential computational advantage of the proposed approach we compared the time required for 1000 evaluations of $y(t)$ in Matlab (a) expressed as in (15) and (b) computed by the Improved Euler method over the interval $[t_0, t_0 + T]$. The parameter values for both cases were kept identical and did not change from one trial to the other. The results are summarized in Table 3.

Table 3. Time for 1000 evaluations of y

Eq. (15)	Improved Euler method	Ratio
2.21311 min	10.43818 min	4.71652

5.2 The Method with RBF Approximation of q

To show feasibility of RBF approximations in this problem we repeated the experiment above but this time with the variable q replaced with its RBF approximation inside the optimisation routine (Nelder–Mead). To produce such approximations we followed steps of Algorithm 1. As the RBF kernel we used the Gaussian function. This transforms (18) into

$$S_q(X) = \sum_{j=1}^{M} w_j e^{-(\alpha\|X-X_{c_j}\|)^2}. \tag{21}$$

Note that the variable q depends only on 3 components of the vector λ, i.e. T_0, V_3, and V_4. And hence all steps of the algorithm related to approximation apply to these 3 relevant components and the variable t only. We considered an extremely sparse setting, in which each of the three parameters have been sampled at 2 points per each relevant sample of t. The values of t where chosen from the grid of 0.002-spaced points in $[0, 15.1692]$ ($N = 7584$ points in the grid). The shape parameter α was set to 0.2107. To see how well $S_q(t_i, \hat{\lambda})$ approximates $q(t_i, \lambda)$ as a function of t_i the following simple criterion has been used:

$$LS = \sum_{i=1}^{N}(q(t_i, \lambda) - S_q(t_i, \hat{\lambda}))^2. \tag{22}$$

In order to judge the efficiency of the approach we run the algorithm 1000 times and recorded empirical errors between λ_i and their estimates $\hat{\lambda}_i$, and computed their L_2 distances as:

$$d(\nu) = \sqrt{\sum_{i=1}^{7}(\lambda_i - \hat{\lambda}_i(\nu))^2}, \tag{23}$$

where $\nu = 1, \cdots, 1000$ is the number of the experiment. Initial guesses for λ were selected randomly in the n-cube $[0, 1] + \lambda_i$, $i = 1, 2, \cdots, 7$, where λ_i are the nominal values. Figure 2 shows histograms of (22), (23) at the initial step of the algorithm. Figure 3 shows histograms of distances between λ and $\hat{\lambda}$ as well as the least square errors (LS) after the application of the Nelder–Mead optimisation routine in which the values of $q(t, \hat{\lambda})$ have been approximated in accordance with Algorithm 1. Note that the histograms are not normalized. We observed a pronounced shift of the histograms to the left, where they concentrate around zero. This contrasts sharply with the initial distributions of errors seen in Fig. 2.

As can be seen from these experiments, RBF approximation is a viable way to further improve scalability and potential of the method.

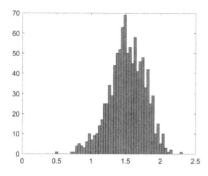
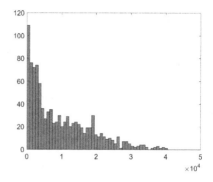

Fig. 2. Histograms of the distributions of $d(\nu)$, $\nu = 1, \cdots, 1000$ (left panel), and least square errors $LS = \sum_{i=1}^{N}(q(t_i, \lambda) - S_q(t_i, \hat{\lambda}))^2$ (right panel) prior to any estimation.

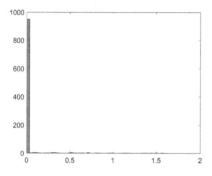
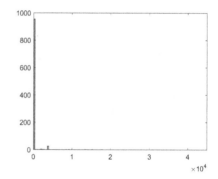

Fig. 3. Histograms of the distributions of $d(\nu)$, $\nu = 1, \cdots, 1000$ (left panel) and $LS = \sum_{i=1}^{N}(q(t_i, \lambda) - S_q(t_i, \hat{\lambda}))^2$ (right panel) after optimisation.

6   Conclusion

The work presented a method for computationally efficient and explicit parameter-dependent representation of periodic solutions of systems of nonlinear ODEs. The method is rooted in the ideas from adaptive observers theory and is an extension of our earlier work [27] in which linear part of the system was supposed to be time-invariant. Here we extended this result to systems with time-varying linear parts. Similar extension can be carried out for other observer structures, including e.g. [17], followed by replacement of condition (7) in Assumption 1 with the requirement of persistency of excitation of relevant terms.

The computational advantage of the method is due to the possible parallel implementation of calculations that the proposed representations offer. In addition to offering scalability and making use of parallel computations, the method offers reduction of dimensionality of the problem due to incorporating linearly parameterized part of the model into internal variables of the proposed representations. These internal variables are uniquely determined by parameters entering the model nonlinearly and are computed as a part of the representation.

An interesting possibility to further improve computational efficiency of the approach to a class of problems emerging in modelling dynamics of neural cells stems from invoking RBF approximations in place of certain integrals in the schemes. Viability of the approach in this setting has been demonstrated with a numerical example.

References

1. Banks, H.T., Robbins, D., Sutton, K.: Theoretical foundations for traditional and generalized sensitivity functions for nonlinear delay differential equations. Math. Biosci. Eng. CRSC-TR12-14 (2012)
2. Bastin, G., Dochain, D.: On-line Estimation and Adaptive Control of Bioreactors. Elsevier, Amsterdam (1990)
3. Bastin, G., Gevers, M.: Stable adaptive observers for nonlinear time-varying systems. IEEE Trans. Autom. Control $33(7)$, 650–658 (1988)
4. Besançon, G.: Remarks on nonlinear adaptive observer design. Syst. Control Lett. $41(4)$, 271–280 (2000)
5. Blelloch, G.E.: Prefix sums and their applications. Technical report CMU-CS-90-190. Carnegie Mellon University (1990)
6. Bock, H.G., Kostina, E., Schlöder, J.P.: Numerical methods for parameter estimation in nonlinear differential algebraic equations. GAMM-Mitteilungen $30(2)$, 376–408 (2007)
7. Buhmann, M.D.: Radial Basis Functions: Theory and Implementations, vol. 12. Cambridge University Press, Cambridge (2003)
8. Capoyleas, V., Rote, G., Woeginger, G.: Geometric clusterings. J. Algorithms $12(2)$, 341–356 (1991)
9. Distefano, J., Cobelli, C.: On parameter and structural identifiability: nonunique observability/reconstructibility for identifiable systems, other ambiguities, and new definitions. IEEE Trans. Autom. Control $25(4)$, 830–833 (1980)
10. Farza, M., MSaad, M., Maatoug, T., Kamoun, M.: Adaptive observers for nonlinearly parameterized class of nonlinear systems. Automatica $45(10)$, 2292–2299 (2009)
11. Hammouri, H., de Morales, L.: Observer synthesis for state-affine systems. In: Proceedings of the 29th IEEE Conference on Decision and Control, pp. 784–785. IEEE (1990)
12. Jain, A.K., Duin, R.P.W., Mao, J.: Statistical pattern recognition: a review. IEEE Trans. Pattern Anal. Mach. Intell. $22(1)$, 4–37 (2000)
13. Jain, A.K., Murty, M.N., Flynn, P.J.: Data clustering: a review. ACM Comput. Surv. (CSUR) $31(3)$, 264–323 (1999)
14. Jing, Z., Chen, L.: The existence and uniqueness of limit cycles in general predator-prey differential equations. Chin. Sci. Bull. 9, 521–523 (1984)
15. Kaufman, L., Rousseeuw, P.J.: Finding Groups in Data: An Introduction to Cluster Analysis, vol. 344. Wiley, Hoboken (2009)
16. Ljung, L.: System Identification - Theory for the User. Prentice Hall, Englewood Cliffs (1987)

17. Loria, A., Panteley, E., Zavala, A.: Adaptive observers for robust synchronization of chaotic systems. IEEE Trans. Circ. Syst. I: Regular Papers **56**(12), 2703–2716 (2009)
18. Marino, R.: Adaptive observers for single output nonlinear systems. IEEE Trans. Autom. Control **35**(9), 1054–1058 (1990)
19. Mohammed, J.A.A., Tyukin, I.: Explicit parameter-dependent representations of periodic solutions for a class of nonlinear systems. IFAC-PapersOnLine **50**(1), 4001–4007 (2017)
20. Morris, C., Lecar, H.: Voltage oscillations in the barnacle giant muscle fiber. Biophys. J. **35**, 193–213 (1981)
21. Nelder, J.A., Mead, R.: A simplex method for function minimization. Comput. J. **7**(4), 308–313 (1965)
22. Park, J., Sandberg, I.W.: Universal approximation using radial-basis-function networks. Neural Comput. **3**(2), 246–257 (1991)
23. Prinz, A., Billimoria, C., Marder, E.: Alternative to hand-tuning conductance-based models: construction and analysis of databases of model neurons. J. Neorophysiol. **90**, 3998–4015 (2003)
24. Soderstrom, T., Stoica, P.: System Identification. Prentice Hall, Englewood Cliffs (1988)
25. Torres, L., Besançon, G., Georges, D., Verde, C.: Exponential nonlinear observer for parametric identification and synchronization of chaotic systems. Math. Comput. Simul. **82**, 836–846 (2012)
26. Tyukin, I.: Adaptation in Dynamical Systems. Cambridge University Press, Cambridge (2011)
27. Tyukin, I.Y., Gorban, A., Tyukina, T., Al-Ameri, J., Korablev, Y.A.: Fast sampling of evolving systems with periodic trajectories. Math. Model. Nat. Phenom. **11**(4), 73–88 (2016)
28. Tyukin, I.Y., Steur, E., Nijmeijer, H., Van Leeuwen, C.: Adaptive observers and parameter estimation for a class of systems nonlinear in the parameters. Automatica **49**(8), 2409–2423 (2013)
29. Zhan, C., Yeung, L.F.: Parameter estimation in systems biology models using spline approximation. BMC Syst. Biol. **5**(1), 1 (2011)

Mathematical Programming

On Vertices of the Simple Boolean Quadric Polytope Extension

Andrei V. Nikolaev[(✉)]

Department of Discrete Analysis, P.G. Demidov Yaroslavl State University,
Yaroslavl, Russia
andrei.v.nikolaev@gmail.com

Abstract. Following the seminal work of Padberg on the Boolean
quadric polytope BQP and its LP relaxation BQP_{LP}, we consider a natural extension: the polytopes $SATP$ and $SATP_{LP}$, with BQP_{LP} being a
projection of $SATP_{LP}$ face (and BQP – projection of $SATP$ face). Various special instances of 3-SAT problem like NAE-3-SAT, 1-in-3-SAT,
weighted MAX-3-SAT, and others can be solved by integer programming over $SATP_{LP}$. We consider the properties of $SATP$ 1-skeleton
and $SATP_{LP}$ fractional vertices. Like BQP_{LP}, the polytope $SATP_{LP}$
has the Trubin-property being quasi-integral (1-skeleton of $SATP$ is a
subset of 1-skeleton of $SATP_{LP}$). However, unlike BQP, not all vertices
of $SATP$ are pairwise adjacent, the diameter of $SATP$ equals 2, and
the clique number of 1-skeleton is superpolynomial in dimension. It is
known that the fractional vertices of BQP_{LP} are half-integral (0, 1 or
1/2 valued). We establish that the denominators of $SATP_{LP}$ fractional
vertices can take any integer values.

Keywords: 3-satisfiability · LP relaxation · 1-skeleton
Fractional vertices

1 Boolean Quadric Polytope

We consider the well-known *Boolean quadric polytope* $BQP(n)$ [20], constructed
from the NP-hard problem of unconstrained Boolean quadratic programming:

$$Q(x) = x^T Q x \rightarrow \max,$$

where vector $x \in \{0, 1\}^n$, and Q is an upper triangular matrix, by introducing
new variables $x_{i,j} = x_i x_j$.

In the standard form $BQP(n)$ can be defined as the convex hull of all integral
solutions of the system

$$x_{i,j}^{1,1} + x_{i,j}^{1,2} + x_{i,j}^{2,1} + x_{i,j}^{2,2} = 1, \tag{1}$$

$$x_{i,j}^{1,1} + x_{i,j}^{1,2} = x_{k,j}^{1,1} + x_{k,j}^{1,2}, \tag{2}$$

$$x_{i,j}^{1,1} + x_{i,j}^{2,1} = x_{i,l}^{1,1} + x_{i,l}^{2,1}, \tag{3}$$

$$x_{i,i}^{1,2} = x_{i,i}^{2,1} = 0, \tag{4}$$

$$x_{i,j}^{1,1} \geq 0, \; x_{i,j}^{1,2} \geq 0, \; x_{i,j}^{2,1} \geq 0, \; x_{i,j}^{2,2} \geq 0, \tag{5}$$

© Springer International Publishing AG, part of Springer Nature 2018
A. Eremeev et al. (Eds.): OPTA 2018, CCIS 871, pp. 155–169, 2018.
https://doi.org/10.1007/978-3-319-93800-4_13

where $1 \leq k \leq i \leq j \leq l \leq n$ [4].

The Boolean quadric polytope arises in many fields of mathematics and physics. Sometimes it is called the *correlation polytope*, since its members can be interpreted as joint correlations of events in some probability space. Also, within the quantum mechanics the Boolean quadric polytope is connected with the representability problem for density matrices of order 2 that render physical properties of a system of particles [12]. Besides, $BQP(n)$ is in one-to-one correspondence via the covariance linear mapping with the well-known *cut polytope* $CUT(n+1)$ of the complete graph on $n+1$ vertices [10].

In recent years, the Boolean quadric polytope has been under the close attention in connection with the problem of estimating the extension complexity. An *extension* of the polytope P is another polytope Q such that P is the image of Q under a linear map. The number of facets of Q is called the *size of an extension*. *Extension complexity* of P is defined as the minimum size over the set of all possible extensions. Fiorini et al. proved that the extension complexity of the Boolean quadric polytope is exponential.

Theorem 1 (see [13] **and** [15]**).** *The extension complexity of $BQP(n)$ and $CUT(n)$ is $2^{\Omega(n)}$.*

Since the polytopes of many combinatorial optimization problems, including stable set, knapsack, 3-dimensional matching, and traveling salesperson, contain a face that is an extension of $BQP(n)$, those polytopes also have an exponential extension complexity. Thus, corresponding problems cannot be solved effectively by linear programming, as any LP formulation will have an exponential number of inequalities [13]. For more details on the special role of the Boolean quadratic polytope see [18].

The system (1)–(5) without integrality constraint describes the *LP relaxation $BQP_{LP}(n)$*. Corresponding cut polytope relaxation is known as the *rooted semimetric polytope $RMET(n)$*.

For any polytope P, we call the collection of its vertices (0-faces) and its edges (1-faces) the 1-*skeleton* of P. Let Q be a polytope that is contained in P. We say that P has the *Trubin-property* (with respect to Q) if 1-skeleton of Q is a subset of 1-skeleton of P [22]. Polytope P with this property is also called *quasi-integral*. If P has the Trubin-property, then all vertices of Q are vertices of P and those facets of Q that define invalid inequalities for P do not create any new adjacencies among the vertices of Q.

Theorem 2 (see [20]**).** *The diameter of the polytope $BQP(n)$ equals 1. The relaxation $BQP_{LP}(n)$ has the Trubin-property with respect to $BQP(n)$.*

As for the fractional vertices in the LP relaxation $BQP_{LP}(n)$, they are completely described and are half-integral.

Theorem 3 (see [20]**).** *Every vertex of $BQP_{LP}(n)$ is $\{0, \frac{1}{2}, 1\}$ valued.*

We note that while the fractional vertices of $BQP_{LP}(n)$ have their denominators bounded by 2, the matrix of the system (1)–(5) is not bimodular. Every nonsingular submatrix of the system has a determinant of $\pm 2^p$ where p is a non-negative integer [20]. Thus, the polynomial time algorithm for bimodular integer linear programming [23], unfortunately, cannot be applied.

2   3-SAT Relaxation Polytope

We consider a more general polytope $SATP(m, n) \subset \mathbb{R}^{6mn}$ (see [4]), obtained as the convex hull of all integral solutions of the system

$$\sum_{k,l} x_{i,j}^{k,l} = 1, \tag{6}$$

$$x_{i,j}^{1,1} + x_{i,j}^{2,1} + x_{i,j}^{3,1} = x_{i,t}^{1,1} + x_{i,t}^{2,1} + x_{i,t}^{3,1}, \tag{7}$$

$$x_{i,j}^{k,1} + x_{i,j}^{k,2} = x_{s,j}^{k,1} + x_{s,j}^{k,2}, \tag{8}$$

$$x_{i,j}^{k,l} \geq 0, \tag{9}$$

where $k = 1, 2, 3$; $l = 1, 2$; $i, s = 1, \ldots m$; $j, t = 1, \ldots n$.

Inequalities (6)–(9) without the integrality constraint define the LP relaxation $SATP_{LP}(m, n)$. Points that satisfy the system can be conveniently represented as a block matrix (Table 1).

Table 1. Fragment of $SATP_{LP}(m, n)$ block matrix

$x_{i,j}^{1,1}$	$x_{i,j}^{1,2}$	$x_{i,t}^{1,1}$	$x_{i,t}^{1,2}$
$x_{i,j}^{2,1}$	$x_{i,j}^{2,2}$	$x_{i,t}^{2,1}$	$x_{i,t}^{2,2}$
$x_{i,j}^{3,1}$	$x_{i,j}^{3,2}$	$x_{i,t}^{3,1}$	$x_{i,t}^{3,2}$
$x_{s,j}^{1,1}$	$x_{s,j}^{1,2}$	$x_{s,t}^{1,1}$	$x_{s,t}^{1,2}$
$x_{s,j}^{2,1}$	$x_{s,j}^{2,2}$	$x_{s,t}^{2,1}$	$x_{s,t}^{2,2}$
$x_{s,j}^{3,1}$	$x_{s,j}^{3,2}$	$x_{s,t}^{3,1}$	$x_{s,t}^{3,2}$

If we consider a face of the polytope $SATP(n, n)$, constructed as follows:

$$\forall i, j : x_{i,j}^{3,1} = x_{i,j}^{3,2} = 0,$$

$$\forall i : x_{i,i}^{1,2} = x_{i,i}^{2,1} = 0,$$

and discard all coordinates with $i > j$ (orthogonal projection), then we get the polytope $BQP(n)$. As a result, we have

Theorem 4. *The extension complexity of $SATP(m, n)$ is $2^{\Omega(\min\{m,n\})}$.*

The polytope $SATP(m, n)$ was introduced in [4] and later studied in [19] as the integer programming formulation of various special instances of 3-SAT problem like not-all-equal-3-SAT, 1-in-3-SAT, weighted MAX-3-SAT, and others. A polynomial time algorithm for some cases of edge constrained bipartite graph coloring, based on the properties of $SATP_{LP}(m,n)$ relaxation, was considered in [19].

The object of research in this paper are the properties of 1-skeleton of $SATP(m,n)$ and the fractional vertices of $SATP_{LP}(m,n)$ relaxation.

3   1-Skeleton

We consider the following 3 characteristics of 1-skeleton: an adjacency relation, a diameter, and a clique number. The adjacency relation together with the local search technique can serve as a basis for optimization algorithms. See, for example, algorithms for the perfect matching, set covering, and other problems [1,9,17]. The study of 1-skeleton's diameter is motivated by its relationship to edge-following algorithms of linear programming such as the simplex method, since the diameter is a lower bound on the number of non-degenerate steps of the algorithm. The clique number of 1-skeleton serves as a lower bound for computational complexity in a class of *direct-type algorithms* based on linear comparisons. In addition, it was found that this characteristic is polynomial for known polynomially solvable problems and is superpolynomial for intractable problems (see, for example, [5–7]).

We encode every integer vertex z of the polytope $SATP(m,n)$ by the vectors $\boldsymbol{row}(z) \in \{0,1\}^m$ and $\boldsymbol{col}(z) \in \{0,1,2\}^n$. We denote by $\boldsymbol{row}_i(z)$ and $\boldsymbol{col}_j(z)$ the i-th and j-th coordinates of the corresponding vectors.

Lemma 1 (see [19]). *Let z be the vertex of the polytope $SATP(m,n)$, then its coordinates are determined by the vectors $\boldsymbol{row}(z) \in \{0,1\}^m$ and $\boldsymbol{col}(z) \in \{0,1,2\}^n$ by the following formulas:*

$$x_{i,j}^{1,1} = \frac{1}{2}(1 - \boldsymbol{row}_i(z))(2 - \boldsymbol{col}_j(z))(1 - \boldsymbol{col}_j(z)),$$

$$x_{i,j}^{1,2} = \frac{1}{2}\boldsymbol{row}_i(z)(2 - \boldsymbol{col}_j(z))(1 - \boldsymbol{col}_j(z)),$$

$$x_{i,j}^{2,1} = (1 - \boldsymbol{row}_i(z))\boldsymbol{col}_j(z)(2 - \boldsymbol{col}_j(z)),$$

$$x_{i,j}^{2,2} = \boldsymbol{row}_i(z)\boldsymbol{col}_j(z)(2 - \boldsymbol{col}_j(z)),$$

$$x_{i,j}^{3,1} = \frac{1}{2}(1 - \boldsymbol{row}_i(z))\boldsymbol{col}_j(z)(1 - \boldsymbol{col}_j(z)),$$

$$x_{i,j}^{3,2} = \frac{1}{2}\boldsymbol{row}_i(z)\boldsymbol{col}_j(z)(1 - \boldsymbol{col}_j(z)).$$

Theorem 5. *Two vertices u and v of the polytope $SATP(m,n)$ are adjacent if and only if one of following conditions is true:*

- *$\boldsymbol{row}(u) \neq \boldsymbol{row}(v)$ and $\boldsymbol{col}(u) \neq \boldsymbol{col}(v)$;*
- *$\exists! i$: $\boldsymbol{row}_i(u) \neq \boldsymbol{row}_i(v)$ and $\boldsymbol{col}(u) = \boldsymbol{col}(v)$;*
- *$\exists! j$: $\boldsymbol{col}_j(u) \neq \boldsymbol{col}_j(v)$ and $\boldsymbol{row}(u) = \boldsymbol{row}(v)$.*

Proof. If the vertices u and v are not adjacent, then the line segment $[u,v]$ intersects with the convex hull of all remaining vertices:

$$\alpha u + \beta v = \sum \lambda_w w,$$

$$\alpha + \beta = \sum \lambda_w = 1,$$

$$\alpha \geq 0, \ \beta \geq 0, \ \lambda_w \geq 0.$$

We consider some vertex w in this convex combination with a positive λ_w. Since u, v and w are zero-one points, equation implies the inequality

$$w \leq u + v. \tag{10}$$

Let the row and col vectors of u and v do not coincide. Since the vertex w is different from u and v, we have

$$\begin{cases} row(w) \neq row(u), \\ col(w) \neq col(v), \end{cases} \quad or \quad \begin{cases} row(w) \neq row(v), \\ col(w) \neq col(u). \end{cases}$$

Without loss of generality, we assume that

$$\exists i : row_i(w) = 0 \neq row_i(u),$$
$$\exists j : col_j(w) = 0 \neq col_j(v).$$

Hence, we have

$$x_{i,j}^{1,1}(w) = \frac{1}{2}(1 - row_i(w))(2 - col_j(w))(1 - col_j(w)) = 1,$$

$$x_{i,j}^{1,1}(u) + x_{i,j}^{1,1}(v) = \frac{1}{2}(1 - row_i(u))(2 - col_j(u))(1 - col_j(u)) +$$

$$+ \frac{1}{2}(1 - row_i(v))(2 - col_j(v))(1 - col_j(v)) = 0.$$

Therefore, the inequality (10) is not satisfied. The remaining cases are treated similarly. Thus, if the row and col vectors of u and v do not coincide, then the vertices u and v are adjacent.

Now let $col(u)$ and $col(v)$ be equal. We suppose that

$$\exists i, j : row_i(u) \neq row_i(v) \text{ and } row_j(u) \neq row_j(v).$$

We consider two vertices w_u and w_v, constructed as follows:

$$col(w_u) = col(w_v) = col(u) = col(v),$$
$$\forall k(k \neq i) : row_k(w_u) = row_k(u), \ row_k(w_v) = row_k(v),$$
$$row_i(w_u) = row_i(v), \ row_i(w_v) = row_i(u).$$

Vertices w_u and w_v are different from u and v, and we have

$$w_u + w_v = u + v,$$

thus, u and v are not adjacent.

Finally, if the vectors $row(u)$ and $row(v)$ differ only in one coordinate, then there are no vertices w other than u and v that satisfy the inequality (10). Consequently, in this case u and v are adjacent.

It remains to study the case $row(u) = row(v)$ that is completely similar to the case $col(u) = col(v)$.

Thus, 1-skeleton of $SATP(m,n)$ is not a complete graph, unlike $BQP(n)$. Still, this graph is very dense.

Theorem 6. *The diameter of $SATP(m,n)$ 1-skeleton equals 2, with the exception of the case $m = n = 1$.*

Proof. First we consider the special case of $m = n = 1$. The system (6)–(9) takes the form

$$\sum_{k,l} x^{k,l} = 1,$$

$$x^{k,l} \geq 0,$$

and defines a standard simplex with all 6 vertices being pairwise adjacent. However, for $m \geq 2$ or $n \geq 2$, by Theorem 5, there are nonadjacent vertices of $SATP(m,n)$.

Let the vertices u and v of the polytope be not adjacent. Let $m \geq 2$. We consider a vertex w, constructed as follows:

$$row(w) \neq row(u), \ row(w) \neq row(v), \tag{11}$$
$$col(w) \neq col(u), \ col(w) \neq col(v). \tag{12}$$

By Theorem 5, we have w being adjacent both to u and v.

Now let $m = 1$ and $n \geq 2$. If $row(u) = row(v)$, then we again consider a vertex w, satisfying (11)–(12), that is adjacent both to u and v. And if $row(u) \neq row(v)$, then we construct a vertex w as follows:

$$row(w) = row(v) \neq row(u),$$
$$col_1(w) \neq col_1(u), \ col_1(w) \neq col_1(v),$$
$$col_j(w) = col_j(v), \ j \geq 2.$$

By Theorem 5, such vertex w is adjacent both to u and v.

Thus, for $m \geq 2$ or $n \geq 2$ the diameter of $SATP(m,n)$ 1-skeleton equals 2.

Theorem 7. *The clique number of $SATP(m,n)$ 1-skeleton is superpolynomial in dimension and bounded from below by $2^{\min\{m,n\}}$.*

Proof. We consider a vertex set W, such that $\forall w \in W$ we have

$$\forall k (k \leq \min\{m,n\}) : \ row_k(w) = col_k(w).$$

All the remaining coordinates of row and col vectors are assumed to be zero. By Theorem 5, each pair of vertices in W is pairwise adjacent, since their row and col vectors do not coincide, and there are exactly $2^{\min\{m,n\}}$ of such vertices.

In the last theorem of this section we will show that the properties of 1-skeleton of the polytope $SATP$ can be transferred to its LP relaxation $SATP_{LP}$.

Theorem 8. $SATP_{LP}$ *has the Trubin-property with respect to* $SATP$.

Proof. Trubin [22] showed that the *relaxation set partitioning polytope*

$$Ax = e, \; x \geq 0, \tag{13}$$

where A is a zero-one matrix, and e is an all unit column, is quasi-integral. The polytope $SATP_{LP}(m, n)$ can be considered as a special case of the relaxation set partitioning polytope. Constraints (6) and (9) already satisfy (13). Each Eq. (7) in pair with the Eq. (6)

$$\begin{cases} x_{i,j}^{1,1} + x_{i,j}^{2,1} + x_{i,j}^{3,1} = x_{i,t}^{1,1} + x_{i,t}^{2,1} + x_{i,t}^{3,1}, \\ x_{i,j}^{1,1} + x_{i,j}^{2,1} + x_{i,j}^{3,1} + x_{i,j}^{1,2} + x_{i,j}^{2,2} + x_{i,j}^{3,2} = 1, \end{cases}$$

can be rewritten in the required form:

$$\begin{cases} x_{i,t}^{1,1} + x_{i,t}^{2,1} + x_{i,t}^{3,1} + x_{i,j}^{1,2} + x_{i,j}^{2,2} + x_{i,j}^{3,2} = 1, \\ x_{i,j}^{1,1} + x_{i,j}^{2,1} + x_{i,j}^{3,1} + x_{i,j}^{1,2} + x_{i,j}^{2,2} + x_{i,j}^{3,2} = 1. \end{cases}$$

Similarly, each Eq. (8) in pair with the Eq. (6)

$$\begin{cases} x_{i,j}^{1,1} + x_{i,j}^{1,2} = x_{s,j}^{1,1} + x_{s,j}^{1,2}, \\ x_{i,j}^{1,1} + x_{i,j}^{1,2} + x_{i,j}^{2,1} + x_{i,j}^{2,2} + x_{i,j}^{3,1} + x_{i,j}^{3,2} = 1, \end{cases}$$

can be rewritten as

$$\begin{cases} x_{s,j}^{1,1} + x_{s,j}^{1,2} + x_{i,j}^{2,1} + x_{i,j}^{2,2} + x_{i,j}^{3,1} + x_{i,j}^{3,2} = 1, \\ x_{i,j}^{1,1} + x_{i,j}^{1,2} + x_{i,j}^{2,1} + x_{i,j}^{2,2} + x_{i,j}^{3,1} + x_{i,j}^{3,2} = 1. \end{cases}$$

Thus, 1-skeleton of $SATP(m, n)$ is a subset of 1-skeleton of $SATP_{LP}(m, n)$. ∎

4 Fractional Vertices

Another area of research are the properties of fractional vertices in the LP relaxation. The structure of fractional vertices is important for the analysis of approximation algorithms based on LP rounding. In particular, if the denominators of the fractional vertices are bounded by some constant d, then LP rounding provides a d-approximation algorithm for the considered problem (d is called the integrality gap). Besides, the fractional vertices are important for constructing new inequalities for branch and cut algorithms [21].

In this section we consider the LP relaxation $SATP_{LP}(m, n)$. It preserves all integral vertices of $SATP(m, n)$, together with their adjacency in 1-skeleton, but as LP relaxation has its own fractional vertices.

Fractional vertices of $BQP_{LP}(n)$ are quite simple with values only from the set $\{0, \frac{1}{2}, 1\}$. Thereby, they can be completely cut off by triangle inequality constraints of the metric polytope $MET(n)$ [20]. The polynomial time algorithm for *integer recognition* problem (determine whether a maximum of a linear objective function is achieved at an integral vertex of a polytope) over $BQP_{LP}(n)$ is based on this fact [4].

The *metric polytope* $MET(n)$ is obtained by augmenting the system (1)–(5) by the triangle inequalities that define the facets of $BQP(3)$. It is the most simple and natural relaxation of the polytope $CUT(n)$, and has many practical applications, such as being a compact LP formulation for the max-cut problem on graphs not contractible to K_5 [3].

Fractional vertices of the metric polytope $MET(n)$ have a much more complicated nature. Their denominators can take any integral values [16], and it is not known if it is possible to cut them off by a polynomial number of additional linear constraints [8]. Furthermore, characteristics of $MET(n)$ fractional vertices were also considered in [2, 11, 14].

Unfortunately, properties of $SATP_{LP}(m, n)$ fractional vertices are similar to $MET(n)$ and not to $BQP_{LP}(n)$, since their denominators can take any integral values as well.

Theorem 9. *The relaxation polytope $SATP_{LP}(n, n)$ has fractional vertices with denominators equal $n + 1$ for all $n \geq 4$.*

Proof. A vertex of the polytope $SATP_{LP}(m, n)$ is a unique solution of the system (6)–(9) with some of the inequalities (9) turned into equations. We construct a required vertex in a few steps.

The basis is the first four blocks as shown in Table 2. We can use the constraints (7)–(8) to establish the relationship between the coordinates:

$$x_{1,2}^{2,2} = x_{1,1}^{3,2} = x_{2,1}^{3,2} = x_{2,2}^{3,2} = x_{1,2}^{3,1}.$$

Hence, for all blocks in the second column

$$x_{i,2}^{2,1} + x_{i,2}^{2,2} = x_{i,2}^{3,1} + x_{i,2}^{3,2}.$$

Table 2. First four blocks of the fractional vertex

$x_{1,1}^{1,1}$	0	$x_{1,2}^{1,1}$	0
$x_{1,1}^{2,1}$	0	0	$x_{1,2}^{2,2}$
0	$x_{1,1}^{3,2}$	$x_{1,2}^{3,1}$	0
$x_{2,1}^{1,1}$	0	$x_{2,2}^{1,1}$	0
$x_{2,1}^{2,1}$	0	$x_{2,2}^{2,1}$	0
0	$x_{2,1}^{3,2}$	0	$x_{2,2}^{3,2}$

Here, we describe the key steps of the construction.

Table 3. Blocks j, j and $j, j+1$ of the fractional vertex

$x_{j,j}^{1,1}$	0	$x_{j,j+1}^{1,1}$	0
$x_{j,j}^{2,1}$	0	0	$x_{j,j+1}^{2,2}$
0	$x_{j,j}^{3,2}$	$x_{j,j+1}^{3,1}$	0

- For all j $(1 \leq j \leq n-1)$ blocks j, j and $j, j+1$ have the form as shown in Table 3. Thus, for all i, j we have

$$x_{i,j}^{3,1} + x_{i,j}^{3,2} = x_{i,j+1}^{2,1} + x_{i,j+1}^{2,2}. \tag{14}$$

- For all k $(3 \leq 2k-1 \leq n)$ there are blocks in the rows $2k-1$ and $2k$ as shown in Table 4. Here, we obtain

$$x_{i,2k-1}^{3,1} + x_{i,2k-1}^{3,2} = x_{i,2k}^{3,1} + x_{i,2k}^{3,2} = x_{i,k}^{2,1} + x_{i,k}^{2,2} + x_{i,k}^{3,1} + x_{i,k}^{3,2} \tag{15}$$

for all blocks in these columns. If $2k > n$, then the row and column $2k$ can be omitted.

- The last part of construction describes the blocks in the rows $n-1$ and n, as shown in Table 5. Hence, for blocks in the first and last columns we have

$$x_{i,1}^{1,1} + x_{i,1}^{1,2} = x_{i,n}^{2,1} + x_{i,n}^{2,2}, \tag{16}$$

$$x_{i,1}^{2,1} + x_{i,1}^{2,2} = x_{i,n}^{3,1} + x_{i,n}^{3,2}. \tag{17}$$

It is possible to make last two rows different from the first two rows since $n \geq 4$.

- We call all of the remaining blocks that were not described in the preceding steps as the *filler blocks*. They are different for the blocks above and below the main diagonal and have the form as shown in Table 6.

Table 4. Fragment of $2k-1$ and $2k$ rows of blocks of the fractional vertex

$x_{2k-1,k}^{1,1}$	0	$x_{2k-1,2k-1}^{1,1}$	0	-	-
0	$x_{2k-1,k}^{2,2}$	$x_{2k-1,2k-1}^{2,1}$	0	-	-
0	$x_{2k-1,k}^{3,2}$	0	$x_{2k-1,2k-1}^{3,2}$	-	-
$x_{2k,k}^{1,1}$	0	-	-	$x_{2k,2k}^{1,1}$	0
0	$x_{2k,k}^{2,2}$	-	-	$x_{2k,2k}^{2,1}$	0
0	$x_{2k,k}^{3,2}$	-	-	0	$x_{2k,2k}^{3,2}$

Now we establish that the system (6)–(9), constructed above, has a unique solution. Let n be odd and equal $2q + 1$. We denote $x_{1,1}^{3,2}$ simply as x. Thereby,

Table 5. Last two rows of blocks of the fractional vertex

0	$x^{1,1}_{n-1,1}$	$x^{1,1}_{n-1,n}$	0
$x^{2,1}_{n-1,1}$	0	0	$x^{2,2}_{n-1,n}$
$x^{3,1}_{n-1,1}$	0	$x^{3,1}_{n-1,n}$	0
$x^{1,1}_{n,1}$	0	$x^{1,1}_{n,n}$	0
0	$x^{2,2}_{n,1}$	$x^{2,1}_{n,n}$	0
$x^{3,1}_{n,1}$	0	0	$x^{3,2}_{n,n}$

Table 6. Form of the filler blocks of the fractional vertex above the main diagonal (a) and beyond the main diagonal (b)

$x^{1,1}_{i,j}$	0
$x^{2,1}_{i,j}$	0
$x^{3,1}_{i,j}$	$x^{3,2}_{i,j}$

(a)

$x^{1,1}_{i,j}$	$x^{1,2}_{i,j}$
$x^{2,1}_{i,j}$	0
0	$x^{3,2}_{i,j}$

(b)

for all i by Eqs. (14)–(15) and induction we get

$$x^{2,1}_{i,2} + x^{2,2}_{i,2} = x,$$

$$x^{3,1}_{i,2} + x^{3,2}_{i,2} = x,$$

$$x^{2,1}_{i,3} + x^{2,2}_{i,3} = x,$$

$$x^{3,1}_{i,3} + x^{3,2}_{i,3} = 2x,$$

$$\dots$$

$$x^{2,1}_{i,2q} + x^{2,2}_{i,2q} = qx,$$

$$x^{3,1}_{i,2q} + x^{3,2}_{i,2q} = qx,$$

$$x^{2,1}_{i,2q+1} + x^{2,2}_{i,2q+1} = qx,$$

$$x^{3,1}_{i,2q+1} + x^{3,2}_{i,2q+1} = (q+1)x.$$

First and last columns are connected by Eqs. (16)–(17), therefore

$$x^{1,1}_{1,1} + x^{1,2}_{1,1} = qx,$$

$$x^{2,1}_{1,1} + x^{2,2}_{1,1} = (q+1)x,$$

$$x^{3,1}_{1,1} + x^{3,2}_{1,1} = x.$$

Since the sum of the coordinates inside a single block is equal to 1, we have

$$qx + (q+1)x + x = 1,$$

and

$$x = \frac{1}{2q+2} = \frac{1}{n+1}. \tag{18}$$

All coordinates of the constructed point are either already directly expressed in terms of x, or can be found using the Eqs. (6)–(8).

Case of n equal $2q$ is considered similarly, the only difference will be that

$$x_{1,1}^{1,1} + x_{1,1}^{1,2} = x_{i,2q}^{2,1} + x_{i,2q}^{2,2} = qx,$$
$$x_{1,1}^{2,1} + x_{1,1}^{2,2} = x_{i,2q}^{3,1} + x_{i,2q}^{3,2} = qx.$$

It remains to verify only that the coordinates of the filler blocks from the Table 6 satisfy the system (6)–(9). We consider the filler blocks above the main diagonal $(i < j)$. Using Eqs. (6)–(8) we can establish that

$$x_{i,j}^{2,1} = \left\lfloor \frac{j}{2} \right\rfloor x,$$

$$x_{i,j}^{3,2} = \left\lfloor \frac{i+1}{2} \right\rfloor x,$$

$$x_{i,j}^{3,1} = \left\lfloor \frac{j+1}{2} \right\rfloor x - \left\lfloor \frac{i+1}{2} \right\rfloor x,$$

$$x_{i,j}^{1,1} = 1 - \left\lfloor \frac{j}{2} \right\rfloor x - \left\lfloor \frac{j+1}{2} \right\rfloor x.$$

Hence, only the inequalities $x_{i,j}^{3,1} \geq 0$ and $x_{i,j}^{1,1} \geq 0$ can be violated. For all $i < j$ we have

$$\left\lfloor \frac{j+1}{2} \right\rfloor \geq \left\lfloor \frac{i+1}{2} \right\rfloor.$$

Therefore, $x_{i,j}^{3,1} \geq 0$. And, since $j \leq n$, we have

$$\left\lfloor \frac{j}{2} \right\rfloor + \left\lfloor \frac{j+1}{2} \right\rfloor < n + 1.$$

Thus, by (18), $x_{i,j}^{1,1} \geq 0$ is satisfied as well.

Now we consider the filler blocks below the main diagonal $(i > j)$:

$$x_{i,j}^{2,1} = \left\lfloor \frac{j}{2} \right\rfloor x,$$

$$x_{i,j}^{3,2} = \left\lfloor \frac{j+1}{2} \right\rfloor x,$$

$$x_{i,j}^{1,2} = \left\lfloor \frac{i+1}{2} \right\rfloor x - \left\lfloor \frac{j+1}{2} \right\rfloor x,$$

$$x_{i,j}^{1,1} = 1 - \left\lfloor \frac{i+1}{2} \right\rfloor x - \left\lfloor \frac{j}{2} \right\rfloor x.$$

Again, for all $i > j$ we have

$$\left\lfloor \frac{i+1}{2} \right\rfloor \geq \left\lfloor \frac{j+1}{2} \right\rfloor,$$

and the inequality $x_{i,j}^{1,2} \geq 0$ is satisfied. And, since $i, j \leq n$:

$$\left\lfloor \frac{j}{2} \right\rfloor + \left\lfloor \frac{i+1}{2} \right\rfloor < n+1.$$

Thus, $x_{i,j}^{1,1} \geq 0$ holds as well.

The constructed system is obtained from (6)–(9) by turning some of inequalities into equations, and it has a unique solution, therefore it defines the fractional vertex of the polytope $SATP_{LP}(n,n)$ with denominator $n+1$.

An example of a fractional vertex for $n = 6$ is shown in Table 7.

Table 7. Fractional vertex of the polytope $SATP_{LP}(6,6)$

$\frac{3}{7}$	0	$\frac{5}{7}$	0	$\frac{4}{7}$	0	$\frac{3}{7}$	0	$\frac{2}{7}$	0	$\frac{1}{7}$	0
$\frac{3}{7}$	0	0	$\frac{1}{7}$	$\frac{1}{7}$	0	$\frac{2}{7}$	0	$\frac{2}{7}$	0	$\frac{3}{7}$	0
0	$\frac{1}{7}$	$\frac{1}{7}$	0	$\frac{1}{7}$	$\frac{1}{7}$	$\frac{1}{7}$	$\frac{1}{7}$	$\frac{2}{7}$	$\frac{1}{7}$	$\frac{2}{7}$	$\frac{1}{7}$
$\frac{3}{7}$	0	$\frac{5}{7}$	0	$\frac{4}{7}$	0	$\frac{3}{7}$	0	$\frac{2}{7}$	0	$\frac{1}{7}$	0
$\frac{3}{7}$	0	$\frac{1}{7}$	0	0	$\frac{1}{7}$	$\frac{2}{7}$	0	$\frac{2}{7}$	0	$\frac{3}{7}$	0
0	$\frac{1}{7}$	0	$\frac{1}{7}$	$\frac{2}{7}$	0	$\frac{1}{7}$	$\frac{1}{7}$	$\frac{2}{7}$	$\frac{1}{7}$	$\frac{2}{7}$	$\frac{1}{7}$
$\frac{2}{7}$	$\frac{1}{7}$	$\frac{5}{7}$	0	$\frac{4}{7}$	0	$\frac{3}{7}$	0	$\frac{2}{7}$	0	$\frac{1}{7}$	0
$\frac{3}{7}$	0	0	$\frac{1}{7}$	$\frac{1}{7}$	0	0	$\frac{2}{7}$	$\frac{2}{7}$	0	$\frac{3}{7}$	0
0	$\frac{1}{7}$	0	$\frac{1}{7}$	0	$\frac{2}{7}$	$\frac{2}{7}$	0	$\frac{1}{7}$	$\frac{2}{7}$	$\frac{1}{7}$	$\frac{2}{7}$
$\frac{2}{7}$	$\frac{1}{7}$	$\frac{5}{7}$	0	$\frac{4}{7}$	0	$\frac{3}{7}$	0	$\frac{2}{7}$	0	$\frac{1}{7}$	0
$\frac{3}{7}$	0	0	$\frac{1}{7}$	$\frac{1}{7}$	0	$\frac{2}{7}$	0	0	$\frac{2}{7}$	$\frac{3}{7}$	0
0	$\frac{1}{7}$	0	$\frac{1}{7}$	0	$\frac{2}{7}$	0	$\frac{2}{7}$	$\frac{3}{7}$	0	$\frac{1}{7}$	$\frac{2}{7}$
0	$\frac{3}{7}$	$\frac{3}{7}$	$\frac{2}{7}$	$\frac{4}{7}$	0	$\frac{2}{7}$	$\frac{1}{7}$	$\frac{2}{7}$	0	$\frac{1}{7}$	0
$\frac{3}{7}$	0	$\frac{1}{7}$	0	0	$\frac{1}{7}$	$\frac{2}{7}$	0	$\frac{2}{7}$	0	0	$\frac{3}{7}$
$\frac{1}{7}$	0	0	$\frac{1}{7}$	0	$\frac{2}{7}$	0	$\frac{2}{7}$	0	$\frac{3}{7}$	$\frac{3}{7}$	0
$\frac{3}{7}$	0	$\frac{3}{7}$	$\frac{2}{7}$	$\frac{4}{7}$	0	$\frac{2}{7}$	$\frac{1}{7}$	$\frac{2}{7}$	0	$\frac{1}{7}$	0
0	$\frac{3}{7}$	$\frac{1}{7}$	0	0	$\frac{1}{7}$	$\frac{2}{7}$	0	$\frac{2}{7}$	0	$\frac{3}{7}$	0
$\frac{1}{7}$	0	0	$\frac{1}{7}$	0	$\frac{2}{7}$	0	$\frac{2}{7}$	0	$\frac{3}{7}$	0	$\frac{3}{7}$

The construction described in Theorem 9 is not working for $n < 4$. However, relaxation polytope $SATP_{LP}(m,n)$ has fractional vertices with denominators $2, 3$ and 4 as well. Some examples are provided in Table 8. It may be noted that Theorem 9 holds for $n = 3$, but not for $n = 1$ and $n = 2$. Polytope $SATP_{LP}(1,1)$ coincide with $SATP(1,1)$ and has only 6 integral vertices. As for the polytope $SATP_{LP}(2,2)$, it has 72 fractional vertices with all of them having denominators equal to 2 (computed by Skeleton software [24]).

Table 8. Fractional vertices with denominators $2, 3$ and 4

$$
\begin{array}{|c|c|c|c|}
\hline
\frac{1}{2} & 0 & 0 & \frac{1}{2} \\ \hline
0 & \frac{1}{2} & \frac{1}{2} & 0 \\ \hline
0 & 0 & 0 & 0 \\ \hline
\end{array}
\quad
\begin{array}{|c|c|c|c|}
\hline
\frac{1}{2} & 0 & \frac{1}{2} & 0 \\ \hline
0 & \frac{1}{2} & 0 & \frac{1}{2} \\ \hline
0 & 0 & 0 & 0 \\ \hline
\end{array}
$$

$$
\begin{array}{|c|c|c|c|}
\hline
0 & \frac{1}{3} & \frac{2}{3} & 0 \\ \hline
\frac{1}{3} & 0 & 0 & \frac{1}{3} \\ \hline
\frac{1}{3} & 0 & 0 & 0 \\ \hline
\end{array}
\quad
\begin{array}{|c|c|c|c|}
\hline
\frac{1}{3} & 0 & \frac{2}{3} & 0 \\ \hline
0 & \frac{1}{3} & 0 & \frac{1}{3} \\ \hline
\frac{1}{3} & 0 & 0 & 0 \\ \hline
\end{array}
\quad
\begin{array}{|c|c|c|c|}
\hline
\frac{1}{3} & 0 & \frac{2}{3} & 0 \\ \hline
\frac{1}{3} & 0 & 0 & \frac{1}{3} \\ \hline
0 & \frac{1}{3} & 0 & 0 \\ \hline
\end{array}
$$

$$
\begin{array}{|c|c|c|c|c|c|}
\hline
\frac{1}{2} & 0 & \frac{1}{2} & 0 & \frac{1}{2} & 0 \\ \hline
\frac{1}{4} & 0 & \frac{1}{4} & 0 & \frac{1}{4} & 0 \\ \hline
0 & \frac{1}{4} & 0 & \frac{1}{4} & 0 & \frac{1}{4} \\ \hline
\end{array}
$$

$$
\begin{array}{|c|c|c|c|c|c|}
\hline
\frac{1}{2} & 0 & \frac{1}{2} & 0 & \frac{1}{2} & 0 \\ \hline
0 & \frac{1}{4} & \frac{1}{4} & 0 & 0 & \frac{1}{4} \\ \hline
\frac{1}{4} & 0 & 0 & \frac{1}{4} & \frac{1}{4} & 0 \\ \hline
\end{array}
$$

$$
\begin{array}{|c|c|c|c|c|c|}
\hline
\frac{1}{2} & 0 & \frac{1}{2} & 0 & 0 & \frac{1}{2} \\ \hline
0 & \frac{1}{4} & 0 & \frac{1}{4} & \frac{1}{4} & 0 \\ \hline
0 & \frac{1}{4} & 0 & \frac{1}{4} & \frac{1}{4} & 0 \\ \hline
\end{array}
$$

5   Conclusions

We have considered $SATP(m, n)$ polytope and its LP relaxation $SATP_{LP}(m, n)$. This polytope is a simple extension of the well-known and important Boolean quadric polytope $BQP(n)$, constructed by adding two additional coordinates per block. The polytope $SATP(m, n)$ is the object of our interest, since it is the integer programming formulation of various special instances of 3-SAT problem like not-all-equal-3-SAT, 1-in-3-SAT, weighted MAX-3-SAT, and others.

We have compared key properties of $BQP(n)$ and $SATP(m, n)$. Like the Boolean quadric polytope, the polytope $SATP(m, n)$ has an exponential extension complexity, and the LP relaxation $SATP_{LP}(m, n)$ is quasi-integral with respect to $SATP(m, n)$. Unlike $BQP(n)$, 1-skeleton of $SATP(m, n)$ is not a complete graph, but is a very dense one, with the diameter equals 2, and the clique number being superpolynomial in dimension. As for the fractional vertices, unlike the Boolean quadric polytope, the denominators of the fractional vertices of the LP relaxation $SATP_{LP}(m, n)$ can take any positive integer values.

Acknowledgments. The research is supported by the grant of the President of the Russian Federation MK-2620.2018.1.

References

1. Aguilera, N.E., Katz, R.D., Tolomei, P.B.: Vertex adjacencies in the set covering polyhedron. Discrete Appl. Math. **218**, 40–56 (2017). https://doi.org/10.1016/j.dam.2016.10.024
2. Avis, D.: On the extreme rays of the metric cone. Can. J. Math. **32**, 126–144 (1980)
3. Barahona, F.: On cuts and matchings in planar graphs. Math. Program. **60**(1–3), 53–68 (1993). https://doi.org/10.1007/BF01580600
4. Bondarenko, V.A., Uryvaev, B.V.: On one problem of integer optimization. Automat. Rem. Contr. **68**(6), 948–953 (2007). https://doi.org/10.1134/S0005117907060021

5. Bondarenko, V.A., Maksimenko, A.N.: Geometricheskie konstruktsii i slozhnost' v kombinatornoy optimizatsii (Geometric constructions and complexity in combinatorial optimization), LKI (2008). (Russian)

6. Bondarenko, V., Nikolaev, A.: On graphs of the cone decompositions for the min-cut and max-cut problems. Int. J. Math. Sci. 2016, article ID 7863650 (2016). https://doi.org/10.1155/2016/7863650

7. Bondarenko, V.A., Nikolaev, A.V.: On the skeleton of the polytope of pyramidal tours. J. Appl. Ind. Math. **12**(1), 9–18 (2018). https://doi.org/10.1134/S1990478918010027

8. Bondarenko, V.A., Nikolaev, A.V., Symanovich, M.E., Shemyakin, R.O.: On a recognition problem on cut polytope relaxations. Automat. Rem. Contr. **75**(9), 1626–1636 (2014). https://doi.org/10.1134/S0005117914090082

9. Chegireddy, C.R., Hamacher, H.W.: Algorithms for finding K-best perfect matchings. Discrete Appl. Math. **18**(2), 155–165 (1987). https://doi.org/10.1016/0166-218X(87)90017-5

10. De Simone, C.: The cut polytope and the Boolean quadric polytope. Discrete Math. **79**(1), 71–75 (1990). https://doi.org/10.1016/0012-365X(90)90056-N

11. Deza, A., Fukuda, K., Pasechnik, D., Sato, M.: On the Skeleton of the metric polytope. In: Akiyama, J., Kano, M., Urabe, M. (eds.) JCDCG 2000. LNCS, vol. 2098, pp. 125–136. Springer, Heidelberg (2001). https://doi.org/10.1007/3-540-47738-1_10

12. Deza, M.M., Laurent, M.: Geometry of Cuts and Metrics. AC, vol. 15. Springer, Heidelberg (1997). https://doi.org/10.1007/978-3-642-04295-9

13. Fiorini, S., Massar, S., Pokutta, S., Tiwary, H.R., De Wolf, R.: Exponential lower bounds for polytopes in combinatorial optimization. J. ACM. **62**(2), 1–17 (2015). https://doi.org/10.1145/2716307

14. Grishukhin, V.P.: Computing extreme rays of the metric cone for seven points. Eur. J. Combin. **13**(3), 153–165 (1992). https://doi.org/10.1016/0195-6698(92)90021-Q

15. Kaibel, V., Weltge, S.: A short proof that the extension complexity of the correlation polytope grows exponentially. Discrete Comput. Geom. **53**(2), 397–401 (2015). https://doi.org/10.1007/s00454-014-9655-9

16. Laurent, M.: Graphic vertices of the metric polytope. Discrete Math. **151**(1–3), 131–153 (1996). https://doi.org/10.1016/0012-365X(94)00091-V

17. Matsui, T., Tamura, S.: Adjacency on combinatorial polyhedra. Discrete Appl. Math. **56**(2–3), 311–321 (1995). https://doi.org/10.1016/0166-218X(94)00092-R

18. Maksimenko, A.N.: A special role of Boolean quadratic polytopes among other combinatorial polytopes. Model. Anal. Inform. Sist. **23**(1), 23–40 (2016). https://doi.org/10.18255/1818-1015-2016-1-23-40

19. Nikolaev, A.: On integer recognition over some boolean quadric polytope extension. In: Kochetov, Y., Khachay, M., Beresnev, V., Nurminski, E., Pardalos, P. (eds.) DOOR 2016. LNCS, vol. 9869, pp. 206–219. Springer, Cham (2016). https://doi.org/10.1007/978-3-319-44914-2_17

20. Padberg, M.: The Boolean quadric polytope: some characteristics, facets and relatives. Math. Program. **45**(1–3), 139–172 (1989). https://doi.org/10.1007/BF01589101

21. Papadimitriou, C.H., Steiglitz, K.: Combinatorial Optimization: Algorithms and Complexity. Dover Publications, Mineola (1998)

22. Trubin, V.: On a method of solution of integer linear programming problems of a special kind. Sov. Math. Dokl. **10**, 1544–1546 (1969)
23. Veselov, S.I., Chirkov, A.J.: Integer program with bimodular matrix. Discrete Optim. **6**(2), 220–222 (2009). https://doi.org/10.1016/j.disopt.2008.12.002
24. Zolotykh, N.Y.: New modification of the double description method for constructing the skeleton of a polyhedral cone. Comp. Math. Math. Phys. **52**(1), 146–156 (2012). https://doi.org/10.1134/S0965542512010162

On Accuracy Estimates for One Regularization Method in Linear Programming

Leonid D. Popov[1,2]($\boxtimes$) (iD)

[1] Krasovskii Institute of Mathematics and Mechanics UB RAS,
Ekaterinburg, Russia
popld@imm.uran.ru
[2] Ural Federal University, Ekaterinburg, Russia

Abstract. In this paper, the alternative duality schemes for mathematical programming problems are considered. These schemes are based on the Lagrange function regularized by Tikhonov in primal and dual variables simultaneously. Earlier such schemes were investigated for convex programming problems only, and the conditions were found which guarantee a convergence of both primal and dual components of the saddle point of the regularized Lagrange function to the optimal sets of primal and dual problems respectively. In addition, some accuracy estimates were obtained for deviation of only one of these components from the normal solution (solution with minimal the Euclidean norm) of the corresponding problem, primal or dual. Unfortunately, these estimates are valid only if primal and dual parameters of regularization have a different order of smallness. In this article, the linear case is investigated in detail. It is shown that for linear programming problem both mentioned sequences converge to the normal solutions of primal and dual programs simultaneously, and it is not essential for such a convergence whether the regularization parameters have a different or the same order of smallness. Also, the alternative accuracy estimates are presented which appear to be more precise and efficient in comparison with the estimates known for a convex case.

Keywords: Linear programs · Auxiliary Lagrangian
Duality in linear programming · Tikhonov-type regularization
Accuracy estimates

1 Introduction

In [1–7], various duality schemes for mathematical programming problems were proposed. These schemes generally use the different modifications of the classical Lagrange function. One of such modifications is the Lagrange function

This work was supported by Russian Science Foundation, grant N 14–11–00109.

© Springer International Publishing AG, part of Springer Nature 2018
A. Eremeev et al. (Eds.): OPTA 2018, CCIS 871, pp. 170–182, 2018.
https://doi.org/10.1007/978-3-319-93800-4_14

symmetrically regularized by Tikhonov in primal and dual variables simultaneously. Properties of such functions for convex programming problems were investigated in [8,9]. In particular, the conditions were found which guarantee a convergence of both primal and dual components of the saddle point of the regularized Lagrange function to the optimal sets of primal and dual problems respectively as the primal and dual parameters of regularization tend to zero. In addition, some accuracy estimates were obtained for deviation of only one of these components from the normal solution (solution with minimal the Euclidean norm) of the corresponding problem, primal or dual. Unfortunately, these estimates are valid only if primal and dual parameters of regularization have a different order of smallness. In this article, the linear case is investigated in detail. It is shown that for the linear programming problem both mentioned sequences converge to the normal solutions of primal and dual programs simultaneously, and it is not essential for such a convergence whether the regularization parameters have a different or the same order of smallness. Also, the alternative accuracy estimates are presented which appear to be more precise and efficient in comparison with the estimates known for a convex case.

2   Symmetrically Regularized Lagrange Function

Consider the linear program

$$\max\{(c, x)\colon Ax \le b, \quad x \ge 0\} \tag{1}$$

and its dual one

$$\min\{(b, y)\colon A^T y \ge c, \quad y \ge 0\}, \tag{2}$$

where vectors $c \in \mathbb{R}^n$, $b \in \mathbb{R}^m$ and matrix $A = (a_{ij})_{m \times n}$ are given, vectors $x \in \mathbb{R}^n$ and $y \in \mathbb{R}^m$ denote primal and dual variables, $(\cdot, \cdot)$ denotes the scalar product.

Assume that problems (1), (2) are solvable and $\bar{x}$ and $\bar{y}$ are their optimal vectors with minimal the Euclidean norm. It's known that these vectors (as any others optimal vectors of problems (1), (2)) together form a saddle point $(\bar{x}, \bar{y})$ of the classical Lagrange function

$$L(x, y) = (c, x) - (y, Ax - b)$$

with respect to domain $\mathbb{R}^n_+ \times \mathbb{R}^m_+$, where $\mathbb{R}^n_+$ and $\mathbb{R}^m_+$ are the non-negative orthants of the corresponding Euclidean spaces [10].

Unfortunately, the classical Lagrange function has some disadvantages from the computational point of view. That is why it is usually replaced by its different modifications in numerical analysis. One of such modifications is the function

$$L_\sigma(x, y) = (c, x) - (y, Ax - b) - \alpha\|x\|^2 + \beta\|y\|^2; \tag{3}$$

it is constructed by means of symmetric regularization of the classical Lagrangian both in primal and dual variables simultaneously. Here $\|\cdot\|$ denotes the Euclidean norm, $\sigma = [\alpha, \beta] \to +0$, $\alpha > 0$, $\beta > 0$ are positive scalar parameters of regularization tending to zero.

Symmetrically regularized function (3) generates a standard pair of maxmin and minmax problems

$$(\mathcal{P}) \quad \max_{x \geq 0} \min_{y \geq 0} L_\sigma(x, y) \quad \left(= \max_{x \geq 0} L_\sigma(x, y(x)), \quad y(x) = \frac{1}{2\beta}(Ax - b)^+ \right),$$

$$(\mathcal{D}) \quad \min_{y \geq 0} \max_{x \geq 0} L_\sigma(x, y) \quad \left(= \min_{y \geq 0} L_\sigma(x(y), y), \quad x(y) = \frac{1}{2\alpha}(c - A^T y)^+ \right),$$

which (after some transformations) may be rewritten as following dual pair of convex programs

$$\max_{x \geq 0} \Phi_\sigma(x), \quad \Phi_\sigma(x) = L_\sigma(x, y(x)) = (c, x) - \frac{1}{4\beta}\|(Ax - b)^+\|^2 - \alpha\|x\|^2; \quad (4)$$

$$\min_{y \geq 0} \Psi_\sigma(y), \quad \Psi_\sigma(y) = L_\sigma(x(y), y) = (b, y) + \frac{1}{4\alpha}\|(c - A^T y)^+\|^2 + \beta\|y\|^2. \quad (5)$$

Here w^+ denotes the Euclidean projection of a vector w onto non-negative orthant of the corresponding Euclidean space.

Since objective functions of modified problems (4), (5) are strongly concave and convex respectively, these problems always are in relation of perfect duality. It means that they are solvable and share the same optimal value. Their optimal vectors x^σ and y^σ are unique and together form a saddle point (x^σ, y^σ) of function (3) with respect to domain $\mathbb{R}^n_+ \times \mathbb{R}^m_+$. Notice that it is just so even if problems (1), (2) are infeasible or ill-posed (see Sect. 4, where the improper linear programs are investigated).

As we already say, some properties of function (3) were described in the paper [8] for the convex case. Being applying to linear programs (1), (2), they allow us to write the following inequalities

$$\bar{v} - \alpha\|\bar{x}\|^2 \leq L_\sigma(x^\sigma, y^\sigma) \leq \bar{v} + \beta\|\bar{y}\|^2, \quad (6)$$

$$\|(Ax^\sigma - b)^+\|^2 \leq 4\beta\left(\alpha\|\bar{x}\|^2 + \beta\|\bar{y}\|^2\right), \quad (7)$$

$$\|(c - A^T y^\sigma)^+\|^2 \leq 4\alpha\left(\alpha\|\bar{x}\|^2 + \beta\|\bar{y}\|^2\right) \quad (8)$$

as well as

$$\bar{v} - \alpha\|\bar{x}\|^2 \leq (c, x^\sigma) \leq \bar{v} + \|\bar{y}\|\,\|(Ax^\sigma - b)^+\|, \quad (9)$$

$$\bar{v} + \beta\|\bar{y}\|^2 \geq (b, y^\sigma) \geq \bar{v} - \|\bar{x}\|\,\|(c - A^T y^\sigma)^+\|, \quad (10)$$

where $\bar{v}$ is the common optimal (finite) value of problems (1), (2). In addition, it is known the estimate

$$\|x^\sigma - \bar{x}\| \leq 1/2\left[\|\bar{y}\|\sqrt{\beta/\alpha} + \sqrt{\alpha\beta}\|[\bar{\lambda}, \bar{\lambda}_0]\|\right], \quad (11)$$

which means that $x^\sigma \to \bar{x}$ if $\beta = o(\alpha)$, $\alpha \to +0$ (parameters $\bar{\lambda}$, $\bar{\lambda}_0$ are discussed below).

The asymmetric estimate is valid

$$\|y^\sigma - \bar{y}\| \leq 1/2 \left[\|\bar{x}\|\sqrt{\alpha/\beta} + \sqrt{\alpha\beta}\|[\bar{\lambda}', \bar{\lambda}'_0]\|\right]$$

if $\alpha = o(\beta)$, $\beta \rightarrow +0$ (parameters $\bar{\lambda}'$, $\bar{\lambda}'_0$ have analogous sense). Thus, only one of these estimates may be using in any concrete numerical process.

Generally, properties (7)–(10) imply convergence of sequences x^σ and y^σ only to the optimal sets of problems (1) and (2) respectively[1] and not allow to concretely specify their limit points. Below it will be established that in linear case the sequences x^σ and y^σ possess more strong properties: they both converge, the first one does to the point $\bar{x}$ and the second one does to the point $\bar{y}$, in spite of any assumptions about the order of smallness of the parameters of regularization α and β. Also, more efficient estimates of deviations $\|x^\sigma - \bar{x}\|$ and $\|y^\sigma - \bar{y}\|$ will be obtained which depend only on $\gamma = \max\{\alpha, \beta\} \rightarrow +0$.

3   Alternative Accuracy Estimates

At first, consider the auxiliary parametric family of linear programs

$$\max\{(c,x)\colon Ax \leq b + u, \ x \geq 0\}, \quad u \in \mathbb{R}^m_+ \text{ is a vector parameter.} \quad (12)$$

Denote by $v(u)$ the optimal value of problem (12) and set

$$\bar{x}(u) = \arg\min\{\|x\|^2\colon Ax \leq b + u, \ (c,x) \geq v(u), \ x \geq 0\}. \quad (13)$$

It is well-known, that $v(u)$ is continuous concave piecewise linear function, finite on $\mathbb{R}^m_+$, and $v(0) = \bar{v}$. Since vector $\bar{y}$ is a sub-gradient of function $v(u)$ computed at zero, we have

$$\bar{v} \leq v(u) \leq \bar{v} + (\bar{y}, u) \leq \bar{v} + \|\bar{y}\|\|u\|. \quad (14)$$

Besides, being optimal vector of convex quadratic program (13) with the strongly convex objective function, the normal solution $\bar{x}(u)$ is a continuous function of right-hand side of constraints of this program [12,13], in our case of u, since $v(u)$ is continuous in u too. Then $\bar{x}(u) \rightarrow \bar{x}(0) = \bar{x}$ as $u \rightarrow +0$.

Let us estimate the norm $\|\bar{x}(u) - \bar{x}\|$. Set

$$\bar{M}(u) = \{x\colon Ax \leq b + u, \ (c,x) \geq v(u), \ x \geq 0\}.$$

From (13) it follows that vector $\bar{x}(u)$ is a projection of zero vector onto the optimal set $\bar{M}(u)$ of problem (12), and vector $\bar{x}$ is a projection of zero vector onto the optimal set $\bar{M}(0)$ of problem (1). Further, according to Hoffman's lemma

---

[1] It follows from well-known Hoffman's lemma [11] estimating the distance of an arbitrary point to the polyhedral set defined by the system of linear inequalities (and the optimal set of a linear programming problem belongs to such a class) in terms of deviations of this point from each inequality separately.

there exists a constant K, depending only on matrix A and vector c, such that Hausdorff's distance $\rho(\bar{M}(u), \bar{M}(0))$ between sets $\bar{M}(0)$ and $\bar{M}(u)$ is evaluated as

$$\rho(\bar{M}(u), \bar{M}(0)) \leq K(\|u\| + |v(u) - \bar{v}|) \leq K(1 + \|\bar{y}\|)\|u\|. \tag{15}$$

Here we use relation (14) too.

Lemma 1. *Let Q and Q' be convex closed subsets of $\mathbb{R}^s$ such that $\rho(Q, Q') \leq \epsilon$, and let $\bar{q}$ and $\bar{q}'$ be two projections of zero vector onto each of these sets. Then we have*

$$\|\bar{q}\| - \epsilon \leq \|\bar{q}'\| \leq \|\bar{q}\| + \epsilon \quad \text{and} \quad \|\bar{q}'\| - \epsilon \leq \|\bar{q}\| \leq \|\bar{q}'\| + \epsilon.$$

Proof. If one of or both projections $\bar{q}$ and $\bar{q}'$ equal to zero, then sought inequalities are evident. Assume that both projections differ from zero. Being a projection of zero vector onto convex closed set, vector $\bar{q}$ satisfies to inequality $(\bar{q}, q - \bar{q}) \geq 0$ for all $q \in Q$. In this inequality, we can replace q by vector $p \in Q$ that is the nearest to vector $\bar{q}' \in Q'$ (with respect to the Euclidean distance). Then we get

$$0 \leq (\bar{q}, p - \bar{q}) = (\bar{q}, p - \bar{q}') + (\bar{q}, \bar{q}' - \bar{q}) \leq \|\bar{q}\|\epsilon + (\bar{q}, \bar{q}' - \bar{q}).$$

It means that $(\bar{q}, \bar{q} - \bar{q}') \leq \epsilon\|\bar{q}\|$, which in turn implies that

$$\|\bar{q}\|^2 - \|\bar{q}\|\|\bar{q}'\| \leq (\bar{q}, \bar{q} - \bar{q}') \leq \epsilon\|\bar{q}\|,$$

or

$$\|\bar{q}\| \leq \|\bar{q}'\| + \epsilon,$$

after dividing of both sides of the previous inequality by $\|\bar{q}\| > 0$.

Now, if we repeat this reasoning with vector $\bar{q}'$ as a projection of zero vector onto set Q', then we obtain the symmetric relation

$$\|\bar{q}'\| \leq \|\bar{q}\| + \epsilon.$$

Finally, transferring ϵ from the right-hand side into the left-hand side of each of these inequalities, we obtain two another sought inequalities. The proof is complete.

Corollary 1. *The estimates hold*

$$\|\bar{x}(u)\| \leq \|\bar{x}\| + K(1 + \|\bar{y}\|)\|u\|, \quad \|\bar{x}\| \leq \|\bar{x}(u)\| + K(1 + \|\bar{y}\|)\|u\|.$$

Now we try to evaluate the norm of vector x^σ in the same way.

Lemma 2. *The estimate is valid*

$$\|x^\sigma\| \leq \|\bar{x}\| + K(1 + \|\bar{y}\|)\|(Ax^\sigma - b)^+\|.$$

Proof. Set $u^\sigma = (Ax^\sigma - b)^+ \in \mathbb{R}_+^m$. From the definition of x^σ, it follows that

$$\Phi_\sigma(x^\sigma) = \max_{x \geq 0} \Phi_\sigma(x) \geq \Phi_\sigma(\bar{x}(u^\sigma)).$$

It other words,

$$(c, x^\sigma) - \frac{1}{4\beta}\|u^\sigma\|^2 - \alpha\|x^\sigma\|^2 \geq v(u^\sigma) - \frac{1}{4\beta}\|(A\bar{x}(u^\sigma) - b)^+\|^2 - \alpha\|\bar{x}(u^\sigma)\|^2.$$

After rearranging of items, we get

$$\alpha\|x^\sigma\|^2 \leq \alpha\|\bar{x}(u^\sigma)\|^2 + \frac{1}{4\beta}\|(A\bar{x}(u^\sigma) - b)^+\|^2 - \frac{1}{4\beta}\|u^\sigma\|^2 + (c, x^\sigma) - v(u^\sigma).$$

Notice that we may omit almost all items on the right-hand side of the last inequality. Indeed, firstly, since $A\bar{x}(u^\sigma) \leq b + u^\sigma$, it follows that

$$0 \leq (A\bar{x}(u^\sigma) - b)^+ \leq u^\sigma,$$

and the inequality holds $\|(A\bar{x}(u^\sigma) - b)^+\| \leq \|u^\sigma\|$. Secondly,

$$(c, x^\sigma) \leq v(u^\sigma),$$

because vector x^σ satisfies to constraints of problem (12) with $u = u^\sigma$.

As a result, we get

$$\alpha\|x^\sigma\|^2 \leq \alpha\|\bar{x}(u^\sigma)\|^2,$$

i.e. $\|x^\sigma\| \leq \|\bar{x}(u^\sigma)\|$. Using Corollary 1, we conclude the proof.

Corollary 2. *The estimate is valid*

$$\|x^\sigma\| \leq \|\bar{x}\| + 2K(1 + \|\bar{y}\|)\left[\beta(\alpha\|\bar{x}\|^2 + \beta\|\bar{y}\|^2)\right]^{1/2}.$$

Here we apply property (7) too.

Now let us investigate another auxiliary linear program

$$\min\{(\bar{x}, x): Ax \leq b, \ (c, x) \geq \bar{v}, \ x \geq 0\} \tag{16}$$

and its dual one

$$\max\{(b, \lambda) + \bar{v}\lambda_0: A^T\lambda + \lambda_0 c \leq \bar{x}, \ \lambda \leq 0, \ \lambda_0 \geq 0\}.$$

Problem (16) may be considered as a result of linearization of quadratic program (13) at point $\bar{x}$ as $u = 0$. It is easy to see that problem (16) is solvable and vector $\bar{x}$ is one of its solutions (by properties of the Euclidean projections). The dual problem is solvable too; we denote by $[\bar{\lambda}, \bar{\lambda}_0]$ one of its solutions (e.g. minimal with respect to the Euclidean norm).

Lemma 3. *For any $x \geq 0$ the estimate is valid*

$$(\bar{x}, x) \geq \|\bar{x}\|^2 - \|\bar{\lambda}\| \|(Ax - b)^+\| - \bar{\lambda}_0(\bar{v} - (c, x))^+.$$

Proof. To begin, include some parameters into the right-hand side of constraints of problem (16) by the following way

$$\min\{(\bar{x}, x): Ax \leq b + u, \;\; (c, x) \geq \bar{v} - \delta, \;\; x \geq 0\}; \tag{17}$$

here $\delta > 0$, $u \in \mathbb{R}_+^m$ are parameters. Using dual prices in the usual manner, estimate the optimal value $v(u, \delta)$ of this problem:

$$v(u, \delta) \geq v(0, 0) + (\bar{\lambda}, u) - \bar{\lambda}_0 \delta \geq \|\bar{x}\|^2 - \|\bar{\lambda}\| \|u\| - \bar{\lambda}_0 \delta.$$

It remains to note that an arbitrary vector $x \geq 0$ is feasible for problem (17) with $u = (Ax - b)^+$ and $\delta = (\bar{v} - (c, x))^+$. Therefore

$$(\bar{x}, x) \geq v\big((Ax - b)^+, (\bar{v} - (c, x))^+\big) \geq \|\bar{x}\|^2 - \|\bar{\lambda}\| \|(Ax - b)^+\| - \bar{\lambda}_0(\bar{v} - (c, x))^+,$$

which was to be proved.

Corollary 3. *The estimate is valid*

$$(\bar{x}, x^\sigma) \geq \|\bar{x}\|^2 - \|\bar{\lambda}\| \Big[\beta\big(\alpha\|\bar{x}\|^2 + \beta\|\bar{y}\|^2\big)\Big]^{1/2} - \alpha\bar{\lambda}_0\|\bar{x}\|^2.$$

Here we apply properties (7), (9) too.

Let's introduce two reduction formulae

$$R(\alpha, \beta) = 2K(1 + \|\bar{y}\|)\Big[\beta\big(\alpha\|\bar{x}\|^2 + \beta\|\bar{y}\|^2\big)\Big]^{1/2},$$

$$H(\alpha, \beta) = \alpha\bar{\lambda}_0\|\bar{x}\|^2 + \|\bar{\lambda}\|\Big[\beta\big(\alpha\|\bar{x}\|^2 + \beta\|\bar{y}\|^2\big)\Big]^{1/2}.$$

Now we are ready to state the first theorem of our investigation.

Theorem 1. *The inequality holds*

$$\|x^\sigma - \bar{x}\| \leq \Big[2\|\bar{x}\|R(\alpha, \beta) + R(\alpha, \beta)^2 + 2H(\alpha, \beta)\Big]^{1/2}.$$

Proof. Corollaries 2 and 3 claim that $x^\sigma \in G = B(\alpha, \beta) \cap Z(\alpha, \beta)$, where $B(\alpha, \beta)$ is a ball with a center at origin and a radius $r = \|\bar{x}\| + R(\alpha, \beta)$, $Z(\alpha, \beta)$ is semi-space defined by linear inequality $(\bar{x}, x) \geq \|\bar{x}\|^2 - H(\alpha, \beta)$. Consequently,

$$\|x^\sigma\| \leq \|\bar{x}\| + 2K(1 + \|\bar{y}\|)\big[\beta\big(\alpha\|\bar{x}\|^2 + \beta\|\bar{y}\|^2\big)\big]^{1/2} = \|\bar{x}\| + R(\alpha, \beta),$$

$$(\bar{x}, x^\sigma) \geq \|\bar{x}\|^2 - \|\bar{\lambda}\|\big[\beta\big(\alpha\|\bar{x}\|^2 + \beta\|\bar{y}\|^2\big)\big]^{1/2} - \alpha\bar{\lambda}_0\|\bar{x}\|^2 = \|\bar{x}\|^2 - H(\alpha, \beta).$$

Hence

$$\|x^\sigma - \bar{x}\|^2 = \|x^\sigma\|^2 + \|\bar{x}\|^2 - 2(\bar{x}, x^\sigma)$$

$$\leq (\|\bar{x}\| + R(\alpha, \beta))^2 + \|\bar{x}\|^2 - 2(\|\bar{x}\|^2 - H(\alpha, \beta))$$

$$= 2\|\bar{x}\| R(\alpha, \beta) + R(\alpha, \beta)^2 + 2H(\alpha, \beta).$$

The proof is complete.

The estimates obtained above have too complicated form. It is more convenient to use more rough ones, but more compact.

Corollary 4. *Let* $\|\sigma\|_\infty = \max\{\alpha, \beta\} < 1$. *Then*

$$\|x^\sigma - \bar{x}\| \leq N_x \|\sigma\|_\infty^{1/2}$$

for some fixed N_x.

Corollary 5. *Let* $\|\sigma\|_\infty = \max\{\alpha, \beta\} \to +0$. *Then* $x^\sigma \to \bar{x}$.

Deducing of these assertions is trivial.

Now it remains to talk over evaluation of $\|y^\sigma - \bar{y}\|$. It is sufficient to note that considered schemes of duality are symmetric (see (4), (5)). Therefore, it is possible to construct the estimates for $\|y^\sigma - \bar{y}\|$ which will be very similar to ones constructed above for $\|x^\sigma - \bar{x}\|$ (with the evident rearranging of parameters of regularization and other constants). Since their structure is clear but very complicated, we don't write them here in detail. Instead, we formulate only summary corollaries, similar to Corollaries 4 and 5.

Corollary 6. *Let* $\|\sigma\|_\infty = \max\{\alpha, \beta\} < 1$. *Then*

$$\|y^\sigma - \bar{y}\| \leq N_y \|\sigma\|_\infty^{1/2}$$

for some fixed N_y.

Corollary 7. *Let* $\|\sigma\|_\infty = \max\{\alpha, \beta\} \to +0$. *Then* $y^\sigma \to \bar{y}$.

Combining Corollaries 5 and 7 into one assertion, we get the second theorem of our investigation.

Theorem 2. *Let* $\|\sigma\|_\infty = \max\{\alpha, \beta\} \to +0$. *Then* $x^\sigma \to \bar{x}$ *and* $y^\sigma \to \bar{y}$.

Thus, in the linear case it is true that both x- and y-components of the saddle points of functions (3) converge to normal solutions of primal and dual problems respectively as $\|\sigma\|_\infty = \max\{\alpha, \beta\} \to +0$ and there are no needs for one of the parameters of regularization to be infinitesimal of smaller order then another.

To conclude this section let us compare the estimate of accuracy known from (11) with similar new estimate stated in Theorem 1. Since estimate (11) is valid only if $\beta = o(\alpha)$, we will enter two infinitesimal positive values α and ϵ

and put $\beta = \alpha\epsilon$. Substituting these formulas into the right-hand side of inequality (11), we get the relation

$$RHS_1 = 1/2\left[\|\bar{y}\|\sqrt{\beta/\alpha} + \sqrt{\alpha\beta}\|[\bar{\lambda}, \bar{\lambda}_0]\|\right] \sim O(\epsilon^{1/2}).$$

But if we substitute the same formulas in the right-hand side of the estimate from Theorem 1 and take into account that $R(\alpha, \alpha\epsilon) \sim O(\alpha\epsilon^{1/2})$ and $H(\alpha, \alpha\epsilon) \sim O(\alpha)$, then we see that

$$RHS_2 = \left[2\|\bar{x}\|R(\alpha, \beta) + R(\alpha, \beta)^2 + 2H(\alpha, \beta)\right]^{1/2} \sim O(\alpha^{1/2}).$$

Thus, new estimates of accuracy received above appear to be more precise compared with well-known ones excepting case when parameter β is infinitesimal of order higher than α^2. Besides, new estimates also are valid whether α and β have an identical order of smallness or even α has the order of smallness, higher than β has.

4   Infeasible Linear Programs

Next, let us consider the case when constraints of primal problem (1) are incompatible while dual problem (2) has at least one feasible vector. Such linear programs are called improper ones of the first kind [10]. In practice, to reduce an improper linear program of the first kind to a solvable one it is enough to correct only the right-hand side of its constraints till they become joint. Of cause, it is reasonable that such data changes must be as minimal as possible.

Denote the feasible set of problem (12) as

$$M(u) = \{x\colon Ax \leq b + u, \quad x \geq 0\}.$$

Also introduce the minimal (with respect to the Euclidean norm) correction vector for right-hand side of the primal program

$$\bar{u} = \arg\min_{M(u)\neq\emptyset} \|u\|. \tag{18}$$

Vector $\bar{u}$ is automatically non-negative and coincides with a projection of zero vector onto convex polyhedral set $\Omega = \{u\colon M(u) \neq \emptyset\}$, which is known as a set of solvability of improper program (1).

Let $\bar{\bar{x}}$ and $\bar{\bar{y}}$ be the normal solutions of the adjusted program

$$\max\{(c, x)\colon Ax \leq b + \bar{u}, \quad x \geq 0\}, \tag{19}$$

and of its dual one respectively

$$\min\{(b + \bar{u}, y)\colon A^T y \geq c, \quad y \geq 0\}.$$

In practice, vector $\bar{\bar{x}}$ may be considered as a generalized solution of the initial infeasible problem. It can provide to decision maker a substantial information about the reasons and sharpness of the contradictions of the model. To find this vector, we can use problem (4) again, and without any changes. Recall that it is generated by regularized Lagrangian

$$L_\sigma(x, y) = (c, x) - (y, Ax - b) - \alpha \|x\|^2 + \beta \|y\|^2$$

and has the form

$$(\mathcal{P}) \quad \max_{x \geq 0} \Phi_\sigma(x), \quad \Phi_\sigma(x) = \max_{x \geq 0} \left[(c, x) - \frac{1}{4\beta} \|(Ax - b)^+\|^2 - \alpha \|x\|^2 \right].$$

Though the initial program is infeasible now, nevertheless, the regularized function $L_\sigma(x, y)$ has a (unique) saddle point with respect to $\mathbb{R}^n_+ \times \mathbb{R}^m_+$ due to its strong concavity in x and strong convexity in y. As usual, let us denote by x^σ the optimal vector of problem (4). In [9] it was already shown that the series of inequalities (analogous to inequalities (7)–(10)) are valid:

$$\|(Ax^\sigma - b - \bar{u})^+\| \leq C_1 \sqrt{\beta}, \tag{20}$$

$$(c, \bar{\bar{x}}) - \alpha \|\bar{\bar{x}}\|^2 \leq (c, x^\sigma) \leq (c, \bar{\bar{x}}) + C_1 \|\bar{\bar{y}}\| \sqrt{\beta}, \tag{21}$$

where

$$C_1 = 2 \left[\sqrt{\beta} \|\bar{\bar{y}}\| + \left(\alpha \|\bar{\bar{x}}\|^2 + \beta \|\bar{\bar{y}}\|^2 \right)^{1/2} \right].$$

Also, for all sufficient small $\alpha > 0$, the estimate is known

$$\|x^\sigma - \bar{\bar{x}}\| \leq \|\bar{\bar{y}}\| (\beta/\alpha)^{1/2},$$

which is similar to estimate (11). It is valid only if $\alpha \to +0$, $\beta = o(\alpha)$.

Let us try to deduce an alternative estimate for $\|x^\sigma - \bar{\bar{x}}\|$.

First, accentuate that now we set $u^\sigma = \bar{u} + (Ax^\sigma - b - \bar{u})^+ \geq \bar{u}$. All other notations are taken from the previous sections. As usual, due to the properties of dual prices, it holds

$$0 \leq v(u^\sigma) - v(\bar{u}) \leq (\bar{\bar{y}}, u^\sigma - \bar{u}) \leq \|\bar{\bar{y}}\| \|u^\sigma - \bar{u}\|.$$

Therefore, by Hoffman's lemma,

$$\rho = \rho\big(\bar{M}(u^\sigma), \bar{M}(\bar{u})\big) \leq K(\|u^\sigma - \bar{u}\| + |v(u^\sigma) - v(\bar{u})|)$$
$$\leq K(1 + \|\bar{\bar{y}}\|) \|(Ax^\sigma - b - \bar{u})^+\|,$$

where ρ is Hausdorf's distance between the sets $\bar{M}(u^\sigma)$ and $\bar{M}(\bar{u})$. Combining this inequality with estimate (20) and using Lemma 1, we get the following assertion.

Corollary 8. *The estimate holds* $\|\bar{x}(u^\sigma)\| \leq \|\bar{x}(\bar{u})\| + K(1 + \|\bar{\bar{y}}\|)C_1\sqrt{\beta}.$

Next we formulate an analogue of Lemma 2.

Lemma 4. *The estimate is valid* $\|x^\sigma\| \le \|\bar{x}(\bar{u})\| + K(1 + \|\bar{\bar{y}}\|)C_1\sqrt{\beta}$.

Proof. Recall that in this section we set $u^\sigma = \bar{u} + (Ax^\sigma - b - \bar{u})^+ \ge \bar{u}$. From the definition of vector x^σ it follows that

$$\Phi_\sigma(x^\sigma) = \max_{x \ge 0} \Phi_\sigma(x) \ge \Phi_\sigma(\bar{x}(u^\sigma)).$$

Consequently,

$$(c, x^\sigma) - \frac{1}{4\beta}\|u^\sigma\|^2 - \alpha\|x^\sigma\|^2 \ge v(u^\sigma) - \frac{1}{4\beta}\|(A\bar{x}(u^\sigma) - b)^+\|^2 - \alpha\|\bar{x}(u^\sigma)\|^2,$$

or, after rearranging of items,

$$\alpha\|x^\sigma\|^2 \le \alpha\|\bar{x}(u^\sigma)\|^2 + \frac{1}{4\beta}\|(A\bar{x}(u^\sigma) - b)^+\|^2 - \frac{1}{4\beta}\|u^\sigma\|^2 + (c, x^\sigma) - v(u^\sigma). \quad (22)$$

We can essentially simplify this relation. Indeed, it holds $A\bar{x}(u^\sigma) \le b + u^\sigma$, which implies $0 \le (A\bar{x}(u^\sigma) - b)^+ \le u^\sigma$, and therefore

$$\|(A\bar{x}(u^\sigma) - b)^+\| \le \|u^\sigma\|.$$

Also, vector x^σ satisfies to constraints of problem (12) with parameter value $u = u^\sigma$. That is why

$$(c, x^\sigma) \le v(u^\sigma).$$

Hence, in really, (22) transforms to

$$\alpha\|x^\sigma\|^2 \le \alpha\|\bar{x}(u^\sigma)\|^2,$$

i.e. $\|x^\sigma\| \le \|\bar{x}(u^\sigma)\|$. This inequality (together with Corollary 1) completes the proof.

Let us turn back to the generalized solution of initial program (1) introduced above as

$$\bar{\bar{x}} = \arg\min\{\|x\|^2 \colon Ax \le b + \bar{u}, \ (c, x) \ge v(\bar{u})\}.$$

Obviously, we have a convex quadratic programming problem again. If we linearize this problem at the point $\bar{\bar{x}}$, then we get a linear program of the form

$$\min\{(\bar{\bar{x}}, x) \colon Ax \le b + \bar{u}, \ (c, x) \ge v(\bar{u})\},$$

which has at least one evident solution $\bar{\bar{x}}$. Also, let us write the dual problem

$$\max\{(b + \bar{u}, \lambda) + v(\bar{u})\lambda_0 \colon A^T\lambda + \lambda_0 c \le \bar{\bar{x}}, \ \lambda \le 0, \ \lambda_0 \ge 0\}$$

and denote by $\bar{\bar{w}} = [\bar{\bar{\lambda}}_0, \bar{\bar{\lambda}}]$ its arbitrary solution, e.g. the solution with minimal the Euclidean norm. By analogy with Lemma 3, it is easy to prove the following assertion.

Lemma 5. *For any $x \geq 0$ the estimate is valid*

$$(\bar{\bar{x}}, x) \geq \|\bar{\bar{x}}\|^2 - \|\bar{\bar{\lambda}}\| \left\| \left(Ax - b - \bar{u}\right)^+ \right\| - \bar{\bar{\lambda}}_0 (v(\bar{u}) - (c, x))^+.$$

Involving properties (20), (21) in our reasoning, we can obtain another assertion

Corollary 9. *The estimate holds*

$$(\bar{\bar{x}}, x^\sigma) \geq \|\bar{\bar{x}}\|^2 - C_1 \sqrt{\beta} \|\bar{\bar{\lambda}}\| - \alpha \bar{\bar{\lambda}}_0 \|\bar{\bar{x}}\|^2.$$

Lemma 4 and Corollary 9 give us the opportunity to carry out just the same reasoning as we use when we prove Theorem 1. Thereby we have the following final theorem.

Theorem 3. *The inequality is true*

$$\|x^\sigma - \bar{\bar{x}}\| \leq \left[2\|\bar{\bar{x}}\| R'(\alpha, \beta) + R'(\alpha, \beta)^2 + 2H'(\alpha, \beta) \right]^{1/2},$$

where

$$R'(\alpha, \beta) = 2K(1 + \|\bar{\bar{y}}\|) C_1 \sqrt{\beta}, \qquad H'(\alpha, \beta) = C_1 \sqrt{\beta} \|\bar{\bar{\lambda}}\| + \alpha \bar{\bar{\lambda}}_0 \|\bar{\bar{x}}\|^2.$$

To end this section let us formulate a simplified estimate.

Corollary 10. *Let $\|\sigma\|_\infty = \max\{\alpha, \beta\} < 1$ and problem (1) be improper of the first kind. Then*

$$\|x^\sigma - \bar{\bar{x}}\| \leq N_x \|\sigma\|_\infty^{1/2}$$

for some fixed N_x.

Remark. As for sequence y^σ, it is non-bounded when constraints of problem (1) are incompatible.

5   Conclusion

In this paper, we consider some alternative schemes of linear programming duality. These schemes are based on the symmetric regularization of the Lagrange function in direct and dual variables simultaneously. We establish that (unlike a non-linear case), the method under consideration converge to the normal (minimum with respect to the Euclidean norm) solutions of the initial primal and dual problems, and it is not essential for such a convergence whether primal and dual regularization parameters have a different or the same order of smallness. Also, the new symmetric estimates of accuracy are given which are more precise and efficient in comparison with ones known for the non-linear case. Besides, the similar estimates are obtained for a primal generalized solution if an origin linear program is improper one of the first kind.

References

1. Tikhonov, A.N., Arsenin, V.Ya.: Methods for Solutions of Ill-Posed Problems. Nauka, Moscow (1979). Wiley, New York (1981)
2. Vasil'ev, F.P.: Optimization Methods, vols. 1, 2. MTsNMO, Moscow (2011). In Russian
3. Rockafellar, R.T.: Augmented Lagrange multiplier functions and duality in nonconvex programming. SIAM J. Control **12**(2), 268–285 (1974)
4. Evtushenko, Ju.: Numerical Optimization Technique. Optimization Software, Inc. Publishing Division, New York (1985)
5. Gol'shtein, E.G., Tret'yakov, N.V.: Modified Lagrange Functions. Theory and Methods of Optimization. Nauka, Moscow (1989). In Rusian
6. Roos, C., Terlaky, T.: Theory and Algorithms for Linear Optimization: An Interior Point Approach. Wiley, New York (1997)
7. Eremin, I.I., Mazurov, Vl.D., Astaf'ev, N.N.: Improper Problems of Linear and Convex Programming. Nauka, Moscow (1983). In Russian
8. Skarin, V.D.: Regularization of the min-max problems occurring in convex programming. USSR Comput. Math. Math. Phys. **17**(6), 65–78 (1977). https://doi.org/10.1016/0041-5553(77)90173-2
9. Skarin, V.D.: On method of regularization for infeasible problems of convex programming. Izv. VUZov. Mathematica **12**, 81–88 (1995). Russian Math. (Iz. VUZ), **39**(12), 78–85 (1995)
10. Eremin, I.I.: Theory of Linear Optimization. Inverse and Ill-Posed Problems Series. VSP, Utrecht, Boston, Koln, Tokyo (2002)
11. Hoffman, A.J.: On approximate solutions of systems of linear inequalities. J. Res. Nat. Bur. Stand. **49**, 263–265 (1952)
12. Guddat, J.: Stability in convex quadratic programming. Mathematische Operationsforschung und Statistik **8**, 223–245 (1976)
13. Dorn, W.S.: Duality in quadratic programming. Q. Appl. Math. **18**, 407–413 (1960)

Binary Solutions to Some Systems
of Linear Equations

Alexandr V. Seliverstov$^{(\boxtimes)}$ (iD)

Institute for Information Transmission Problems of the Russian Academy of Sciences
(Kharkevich Institute), Bolshoy Karetny per. 19, build.1, Moscow 127051, Russia
slvstv@iitp.ru

Abstract. A point is called binary if its coordinates are equal to either
zero or one. It is well known that it is hard to find a binary solution
to the system of linear equations whose coefficients are integers with
small absolute values. The aim of the article is to propose an effective
probabilistic reduction from the system to the unique equation when
there is a small difference between the number of binary solutions to
the first equation and the number of binary solutions to the system.
There exist nontrivial examples of linear equations with small positive
coefficients having a small number of binary solutions in high dimensions.

Keywords: Subset sum · Linear equation · Probabilistic algorithm
Computational complexity

1 Introduction

Let us consider a system of linear equations over integers. The problem of the
existence of a $(0, 1)$-solution to the system is NP-complete [1]. The $(0, 1)$-solution
is also referred to as either binary or Boolean one.

In case of the unique equation, one can either find some $(0, 1)$-solution or
prove the absence of such solutions, using dynamic programming [2–6]. Moreover, the number of $(0, 1)$-solutions to the linear equation over integers can be
computed in pseudopolynomial time [3]. On the other hand, the problem is to
solve the system that consists of the linear equation and the set of quadratic
equations $x_1^2 = x_1, \ldots, x_n^2 = x_n$. If there is no solution, then a direct proof of
the insolvability of the system by means of Hilbert's Nullstellensatz requires to
produce polynomials of high degree [7]. All known methods for solving systems
of algebraic equations require at least exponential time in general case [8,9].
There exists a one-to-one correspondence between the $(0, 1)$-solutions and singular points of the effectively computed cubic hypersurface [10]. Some singular
points can be found by means of the method described in [11]. Another approach
to the problem is based on L-class enumeration algorithm [12].

This research has been supported by the Russian Science Foundation, Project No.
14–50–00150.

© Springer International Publishing AG, part of Springer Nature 2018
A. Eremeev et al. (Eds.): OPTA 2018, CCIS 871, pp. 183–192, 2018.
https://doi.org/10.1007/978-3-319-93800-4_15

There is also the related optimization problem to find the maximum of the linear functional on the set of $(0, 1)$-points satisfying a unique inequality. It is called the knapsack problem. There are well known both fully polynomial time approximation scheme and pseudopolynomial time algorithm for solving the problem. The obstacle for solving the optimization problem is a large number of values of the linear functional at different $(0, 1)$-points. If all coefficients are small positive integers, then the linear functional is bounded. Thus, the set of its values at $(0, 1)$-points is small. Howbeit, the NP-complete problem seems insolvable in polynomial time. Moreover, the polynomial hierarchy is infinite relative to a random oracle with probability one [13].

2   Preliminaries

The running time of the algorithm is the number of arithmetic operations $(+, -,$ and $\times)$ as well as of verifications of two binary predicates $=$ and $<$ over integers. The $\tilde{O}$ notation suppresses a factor that is polylogarithmic in the input size. The O^* notation suppresses a factor that is polynomial in the input size.

The symbol $\mathbf{x}$ denotes the integer sequence $(x_1, \ldots, x_n)$. Both k and j are integer so that $k \leq m$ and $j \leq n$, where m is an integer.

The number of $(0, 1)$-solutions to the linear equation $\beta + \alpha_1 x_1 + \ldots + \alpha_n x_n = 0$ over integers is equal to the coefficient of the monomial $t^{-\beta}$ of the univariate Laurent polynomial

$$F(t) = \prod_{j=1}^{n} (1 + t^{\alpha_j})$$

In case the j-th coefficient α_j is negative, one can make the linear transformation $x_j \mapsto 1 - x_j$. Thus, without loss of generality, one can assume that all coefficients α_j are positive, that is, the Laurent polynomial $F(t)$ is a polynomial.

Proposition 1 (Smolev [3]). *The number of $(0, 1)$-solutions to the linear equation $\beta + \alpha_1 x_1 + \ldots + \alpha_n x_n = 0$ over integers can be computed in pseudopolynomial time $O(n^3 a)$, where $a = \max_j |\alpha_j|$.*

So, the counting problem seems to be as hard as the recognition problem, that is, whether there exists a $(0, 1)$-solution to the linear equation. If all coefficients α_j are positive, the recognition problem coincides with the subset sum problem.

Remark 1. The subset sum problem can be solved in exponential time $O^*(2^{n/2})$ as well as in exponential space $O^*(2^{n/2})$ according to [14]. On the other hand, it can be solved in probabilistic time $O^*(2^{0.86n})$ and in polynomial space [15].

The running time of the algorithm solving the subset sum problem by means of dynamic programming is bounded by $O(n^2 a \log_2(na))$, where $a = \max_j |\alpha_j|$. Furthermore, in case the coefficients α_k are large, if the difference between $\max_k \alpha_k$ and $\min_k \alpha_k$ is bounded by a polynomial in n, then the subset sum problem can be solved in polynomial time [3].

There are some ways to improve the upper bound. In accordance with [5], in case $n < |\beta|$, the problem can be solved in pseudopolynomial time $\tilde{O}(\sqrt{n}|\beta|)$. In accordance with [6], the problem can be solved by a probabilistic algorithm in pseudopolynomial time $O(n + |\beta| \log_2 |\beta| \log_2^3(n/\varepsilon) \log_2 n)$ with error probability at most ε. On the other hand, the subset sum problem can be solved in polynomial space $\tilde{O}(n^2)$ and in pseudopolynomial time $\tilde{O}(n^3|\beta| \log_2 |\beta|)$ according to [16]. There exists another space-efficient algorithm [17].

Let us consider linear forms $\alpha_1 x_1 + \cdots + \alpha_n x_n$ over integers, where the greatest common divisor $\mathrm{GCD}(\alpha_1, \ldots, \alpha_n) = 1$. The greatest coefficient that appears in such linear forms vanishing on a set of $n - 1$ linearly independent $(0,1)$-points is at most $2^{-n}(\sqrt{n+1})^{n+1}$. The upper bound is based on the inequality for determinants [18]. It is almost tight [19,20]. So, the distribution of the number of $(0,1)$-solutions as a function in β is complicated [21].

Proposition 2 [1,22]. *Given the system of m linear equations $\beta_k + \alpha_{k1}x_1 + \cdots + \alpha_{kn}x_n = 0$ over integers. The set of $(0,1)$-solutions to the system coincides with the set of $(0,1)$-solutions to the unique equation*

$$\sum_{k=1}^{m} \gamma_k \left(\beta_k + \sum_{j=1}^{n} \alpha_{kj}x_j \right) = 0,$$

where integers $\gamma_k = (an + b + 1)^{k-1}$, $a = \max_{k,j} |\alpha_{kj}|$, and $b = \max_k |\beta_k|$.

Remark 2. On the other hand, in accordance with Proposition 1 as well as Remark 1, if all coefficients of the unique linear equation belong to a small segment near zero, then the equation can be solved by dynamic programming. Therefore, it is important to look for the coefficients γ_k as small as possible.

Propositions 2 and 1 together provide an algorithm whose running time is exponential in the number m. Let us compare the algorithm with what is obtained as a result of elimination of m variables. In this case, the absolute values of the coefficients of the resulting linear equation can rapidly increase during the process of elimination. This method allows to quickly find all $(0,1)$-solutions to the system only under the condition $n - m = O(\log_2 n)$.

Proposition 3. *There exists an algorithm that accepts the system of m independent linear equations $\beta_k + \alpha_{k1}x_1 + \cdots + \alpha_{kn}x_n = 0$ if and only if it has some $(0,1)$-solution. The running time of the algorithm is bounded by $O(nm^2 + nm2^{n-m})$.*

Proof. Elimination of m variables produces a linear equation that depends on at most $n - m$ variables. Therefore, it suffices to go over all $(0,1)$-points of $(n-m)$-dimensional space and to verify for each of them whether it corresponds to the $(0,1)$-solution to the input system. □

So, the most difficult case is when $n \approx 2m$.

Remark 3. If some linear equation of the system has small coefficients and a small number of $(0, 1)$-solutions, then one can compute the list of all $(0, 1)$-solutions to the equation by means of a binary search tree. Next, one can check step by step whether a $(0, 1)$-solution from the list is the solution to the system. But the task is more difficult, when there are sufficiently many $(0, 1)$-solutions to each linear equation.

3   Main Results

Theorem 1. *Given the positive number ε and the system of $m \geq 2$ linear equations $\ell_k(\mathbf{x}) = 0$ over integers, where $\ell_k(\mathbf{x}) = \beta_k + \alpha_{k1}x_1 + \cdots + \alpha_{kn}x_n$. Assume the first linear equation has at most μ redundant $(0, 1)$-solutions, which do not satisfy the system. If all random integers $\eta_2, \ldots, \eta_m$ are independent and uniformly distributed over the set from zero up to the number $N = \lceil \mu/\varepsilon \rceil$, then the probability that each $(0, 1)$-solution to the linear equation*

$$(Nm(an + b) + 1)\ell_1(\mathbf{x}) + \sum_{k=2}^{m} \eta_k \ell_k(\mathbf{x}) = 0$$

satisfies the system is at least $1 - \varepsilon$, where $a = \max_{k,j} |\alpha_{kj}|$ and $b = \max_k |\beta_k|$.

Proof. If either the first equation has no $(0, 1)$-solution or each $(0, 1)$-solution to the first equation satisfies the whole system, then the desired result is obvious. Else let us define a subset of the set of all $(0, 1)$-points

$$\mathcal{S} = \{\mathbf{x} \in \{0, 1\}^n : \ell_1(\mathbf{x}) = 0 \wedge (\exists k \leq m)\ell_k(\mathbf{x}) \neq 0\}.$$

The cardinality of the set $\mathcal{S}$ is at most μ. Let us define the polynomial

$$f(y_2, \ldots, y_m) = \prod_{\mathbf{x} \in \mathcal{S}} \left(\sum_{k=2}^{m} \ell_k(\mathbf{x})y_k \right)$$

In particular, if the set $\mathcal{S}$ is empty, then one can set $f = 1$. If a sequence $\gamma_2, \ldots, \gamma_m$ increases sufficiently fast, then $f(\gamma_2, \ldots, \gamma_m)$ does not vanish, consequently, the polynomial f does not vanish identically. Note that $\deg f \leq \mu$.

Let random integers η_k be independent and each η_k is uniformly distributed over the set $\{0, \ldots, N\}$. In accordance with the Schwartz–Zippel lemma [23], the probability of vanishing $f(\eta_2, \ldots, \eta_m)$ is at most ε.

In case $f(\eta_2, \ldots, \eta_m) \neq 0$, to prove that the system has no redundant $(0, 1)$-solution, it is sufficient to prove that there exists no redundant $(0, 1)$-solution to the following system of two linear equations

$$\begin{cases} \ell_1(\mathbf{x}) = 0 \\ \eta_2\ell_2(\mathbf{x}) + \cdots + \eta_m\ell_m(\mathbf{x}) = 0 \end{cases}$$

In turn, a $(0,1)$-point is the solution to the system if and only if it satisfies the unique linear equation

$$(Nm(an + b) + 1)\ell_1(\mathbf{x}) + \sum_{k=2}^{m} \eta_k \ell_k(\mathbf{x}) = 0.$$

In particular, if $\mu = 0$, then the equation coincides with the first equation. $\quad\square$

Remark 4. The number N can be replaced by another large number. So, without loss of generality one can assume $N = 2^{\nu} - 1$, where ν is integer. In this case, random numbers can be identified with sequences of independent random bits. There exist other methods for calculating random variables by coin tossing, cf. [24].

Remark 5. Of course, instead of the first equation one can use the sum $h(\mathbf{x})$ of both the first equation and a linear combination of all other equations having small coefficients. But this $h(\mathbf{x})$ must be explicitly defined.

Next, let us consider a Las Vegas algorithm what uses random integers while it is running, but always either returns the correct answer or never halts.

Theorem 2. *There exists a zero-error probabilistic algorithm such that for each integer $\mu \geq 0$ and for each system of $m \geq 2$ linear equations $\ell_k(\mathbf{x}) = 0$ over integers, where $\ell_k(\mathbf{x}) = \beta_k + \alpha_{k1}x_1 + \cdots + \alpha_{kn}x_n$, if the first linear equation has at most μ redundant $(0,1)$-solutions, which do not satisfy the system, then the algorithm returns the linear equation $h(\mathbf{x}) = 0$ over integers, where*

$$h(\mathbf{x}) = (2\mu m(an + b) + 1)\ell_1(\mathbf{x}) + \gamma_2 \ell_2(\mathbf{x}) + \cdots + \gamma_m \ell_m(\mathbf{x})$$
$$a = \max_{k,j} |\alpha_{kj}|$$
$$b = \max_k |\beta_k|$$
$$(\forall k)\gamma_k \leq 2\mu$$

so that each $(0,1)$-solution to the equation $h(\mathbf{x}) = 0$ is the solution to the system. In the case, the running time of the algorithm is pseudopolynomial in expectation. If the condition for μ is false, then the algorithm either returns an equation $h(\mathbf{x}) = 0$ or never halts.

Proof. Let $\eta_2, \ldots, \eta_m$ be independent random integers from zero to 2μ.
   At first, the algorithm chooses these random integers, sets

$$h(\mathbf{x}) = (2\mu m(an + b) + 1)\ell_1(\mathbf{x}) + \eta_2 \ell_2(\mathbf{x}) + \cdots + \eta_m \ell_m(\mathbf{x}),$$

and computes the number λ_0 of $(0,1)$-solutions to the equation $h(\mathbf{x}) = 0$ by means of Proposition 1. For each $1 \leq k \leq m$, it computes the number λ_k of $(0,1)$-solutions to the system of two equations $h(\mathbf{x}) = 0$ and $\ell_k(\mathbf{x}) = 0$ by means of Propositions 2 and 1.
   If for all k the equation $\lambda_0 = \lambda_k$ holds, then the algorithm returns the current equation $h(\mathbf{x})$, where for all k the coefficients $\gamma_k = \eta_k$.

Otherwise the algorithm repeats the same computation with new choice of random integers $\eta_2, \ldots, \eta_m$.

If the number μ satisfies the condition, then the probability that the algorithm returns a correct answer is at least $\frac{1}{2}$ at each round according to Theorem 1. The probability of there is no correct answer in a long series of repeats is small. Thus, the expected running time is almost as small as the running time of one round of the algorithm.   □

Theorem 3. *Given the system of m linear equations $\ell_k(\mathbf{x}) = 0$ over integers, where $m > r > 0$. Assume the subsystem of equations $\ell_1(\mathbf{x}) = 0, \ldots, \ell_r(\mathbf{x}) = 0$ has at most μ redundant $(0,1)$-solutions, which do not satisfy the system. There exist integers $\gamma_{r+1}, \ldots, \gamma_m$ belonging to the segment from zero up to the integer μ such that each $(0,1)$-solution to the new system of linear equations $\ell_1(\mathbf{x}) = 0, \ldots, \ell_r(\mathbf{x}) = 0$, and $\gamma_{r+1}\ell_{r+1}(\mathbf{x}) + \cdots + \gamma_m\ell_m(\mathbf{x}) = 0$ satisfies the initial system.*

Proof. If either the considered subsystem has no $(0,1)$-solution or each $(0,1)$-solution to the subsystem satisfies the whole system, then the desired result is obvious. Else let us define a subset of the set of all $(0,1)$-points

$$ \mathcal{S} = \{\mathbf{x} \in \{0,1\}^n : \ell_1(\mathbf{x}) = 0 \wedge \cdots \wedge \ell_r(\mathbf{x}) = 0 \wedge (\exists k \leq m)\ell_k(\mathbf{x}) \neq 0\}. $$

The cardinality of the set $\mathcal{S}$ is at most μ. Let us define the polynomial

$$ f(y_{r+1}, \ldots, y_m) = \prod_{\mathbf{x} \in \mathcal{S}} \left(\sum_{k=r+1}^{m} \ell_k(\mathbf{x}) y_k \right) $$

In particular, if the set $\mathcal{S}$ is empty, then one can set $f = 1$. If a sequence $\gamma_{r+1}, \ldots, \gamma_m$ increases sufficiently fast, then $f(\gamma_{r+1}, \ldots, \gamma_m)$ does not vanish, consequently, the polynomial f does not vanish identically. On the other hand, the inequality $\deg f \leq \mu$ holds. In accordance with the Schwartz–Zippel lemma [23], there exist desired integers $\gamma_{r+1}, \ldots, \gamma_m$ belonging to the segment from zero up to the integer μ.   □

4   Discussion

In case a correct value for μ is known, either Theorems 1 or 2 together with Proposition 1 provide the probabilistic algorithm to enumerate $(0,1)$-solutions to the system of linear equations over integers because each solution to the system satisfies all linear combinations of the equations. The first algorithm halts in one-sided error polynomial time. The second algorithm does not make errors.

If a $(0,1)$-solution exists, then it can be found by binary search. Moreover, all $(0,1)$-solutions can be listed in this way. Any substitution for a variable by either zero or one does not increase the number of solutions. Thus, the reduction of dimension require at most $2n$ steps. If all coefficients α_{kj} are nonnegative, then the search of $(0,1)$-solutions to the system can be improved by means of new algorithms for the subset sum problem, which are listed in Remark 1.

The algorithms can be useful for small both a and μ. The restriction on both values $a = \max_{k,j} |\alpha_{kj}|$ and $b = \max_k |\beta_k|$ is not crucial. The recognition problem of the existence of a $(0,1)$-solution to the system is NP-complete in case $a = 1$ without any restriction on the number of solutions, that is, $\mu = 2^n$. The reduction is obvious [1]. Thus, the linear system in n variables can be reduced to the another linear system in $O(n \log_2(ab))$ variables such that new coefficients have small absolute values. Furthermore, in case a is small, the running time of the algorithms depends weakly on b because without loss of generality one can assume the inequality $b \leq an$ holds. Otherwise the system has no $(0,1)$-solution. But the upper bound on the value μ is crucial.

If the first equation of the system has a small number of $(0,1)$-solutions, then μ can be chosen small too. But in the case, one can to check all these $(0,1)$-solutions by means of the deterministic algorithm. Nontrivial case is when each equation has many $(0,1)$-solutions, but there is a small difference between the number of $(0,1)$-solutions to the first equation and the number of $(0,1)$-solutions to the system.

There exist at most 2^{n-m} binary solutions to the system of m linearly independent linear equations in n variables. In accordance with Proposition 1, the number λ_1 of $(0,1)$-solutions to the first equation can be found in pseudopolynomial time. So, there is the lower bound on the value $\mu \geq \lambda_1 - 2^{n-m}$. Another way to obtain the lower bound on the value μ is to compute the upper bound on the dimension of the affine hull of $(0,1)$-solutions to the first equation of the system.

Note that μ can be a rough upper bound on the difference between the total number of $(0,1)$-solutions to the first equation and the number of $(0,1)$-solutions to the system. On the other hand, it is hard to compute this difference. Otherwise, it would be easy to calculate the number of $(0,1)$-solutions to the system, that is, to solve the hard counting problem.

Of course, if all absolute values of the coefficients α_j are small integers, then there exists a number β such that the linear equation $\beta + \alpha_1 x_1 + \cdots + \alpha_n x_n = 0$ has at least $2^n/(1 + na)$ binary solutions, where $a = \max_j |\alpha_j|$. Let us consider examples of linear equations with small positive coefficients having a few $(0,1)$-solutions. In particular, if the first equation of the system coincides with one of exemplified equations, then one can use a small value of μ.

Example 1. If all the coefficients α_k are strictly positive, then there exists exactly one $(0,1)$-solution to the equation

$$\sum_{j=1}^{n} \alpha_j x_j = \sum_{j=1}^{n} \alpha_j,$$

that is, $(1, \ldots, 1)$. Moreover, if the inequality $\alpha_1 + \cdots + \alpha_{n-1} < \alpha_n$ holds, then the equation

$$\sum_{j=1}^{n} \alpha_j x_j = \sum_{j=1}^{n-1} \alpha_j$$

has exactly one $(0,1)$-solution, that is, $(1, \ldots, 1, 0)$.

Example 2. If $n = p+q$, where $p \neq q$ and both numbers p and q are prime, then there exist exactly two $(0,1)$-solutions to the equation

$$\sum_{j=1}^{p} qx_j + \sum_{j=p+1}^{p+q} px_j = pq.$$

These antipodal points are $(1,\ldots,1,0\ldots,0)$ and $(0,\ldots,0,1,\ldots,1)$, where the number of zeros is equal to either p or q. The equations $x_1 = x_2 = \cdots = x_p$ hold because

$$q\sum_{j=1}^{p} x_j \equiv 0 \pmod{p}.$$

The equations $x_{p+1} = x_{p+2} = \cdots = x_n$ hold because

$$p\sum_{j=p+1}^{p+q} x_j \equiv 0 \pmod{q}.$$

The maximum of the linear form over the set $\{0,1\}^n$ is equal to $2pq$.

Example 3. If $n = p+q+1$, where $p \neq q$ and both numbers p and q are prime, then there are exactly three $(0,1)$-solutions to the equation

$$\sum_{j=1}^{p} qx_j + \sum_{j=p+1}^{p+q} px_j + pqx_n = pq.$$

These points are $(1,\ldots,1,0\ldots,0,0)$, $(0,\ldots,0,1,\ldots,1,0)$, and $(0,\ldots,0,1)$. The maximum of the linear form over the set $\{0,1\}^n$ is equal to $3pq$.

Example 4. If $n = p+q+r$, where $p < q < r$ and the numbers p, q, and r are prime, then there are exactly three $(0,1)$-solutions to the equation

$$\sum_{j=1}^{p} qrx_j + \sum_{j=p+1}^{p+q} prx_j + \sum_{j=p+q+1}^{p+q+r} pqx_j = pqr.$$

The maximum of the linear form over the set $\{0,1\}^n$ is equal to $3pqr$.

In this way, one can construct other examples with arbitrary given number of $(0,1)$-solutions for almost all n. Linear transformations of coordinates of the type $x_j \mapsto 1 - x_j$ allow constructing other examples with coefficients of different signs.

The abundance of such examples allows to hope that the discussed algorithm can find practical application, in particular, in bioinformatics and economics [25].

Theorem 3 provides an improvement of the Proposition 2. If there exists a subsystem with a small number of redundant $(0,1)$-solutions, which do not satisfy the system, then one can reduce the number of equations without a considerable increment of absolute values of its coefficients. Unfortunately, it requires guessing this subsystem. Assume the initial system has no $(0, 1)$-solution. At first, it can be reduced to the new system according to Theorem 3. Next, it can be reduced to the unique equation according to Proposition 2. At last, one can count the number of $(0, 1)$-solution according to Proposition 1. So, this particular instance of the $coNP$-complete problem can be solved by the non-deterministic algorithm. Of course, if the hypothesis $NP \neq coNP$ holds, then the running time of the algorithm must be sufficiently large in some cases.

The same result is also applicable to the case of $(-1, 1)$-solutions, that is, solutions to the set partition problem.

Acknowledgements. The author would like to thank the anonymous reviewers for useful comments.

References

1. Schrijver, A.: Theory of Linear and Integer Programming. Wiley, New York (1986)
2. Dantzig, G.B.: Discrete-variable extremum problems. Oper. Res. **5**(2), 266–277 (1957)
3. Smolev, V.V.: On an approach to the solution of a Boolean linear equation with positive integer coefficients. Discrete Math. Appl. **3**(5), 523–530 (1993). https://doi.org/10.1515/dma.1993.3.5.523
4. Tamir, A.: New pseudopolynomial complexity bounds for the bounded and other integer Knapsack related problems. Oper. Res. Lett. **37**(5), 303–306 (2009). https://doi.org/10.1016/j.orl.2009.05.003
5. Koiliaris, K., Xu, C.: A faster pseudopolynomial time algorithm for subset sum. In: SODA 2017 Proceedings of the Twenty-Eighth Annual ACM-SIAM Symposium on Discrete Algorithms, pp. 1062–1072. Society for Industrial and Applied Mathematics, Philadelphia (2017)
6. Bringmann, K.: A near-linear pseudopolynomial time algorithm for subset sum. In: SODA 2017 Proceedings of the Twenty-Eighth Annual ACM-SIAM Symposium on Discrete Algorithms, pp. 1073–1084. Society for Industrial and Applied Mathematics, Philadelphia (2017)
7. Margulies, S., Onn, S., Pasechnik, D.V.: On the complexity of Hilbert refutations for partition. J. Symbolic Comput. **66**, 70–83 (2015). https://doi.org/10.1016/j.jsc.2013.06.005
8. Chistov, A.L.: An improvement of the complexity bound for solving systems of polynomial equations. J. Math. Sci. **181**(6), 921–924 (2012). https://doi.org/10.1007/s10958-012-0724-4
9. Jeronimo, G., Sabia, J.: Sparse resultants and straight-line programs. J. Symbolic Comput. **87**, 14–27 (2018). https://doi.org/10.1016/j.jsc.2017.05.005
10. Latkin, I.V., Seliverstov, A.V.: Computational complexity of fragments of the theory of complex numbers. Bull. Karaganda Univ. Math. **1**, 47–55 (2015). (In Russian). http://vestnik.ksu.kz
11. Seliverstov, A.V.: On tangent lines to affine hypersurfaces. Vestnik Udmurtskogo Universiteta. Matematika. Mekhanika. Komp'yuternye Nauki **27**(2), 248–256 (2017). (In Russian). https://doi.org/10.20537/vm170208

12. Kolokolov, A.A., Zaozerskaya, L.A.: Finding and analysis of estimation of the number of iterations in integer programming algorithms using the regular partitioning method. Russian Math. (Iz. VUZ) **58**(1), 35–46 (2014). https://doi.org/10.3103/S1066369X14010046

13. Håstad, J., Rossman, B., Servedio, R.A., Tan, L.-Y.: An average-case depth hierarchy theorem for Boolean circuits. J. ACM **64**(5), 35 (2017). https://doi.org/10.1145/3095799

14. Horowitz, E., Sahni, S.: Computing partitions with applications to the knapsack problem. J. ACM **21**(2), 277–292 (1974). https://doi.org/10.1145/321812.321823

15. Bansal, N., Garg, S., Nederlof, J., Vyas, N.: Faster space-efficient algorithms for subset sum and k-sum. In: STOC 2017 Proceedings of the 49th Annual ACM SIGACT Symposium on Theory of Computing, pp. 198–209 ACM, New York (2017). https://doi.org/10.1145/3055399.3055467

16. Lokshtanov, D., Nederlof, J.: Saving space by algebraization. In: STOC 2010 Proceedings of the Forty-second ACM Symposium on Theory of Computing, pp. 321–330. ACM, New York (2010). https://doi.org/10.1145/1806689.1806735

17. Gál, A., Jang, J.-T., Limaye, N., Mahajan, M., Sreenivasaiah, K.: Space-efficient approximations for Subset Sum. ACM Trans. Comput. Theory **8**(4), 16 (2016). https://doi.org/10.1145/2894843

18. Williamson, J.: Determinants whose elements are 0 and 1. Am. Math. Monthly **53**(8), 427–434 (1946)

19. Alon, N., Vũ, V.H.: Anti-Hadamard matrices, coin weighing, threshold gates and indecomposable hypergraphs. J. Comb. Theory A **79**(1), 133–160 (1997). https://doi.org/10.1006/jcta.1997.2780

20. Babai, L., Hansen, K.A., Podolskii, V.V., Sun, X.: Weights of exact threshold functions. In: Hliněný, P., Kučera, A. (eds.) MFCS 2010. LNCS, vol. 6281, pp. 66–77. Springer, Heidelberg (2010). https://doi.org/10.1007/978-3-642-15155-2_8

21. Gorbunov, K.Yu., Seliverstov, A.V., Lyubetsky, V.A.: Geometric relationship between parallel hyperplanes, quadrics, and vertices of a hypercube. Probl. Inform. Transm. **48**(2), 185–192 (2012). https://doi.org/10.1134/S0032946012020081

22. Williams, R.: New algorithms and lower bounds for circuits with linear threshold gates. In: STOC 2014 Proceedings of the Forty-sixth Annual ACM Symposium on Theory of Computing, pp. 194–202. ACM, New York (2014). https://doi.org/10.1145/2591796.2591858

23. Schwartz, J.T.: Fast probabilistic algorithms for verification of polynomial identities. J. ACM **27**(4), 701–717 (1980). https://doi.org/10.1145/322217.322225

24. Bacher, A., Bodini, O., Hwang, H.-K., Tsai, T.-H.: Generating random permutations by coin-tossing: classical algorithms, new analysis, and modern implementation. ACM Trans. Algorithms **13**(2), 24 (2017). https://doi.org/10.1145/3009909

25. Beresnev, V.L., Melnikov, A.A.: An upper bound for the competitive location and capacity choice problem with multiple demand scenarios. J. Appl. Ind. Math. **11**(4), 472–480 (2017). https://doi.org/10.1134/S1990478917040020

Variant of the Cutting Plane Method with Approximation of the Set of Constraints and Auxiliary Functions Epigraphs

I. Ya. Zabotin and K. E. Kazaeva[✉]

Kazan (Volga Region) Federal University,
18, Kremlyovskaya street, Kazan 420008, Russia
iyazabotin@mail.ru, k.e.kazaeva@gmail.com

Abstract. We propose a method of solving a convex programming problem, which is based on the ideas of cutting plane methods and the method of penalty functions. To construct each approximation, the method uses an operation of immersing the epigraph of auxiliary function into a polyhedral set. The auxiliary function is constructed as the sum of the objective function and the external penalty function of the constraint area. In addition, an admissible set of the original problem is immersed in the polyhedron simultaneously. In connection with this, the problem of constructing an iterative point is a linear programming problem, in which constraints are polyhedrons approximating the epigraph of auxiliary function and the admissible set. Both next approximating sets are based on the previous ones by cutting off the iterative point from them by hyperplanes. The convergence of the method is proved. We describe its algorithms. One of them can be the implementation of the method of penalty functions.

Keywords: Conditional minimization · Algorithm · Epigraph
Penalty function · Approximating set · Cutting hyperplane
Iterative point · Convergence · Subgradient

1 Introduction

Cutting plane methods form a well-known class of mathematical programming problem solving methods (e.g., [1–7]). From a practical point of view, their convenience lies in the fact that at each step it is possible to estimate the closeness of the current value of the objective function to its optimal value.

The methods of this class are characterized by the fact that in order to find iterative points they use the operation of immersing the epigraph of the objective function or the region of constraints of the original problem into polyhedral sets. At the same time, there is a small group of cutting methods, which simultaneously use the approximation of the epigraph and the set of constraints (e.g., [8–10]). The method proposed here belongs to this group. It differs in that,

© Springer International Publishing AG, part of Springer Nature 2018
A. Eremeev et al. (Eds.): OPTA 2018, CCIS 871, pp. 193–204, 2018.
https://doi.org/10.1007/978-3-319-93800-4_16

on each iteration, it uses the approximation of the epigraph not of the objective function, but of some auxiliary function, which is built on the basis of an external penalty of the constraint area.

As is known, in the method of penalty functions, the properties of auxiliary functions could deteriorate significantly with increasing number of steps. Wherein the minimization problems of auxiliary functions have to be solved with accuracy increasing from step to step. The iterative point of the basic sequence is fixed, and the penalty function changes only after reaching the specified accuracy. In the proposed method, it is not necessary to solve the problems of minimizing such auxiliary functions. Penalties, and hence auxiliary functions, can be changed at each iteration of the method.

At the same time, the proposed method allows for an approximate solution of the above auxiliary problems. In this regard, one of its algorithms can serve as a realization of the method of penalty functions. The peculiarity of this realization is as follows. Due to the approximating procedures of the epigraph and the area of constraints, it is possible to estimate the closeness of the current value of the auxiliary function to its optimal value at each step.

2   Problem Setting

We solve the problem
$$\min \{f(x) : x \in D\}, \tag{1}$$

where $f(x)$ is a convex function defined in the n-dimensional Euclidean space R_n, $D \subset R_n$ is a convex closed set, int $D \neq \emptyset$.

Let $f^* = \min \{f(x) : x \in D\} > -\infty$, $X^* = \{x \in D : f(x) = f^*\}$, $x^* \subset X^*$, epi$(g, G) = \{(x, \gamma) \in R_{n+1} : x \in G, \gamma \geqslant g(x)\}$, where $G \subset R_n$, $g(x)$ is the function defined in R_n, $W(z, Q)$ — the bunch of normalized generally support vectors for the set Q at the point z, $K = \{0, 1, \ldots\}$.

3   The Cutting Plane Method

The proposed method of solving the problem (1) generates a sequence of approximations $\{x_k\}$, $k \in K$, by the following rule.

Points
$$v' \in \text{int epi}(f, D), \quad v'' \in \text{int } D$$

and a convex penalty function $P_0(x)$ such as
$$P_0(x) = 0 \,\forall x \in D, \; P_0(x) > 0 \,\forall x \notin D$$

are selected.

A convex bounded closed set $D_0 \subset R_n$ containing the point x^* is constructed. Set numbers $\bar{\gamma}$, q and Δ_0 such that $\bar{\gamma} \leqslant f_0^* = \min \{f(x) : x \in D_0\}$, $q \geqslant 1$, $\Delta_0 > 0$. We set
$$F_0(x) = f(x) + P_0(x), \quad M_0 = R_{n+1},$$

$i = 0$, $k = 0$.

1. We find a solution $u_i = (y_i, \gamma_i)$ where $y_i \in R_n$, $\gamma_i \in R_1$ of the problem

$$\min \{\gamma : x \in D_i, \ (x, \gamma) \in M_i, \ \gamma \geqslant \bar{\gamma}\}. \tag{2}$$

If

$$y_i \in D, \quad \gamma_i \geqslant f(y_i), \tag{3}$$

then $y_i \in X^*$, and the process is over.

2. In the interval (v', u_i) we select a point $v'_i \notin \operatorname{int} \operatorname{epi}(F_i, R_n)$, so that for the point $z'_i = u_i + q'_i (v'_i - u_i)$ with some $q'_i \in [1, q]$ the inclusion

$$z'_i \in \operatorname{epi}(F_i, R_n)$$

is performed.

3. A finite set $A_i \subset W(v'_i, \operatorname{epi}(F_i, R_n))$ is chosen and let

$$M_{i+1} = M_i \cap \{u \in R_{n+1} : \langle a, u - v'_i \rangle \leqslant 0 \ \forall a \in A_i\}. \tag{4}$$

4. If $y_i \in D$, then it is assumed that $D_{i+1} = D_i$, $z''_i = y_i$, and proceed to step 7. Otherwise, go to step 5.

5. In the interval (v'', y_i) select a point $v''_i \notin \operatorname{int} D$, such that $z''_i = y_i + q''_i (v''_i - y_i) \in D$ for some $q''_i \in [1, q]$.

6. A finite set $B_i \subset W(v''_i, D)$ is chosen. Let

$$D_{i+1} = D_i \cap \{x \in R_n : \langle b, x - v''_i \rangle \leqslant 0 \ \forall b \in B_i\}. \tag{5}$$

7. If

$$F_i(y_i) - \gamma_i > \Delta_k, \tag{6}$$

then set

$$P_{i+1}(x) = P_i(x), \quad F_{i+1}(x) = F_i(x),$$

and go to step 9. Otherwise, proceed to step 8.

8. We choose convex penalty function $P_{i+1}(x)$ with the condition that

$$P_{i+1}(x) = 0 \ \forall x \in D, \quad P_{i+1}(x) \geqslant P_i(x) \ \forall x \notin D.$$

Let $F_{i+1}(x) = f(x) + P_{i+1}(x)$, $i_k = i$,

$$x_k = y_{i_k}, \quad \sigma_k = \gamma_{i_k}. \tag{7}$$

Set $\Delta_{k+1} > 0$ and go to step 9 with the value of k increased by one.

9. The value of i is incremented by one and we go to step 1.

Let us make some remarks concerning the method.

The ways of specifying the functions $P_i(x)$, $i \in K$, can be found, for example, by references [11–13]. In particular, if

$$D = \{x \in R_n : f_j(x) \leqslant 0, \ j = 1, \ldots, m\},$$

where functions $f_j(x)$, $j = 1, \ldots, m$, are convex, then we could let

$$P_i(x) = \alpha_i P(x),$$

where

$$P(x) = \sum_{j=1}^{m} (\max\{f_j(x), \ 0\})^p,$$

$p \geqslant 1$, $\alpha_{i+1} \geqslant a_i > 0$, $i \in K$.

If the set D_0 is chosen as a polyhedron, then taking into account the above conditions (4), (5), problems (2) are linear programming problems for all $i \in K$.

Note that for $i = 0$ any point $(y_0, \bar{\gamma})$, such that $y_0 \in D$, is a solution of the problem (2).

The initial immersion of the set M_0 need not be set to coincide with R_{n+1}. It could be chosen in the form of any polyhedral set with a condition that the following inclusion holds

$$\text{epi}(F_0, R_n) \subset M_0.$$

The set M_0 could be defined using single inequality function subgradient as, for example, this is done in the description of the cutting-plane method by reference [7]. In the case of a choice of M_0 different from R_{n+1}, the restriction $\gamma \geq \bar{\gamma}$ in problem (2) can be removed in connection with the condition of boundedness of the initial approximating set D_0.

Note that on steps 2 and 5 of the method the points v_i', v_i'' can be chosen as points of intersection of segments $[v', u_i]$ and $[v'', y_i]$ with the boundaries of sets epi (F_i, R_n) and D, respectively. Then we can assume $q_i' = q_i'' = 1$, $z_i' = v_i'$, $z_i'' = v_i''$. The conditions for choosing v_i', v_i'' points in the method in fact allows us to find the mentioned points approximately.

Generalized support vectors from the sets $W(v_i', \text{epi}(F_i, R_n))$ and $W(v_i'', D)$ can be constructed using the subgradients of corresponding functions $F_i(x)$ and functions that define D on steps 3, 6 of the method. For example, if $v_i' = (p_i, \alpha_i)$, where $p_i \in R_n$, $\alpha_i \in R_1$ and c is a subgradient of the function $F_i(x)$ at a point p_i, then the vector $a = (c, -1)$ is a generalized support vector to the set epi (F_i, R_n) at the point v_i' [14].

Lemma 1. *The point (x^*, f^*) satisfies the constraints of the problem (2) for all $i \in K$.*

Proof. Since $x^* \in D_0$ and $D_{i+1} = D_i$ or D_{i+1} has the form (5), then the inclusion $x^* \in D_i$ holds for all $i \in K$. Moreover, $\bar{\gamma} \leqslant f_0^* \leqslant f^*$. Therefore, in order to justify the lemma, it suffices to show that

$$(x^*, f^*) \in M_i \tag{8}$$

for all $i \in K$.

The inclusion (8), for $i = 0$, is valid, since equality $M_0 = R_{n+1}$ is satisfied. Now we assume that the inclusion (8) holds for $i = l \geqslant 0$. Lets show the fulfillment of (8) for $i = l + 1$, then the lemma will be proved. Indeed, since

$x^* \in D$, and taking into account the assignment of functions $P_i(x)$, $i \in K$, then the equality $f(x^*) = f^* = F_i(x^*)$, $i \in K$, holds and

$$(x^*, f^*) \in \mathrm{epi}(F_i, R_n), \quad i \in K.$$

Consequently, inequalities

$$\langle a, (x^*, f^*) - v'_l \rangle \leqslant 0$$

hold for all $a \in A_l$. Taking into account the inductive hypothesis, according to (4) $(x^*, f^*) \in M_{l+1}$. The lemma is proved.

Since $x^* \in D_i$, $i \in K$, then (8) is fulfilled for all $i \in K$ and, in addition, the inequality $f^* \geq \bar{\gamma}$ holds, then the points (x^*, f^*) satisfy the constraints of problem (2) for all $i \in K$. Hence follows

Lemma 2. *For each $i \in K$, the inequality (9) is valid.*

$$\gamma_i \leqslant f^*. \tag{9}$$

With the use of Lemma 2, we justify the criterion of optimality laid down in paragraph 1 of the method.

Theorem 1. *Suppose that for some $i \in K$ the relation (3) is satisfied. Then y_i^* is a solution of problem (1).*

Proof. From condition (3), taking into account (9), inequality $f^* \leqslant f(y_i) \leqslant \gamma_i \leqslant f^*$ holds. Thus we obtain the equality $f(y_i) = f^*$, which proves the theorem.

4   Investigation of the Method's Convergence

We proceed to the justification of the convergence of the proposed method. First of all, we note that the sequence $\{u_i\}$, $i \in K$, is bounded. This follows from the inclusions $y_i \in D_0$, $i \in K$, inequalities (9), and the last constraint of the problem (2). Thus, sequences $\{v'_i\}$, $\{z'_i\}$, $\{z''_i\}$, $i \in K$, are also bounded.

Lemma 3. *Let $\{u_i\}$, $i \in K' \subset K$, is a convergent subsequence of the sequence $\{u_i\}$, $i \in K$. Then the following equalities hold:*

$$\lim_{i \in K'} \|v'_i - u_i\| = 0. \tag{10}$$

Proof. By the choice of the points v'_i, $i \in K$, there is such $\mathcal{T}_i \in [0,1)$ for each $i \in K$ that

$$v'_i = u_i + \mathcal{T}_i(v' - u_i). \tag{11}$$

Fix numbers i', $i'' \in K'$ so that $i'' > i'$. According to condition (4), $M_{i''} \subset M_{i'}$, and this means that

$$A_{i'} \subset W(v'_i, M_{i''}).$$

Then, in view of inclusion $u_{i''} \in M_{i''}$ we have $\langle a, u_{i''} - v'_{i'} \rangle \leqslant 0$ for all $a \in A_{i'}$. From this and (11) follows the inequality

$$\langle a, u_{i'} - u_{i''} \rangle \geqslant T_{i'} \langle a, u_{i'} - v' \rangle \quad \forall a \in A_{i'}. \tag{12}$$

As $v' \in \operatorname{int} \operatorname{epi}(f, D)$ and $v'_i \notin \operatorname{int} \operatorname{epi}(F_i, R_n)$, and then $v'_i \notin \operatorname{int} \operatorname{epi}(f, D)$, so by Lemma 1 from [15] there is a such number $\delta > 0$ that $\langle a, v' - v'_i \rangle \leqslant -\delta$ for all $a \in A_i$, $i \in K$. Hence, taking into account the equality (11) and inequalities $0 \leqslant T_i \leqslant 1$, we have $\langle a, u'_i - u''_i \rangle \leqslant -\delta$ for all $a \in A_i$, $i \in K$. Then, from the proposition (12) we obtain $\langle a, u_{i'} - u_{i''} \rangle \geqslant T_{i'} \delta$ for all $a \in A_{i'}$ or

$$\| u_{i'} - u_{i''} \| \geqslant T_{i'} \delta,$$

because $\|a\| = 1$ for all $a \in A_{i'}$. From the last inequality and from the convergence of the sequence $\{u_i\}$, $i \in K'$, the limiting relation $T_i \to 0$, $i \in K'$, holds. Then the equality (11) and the boundedness of the sequence $\{\|v' - u_i\|\}$, $i \in K'$, imply the equality (10). The lemma is proved.

Corollary. When the conditions of Lemma 3 are satisfied, the following equality holds.

$$\lim_{i \in K'} \| z'_i - u_i \| = 0. \tag{13}$$

Proof. According to the conditions in step 2 of the method $z'_i - u_i = q'_i (v'_i - u_i)$, $i \in K$. Thus, the assertion (13) follows from (10) and the boundedness of the sequence $\{q'_i\}$, $i \in K$.

The consequence to Lemma 3 enables us to prove further that, together with the sequence $\{u_i\}$, $i \in K$, the method will also construct sequences $\{x_k\}$, $\{\sigma_k\}$, $k \in \bar{K}$.

Lemma 4. *For each $k \in \bar{K}$ there is an index $i_k \in K$ which satisfies (7).*

Proof. Let prove the derivation of the lemma by induction.

(1) Suppose that $k = 0$. Show the existence of such index $i_0 \geqslant 0$ for which the inequality $F_{i_0}(y_{i_0}) - \gamma_{i_0} \leqslant \Delta_0$ holds. Then, in view of step 8 of the method,

$$x_0 = y_{i_0}, \quad \delta_0 = \gamma_{i_0}$$

and the assertion for $k = 0$ will be proved.

Suppose the contrary, that is

$$F_i(y_i) - \gamma_i > \Delta_0 \quad \forall i \in K. \tag{14}$$

In this case, according to the step 7 of the method, we have the equalities $P_i(x) = P_0(x)$, $F_i(x) = F_0(x)$ for all $i > 0$. By virtue of (14)

$$F_0(y_i) - \gamma_i > \Delta_0 \quad \forall i \in K. \tag{15}$$

From the sequence $\{u_i\}$, $i \in K$, we distinguish a convergent subsequence $\{u_i\}$, $i \in K' \subset K$, and let $u' = (y', \gamma')$ be its limit point. Then, in view of condition (15) and the continuity of the function $F_0(x)$,

$$F_0(y') - \gamma' \geqslant \Delta_0. \tag{16}$$

By the consequence to Lemma 3, the equality (13) holds for the selected subsequence $\{u_i\}$, $i \in K'$. From (13) and the inclusion $z'_i \in \operatorname{epi}(F_0, R_n)$, $i \in K'$, follows $u' \in \operatorname{epi}(F_0, R_n)$, i.e.

$$F_0(y') \leqslant \gamma'. \tag{17}$$

Contrariwise, according to the specification of sets A_i, $i \in K$, and method (4) for constructing sets M_i, $i \in K$, the inclusion takes place $\operatorname{epi}(F_0, R_n) \subset M_i$, $i \in K$, and it means that the point $(y_i, F_0(y_i))$ is a permissible solution of the problem (2) for every $i \in K$. So the inequality $\gamma_i \leqslant F_0(y_i)$ holds for the solution (y_i, γ_i) of the problem (2) for all $i \in K$. Passing to the limit in $i \in K'$ in this inequality we obtain $\gamma' \leqslant F_0(y')$. This and (17) imply the equality $F_0(y') = \gamma'$ which contradicts (16). Thus, the existence of the number i_0 for which the inequality $F_{i_0}(y_{i_0}) - (\gamma_{i_0}) \leq \Delta_0$ is proved unjustly, and for $k = 0$ the assertion of the Lemma is valid.

(2) Now assume that the equalities (7) are satisfied for some fixed $k \geqslant 0$. We show the existence of a number $i_{k+l} \geqslant i_k + 1 = r$ such that the equalities hold

$$x_{k+1} = y_{i_{k+1}}, \quad \sigma_{k+1} = \gamma_{i_{k+1}}, \tag{18}$$

and then the lemma will be proved.

Assume the opposite, i.e. let the inequality (6) hold for all $i \geqslant r$. It means that $P_i(x) = P_r(x)$, $F_i(x) = F_r(x)$, and by virtue of the inequality (6)

$$F_r(y_i) - \gamma_i > \Delta_k \quad \forall i \in K, \quad i \geqslant r.$$

From the points u_i, $i \in K$, $i \geqslant r$, we select a convergent subsequence $\{u_i\}$, $i \in K'' \subset K$. Then for its limit point $u'' = (y'', \gamma'')$ we have the inequality

$$F_r(y'') - \gamma'' \geqslant \Delta_k. \tag{19}$$

As in the first part of the proof, for points (y'', γ'') it is not difficult to obtain firstly the inequality $F_r(y'') \leqslant \gamma''$, and then the inequality $\gamma'' \leqslant F_r(y'')$. Consequently, $F_r(y'') = \gamma''$. The last equality contradicts inequality (19) in view of the choice of Δ_k. Thus, the existence of the number i_{k+1} for which (18) holds is proved. The rationale for the Lemma is complete.

Lemma 4 proves the existence of sequences $\{x_k\}$, $\{\sigma_k\}$. To study their convergence, we give below three more auxiliary assertions.

Lemma 5. *Any limit point of the sequence $\{y_i\}$, $i \in K$, belongs to the set D.*

Proof. Let $\{y_i\}$, $i \in K' \subset K$, is a convergent subsequence of the sequence $\{y_i\}$, $i \in K$, and $\bar{y}$ is its limit point. We should show the inclusion

$$\bar{y} \in D, \tag{20}$$

then the lemma will be proved. If $y_i \in D$ holds for an infinite number of indexes $i \in K'$, then the inclusion (20) is is satisfied because the set D is closed. Therefore, we suppose that $y_i \notin D$ for all $i \in K'$, starting with some number i'.

Fix numbers $l, p \in K'$, so that $p > l \geqslant i'$. Since $D_p \subset D_l$, $y_p \in D_p$, and any element of the set B_l is a generalized support for the set D_p at the point v_l'' then

$$\langle b, y_p - v_l'' \rangle \leqslant 0 \; \forall b \in B_l.$$

Note that

$$v_i'' = y_i + \beta_i \left(v'' - y_i \right), \quad i \in K, \tag{21}$$

where $\beta \in [0, 1)$. Then it follows from the last inequality that

$$\langle b, y_l - y_p \rangle \geqslant \beta_l \langle b, y_l - v'' \rangle \quad \forall b \in B_l. \tag{22}$$

As $v'' \in \operatorname{int} D$, but $v_i'' \notin \operatorname{int} D$ there is a number [15] $\delta > 0$ such that

$$\langle b, v'' - v_i'' \rangle \leqslant -\delta \quad \forall b \in B_i, \; i \in K'.$$

Hence, taking (21) into account, we get the inequality $\langle b, v'' - y_i \rangle \leqslant -\delta$ for all $b \in B_i$, $i \in K'$. Then it follows from (22) that $\langle b, y_l - y_p \rangle \geqslant \beta_l \delta$ for all $b \in B_l$, or

$$\| y_l - y_p \| \geqslant \beta_l \delta,$$

because $\| b \| = 0 \; \forall b \in B_l$. Since the sequence $\{ y_i \}$, $i \in K'$, is convergent, from the last inequality we have $\beta_i \to 0$, $i \in K'$. Hence, adding condition (21), we obtain the equality

$$\lim_{i \in K'} \| v_i'' - y_i \| = 0. \tag{23}$$

Further, according to step 5 of the method $z_i'' - y_i = q_i'' \left(v_i'' - y_i \right)$, $q_i'' \in [1, q]$, $i \in K'$. Therefore, by virtue of (23)

$$\lim_{i \in K'} \| z_i'' - y_i \| = 0.$$

But $z_i'' \in D$, $i \in K'$. Hence, taking into account the fact that the set D is closed, (20) follows. The lemma is proved.

Lemma 6. *Let $\{ z_i' \}$, $i \in K' \subset K$, is a convergent subsequence of the sequence $\{ z_i' \}$, $i \in K$, and z' is its limit point. Then*

$$z' \in \operatorname{epi} \left(f, R_n \right). \tag{24}$$

Proof. We assume that $z_i' = (\omega_i, \eta_i)$, $i \in K$, $z' = (\omega', \eta')$, where $\omega_i, \omega' \in R_n$, $\eta_i, \eta' \in R_1$, i.e.

$$\omega_i \to \omega' \text{ and } \eta_i \to \eta' \text{ for } i \to \infty, \quad i \in K'.$$

According to step 2 of the method, we have we have inclusions $z' \in \operatorname{epi} \left(F_i, R_n \right)$ and hence inequalities $F_i \left(\omega_i \right) \leqslant \eta_i$ or $f \left(\omega_i \right) + P_i \left(\omega_i \right) \leqslant \eta_i$ for all $i \in K$. But $P_i \left(\omega_i \right) \geqslant 0$, $i \in K$. Then $f \left(\omega_i \right) \leqslant \eta_i$, $i \in K'$, which means that $f \left(\omega' \right) \leqslant \eta'$. Consequently, the inclusion (24) is proved.

Lemma 7. *If $\bar{u}$ is a limit point of the sequence $\{u_i\}$, $i \in K$, then*

$$\bar{u} \in \mathrm{epi}\,(f, D)\,. \tag{25}$$

Proof. Let $\bar{u} = (\bar{y}, \bar{\gamma})$ be a limit point of a convergent subsequence $\{u_i\}$, $i \in K' \subset K$. Select a convergent subsequence $\{z_i'\}$, $i \in K'' \subset K'$, from the sequence $\{z_i'\}$, $i \in K'$. Let z' is its limit point. Then, by Lemma 6, the inclusion (24) holds for this point. On the other hand, according to the consequence to Lemma 3, $\lim\limits_{i \in K''} \|z_i' - u_i\| = 0$. Hence, $\bar{u} = (\bar{y}, \bar{\gamma}) \in \mathrm{epi}\,(f, R_n)$ and

$$f\,(\bar{y}) \leqslant \bar{\gamma}.$$

At the same time, according to Lemma 5, $\bar{y} \in D$. Thus, taking into account the last inequality, we have the inclusion (25). The lemma is proved.

Theorem 2. *Let $\{(x_k, \sigma_k)\}$, $k \in K_1 \subset K$, be a convergent subsequence of the sequence $\{(x_k, \sigma_k)\}$, $k \in K$, and $\bar{u} = (\bar{x}, \bar{\sigma})$ is its limit point. Then*

$$\bar{x} \in X^*, \quad \bar{\sigma} = f^*.$$

Proof. We recall that according to (7)

$$(x_k, \sigma_k) = (y_{i_k}, \gamma_{i_k}) = u_{i_k}, \ k \in K.$$

Then, by the hypothesis of the theorem, $u_{i_k} \to \bar{u}$, $k \in K_1$ and for the point $\bar{u} = (\bar{x}, \bar{\sigma})$, by Lemma 7, we have the inclusion (25). It means that $\bar{x} \in D$, $f\,(\bar{x}) \leqslant \bar{\sigma}$. But by the condition (9) $\bar{\sigma} \leqslant f^*$. Hence, $f\,(\bar{x}) \leqslant \bar{\sigma} \leqslant f^* \leqslant f\,(\bar{x})$, i.e. $f\,(\bar{x}) = \bar{\sigma} = f^*$. The theorem is proved.

5   Conclusion

Earlier, the authors proposed cutting-plane methods [16,17] for the solution of problem (1), in which a similar idea of immersing the epigraphs of auxiliary functions into polyhedral sets is used. The method proposed in this paper differs from the methods [16,17] in that it additionally implements the procedure for approximating the set of constraints of the original problem by polyhedron. Numerical calculations have shown that the successive approximation of the set D by the sets D_i improves the rate of convergence in comparison with another version of the method, where $D_i = D_0 \, \forall\, i > 0$.

In addition, according to the calculations for large values of i the iterative points y_i often belong to D. On such iterations, it is possible to estimate the proximity of the current value $f(y_i)$ to the optimal value f^*, since, taking into account condition (9), inequality holds

$$\gamma_i \leqslant f^* \leqslant f\,(y_i)\,.$$

Suppose that the function $f(x)$ is strongly convex on the set D with the constant of strong convexity μ. Then the points $x_k \in D$ satisfy the inequality

$$\frac{1}{2}\mu\|x_k - x^*\|^2 \leqslant f(x_k) - f^*$$

([11], p. 207). Since, according to (7), (9), $\sigma_k \leqslant f^*$, then $\frac{1}{2}\mu\|x_k - x^*\|^2 \leqslant f(x_k) - \sigma_k$. Hence, for the indicated x_k the estimate follows

$$\|x_k - x^*\| \leqslant \sqrt{\frac{2\varepsilon_k}{\mu}},$$

where $\varepsilon_k = f(x_k) - \sigma_k$.

Below we make a few more remarks concerning the choice of functions $P_i(x)$ in the proposed method.

Note that, under condition

$$F_i(y_i) - \gamma_i \leqslant \Delta_k$$

for the point u_i, method makes it possible to set $P_{i+1}(x) = P_i(x)$ at the step 8 for all $x \in R_n$. Taking into account the point 7 of the method, this means that it is permissible to assume

$$P_i(x) = P_0(x), \quad F_i(x) = F_0(x) \quad \forall i \in K \tag{26}$$

regardless of the condition (6) in the method. Thus, the method admits such an implementation, where the penalty, and therefore the auxiliary functions, do not change at all during the computational process. In this case, the sets M_i approximate the epigraph of the initial function $F_0(x)$ for all $i \in K$. The possibility of specifying functions $P_i(x)$, $F_i(x)$ in the form (26) essentially distinguishes the proposed method from both the penalty method and the cutting-plane methods [16, 17].

The point is that the theorem of convergence of the method is proved without any additional requirements to the sequence $\{\Delta_k\}$, except for its positivity. In particular, it is permissible to set

$$\Delta_k = \Delta > 0 \forall k \in K.$$

If we assume that Δ is arbitrarily large, then $F_i(y) - \gamma_i < \Delta_k$ for all $i \in K$, $k \in K$. In this case, for each $i \in K$, step 8 of the method is performed, $x_k = y_i$, $\sigma_k = \gamma_i$, and the penalty function $P_{i+1}(x)$ can be chosen with conditions

$$P_{i+1}(x) = 0 \quad \forall x \in D, \qquad P_{i+1}(x) > P_i(x) \quad \forall x \notin D.$$

Indeed, the functions $P_{i+1}(x)$ and $F_{i+1}(x)$ will differ, respectively, from $P_i(x)$ and $F_i(x)$ for all $i \in K$.

Now let the numbers Δ_k be chosen in the method in such a way that

$$\Delta_k \to 0 \quad k \to \infty,$$

and the functions $P_i(x)$ are given with the additional condition

$$P_i(x) \to +\infty, \quad i \to \infty, \quad \forall x \notin D.$$

Let $F_i^* = \min\{F_i(x) : x \in D_i\}$. Since the inclusions $y_i \in D_i$, $M_i \supset \mathrm{epi}(F_i, R_n)$, $i \in K$, are fulfilled, the inequalities

$$\gamma_i \leqslant F_i^* \leqslant F_i(y_i)$$

are valid. Consequently, the condition (6) can serve as a test of the closeness of the values $F_i(y_i)$ and F_i^* As long as, (6) is satisfied for fixed Δ_k, the functions $P_i(x)$, $F_i(x)$ do not change. According to Lemma 4, an inequality $F_i(y_i) - \gamma_i \leqslant \Delta_k$ holds for some $i = i_k$, and the point $y_{i_k} = x_k$ can be regarded as an approximate solution of problem $\min\{F_{i_k}(x) : x \in D_{i_k}\}$. Such an algorithm of the proposed method can be adopted as an implementation of the classical method of penalty functions.

References

1. Bulatov, V.P.: Embedding methods in optimization problems. Nauka, Novosibirsk, vol. 161 (1977). (in Russian)
2. Bulatov, V.P., Khamisov, O.V.: Cutting methods in E^{n+1} for global optimization of a class of functions. Zh. Vychisl. Mat. Mat. Fiz. **47**(11), 1830–1842 (2007). (in Russian)
3. Zabotin, I.Ya., Yarullin, R.S.: A cutting-plane method based on epigraph approximation with discarding the cutting planes. Autom. Remote Control **76**(11), 1966–1975 (2015)
4. Zabotin, I.Ya., Yarullin, R.S.: A cutting-plane method without inclusions of approximating sets for conditional minimization. Lobachevskii J. Math. **36**(2), 132–138 (2015)
5. Nesterov, Yu.E.: Introduction to convex optimization. Moscow, MCCME, 274 (2010). (in Russian)
6. Nurminski, E.A.: A separating plane algorithm with limited memory for convex nonsmooth optimization. Vychisl. Metody Programm. **7**(1), 133–137 (2006). (in Russian)
7. Polyak, B.T.: Introduction to Optimization, 438 pp. Optimization Software Inc., New York (1987)
8. Demyanov, V.F., Vasilev, L.V.: Nondifferentiable Optimization, 384 pp. Nauka, Moscow (1981). (in Russian)
9. Zabotin, I.Y., Shulgina, O.N., Yarullin, R.S.: A minimization method with approximation of feasible set and epigraph of objective function. Russ. Math. (Iz. VUZ) **60**(11), 78–81 (2016)
10. Zabotin, I.Y., Shulgina, O.N., Yarullin, R.S.: A minimization algorithm with approximation of an epigraph of the objective function and a constraint set. In: Proceedings of the DOOR, Vladivostok, Russia, CEUR-WS, 19–23 September 2016, vol. 1623, pp. 321–324 (2016)
11. Vasilev, F.P.: Optimization methods, 620 p. MCCME, Moscow (2011). (in Russian)
12. Karmanov, V.G.: Mathematical Programming, 288 p. Nauka, Moscow (1986). (in Russian)

13. Konnov, I.V.: Nonlinear Optimization and Variational Inequalities, 508 pp. Kazan (2013). (in Russian)
14. Zabotin, I.Ya., Shulgina, O.N., Yarullin, R.S.: A cutting method and construction of mixed minimization algorithms on its basis. Uchenye Zapiski Kazanskogo Universiteta., Ser. Fiz.-Matem. Nauki **156**(4), 14–24 (2014). (in Russian)
15. Zabotin, I.Ya.: Some embedding-cutting algorithms for mathematical programming problems. Izv. Irkutsk. Gos. Univ., Ser. Matem. **4**(2), 91–101 (2011). (in Russian)
16. Zabotin, I.Ya., Kazaeva, K.E.: One cutting plane algorithm using auxiliary functions. In: IOP Conference Series: Materials Science and Engineering (MSE), vol. 158(1) (2016)
17. Zabotin, I.Ya., Kazaeva, K.E.: Cutting-plane method with embedding of epigraphs of auxiliary functions. In: Constructive Nonsmooth Analysis and Related Topics (dedicated to the memory of V.F. Demyanov) (St. Petersburg, Russia, 22–27 May). IEEE (2017). https://doi.org/10.1109/CNSA.2017.7974033

Game Theory and Economical Applications

Sorger Game Under Uncertainty: Discrete Case

Natalia V. Adukova[1(✉)] and Konstantin N. Kudryavtsev[1,2]

[1] South Ural State University, Chelyabinsk, Russia
`adnatasha94@mail.ru`, `kudrkn@gmail.com`
[2] Chelyabinsk State University, Chelyabinsk, Russia

Abstract. At the present time, the discrete-time models are not given enough attention. But these models are more realistic than the continuous models, because the allocation of funds is discrete. In the paper a new discrete model of optimal advertising is proposed. This model takes into account the uncertainties. These uncertainties are caused by acts of a set of the small companies. The companies' problem is to maximize their market share taking into account the reaction of competitors. The problem is a discrete multistep optimal control problem. For this model the optimal control problem is solved explicitly. The Bellman method of dynamic programming is used to construct the guaranteed equilibrium.

Keywords: Strong guaranteed equilibrium
Multistage positional two-player game
Optimal control of advertising

1 Introduction

Many industries are characterized by firms competing for the market share mainly on the basis of advertising. The markets for cola drinks, beer and cigarettes are some examples that have been studied in [1, 2]. The goal of a firm's advertising is to increase its market share while the advertising of competitors act to reduce the market share of the firm. The interaction of companies is usually considered as a differential game, were the each company tries to increase its market share taking into account the reaction of competitors. The strategies of companies are the allocation of funds for advertising.

However, the financing of the advertising action is discrete. Therefore the multistage discrete games are more realistic.

One of the first models that describes an influence of advertising costs on the market share of a monopolistic firm was proposed by Vidale and Wolfe in 1957 [3]. In the model, the dynamics of the sales rate $s(t)$ is given by the following differential equation

$$\dot{s}(t) = \gamma u(t)[m - s(t)] - \delta s(t), \quad s(t_0) = s_0, \tag{1}$$

© Springer International Publishing AG, part of Springer Nature 2018
A. Eremeev et al. (Eds.): OPTA 2018, CCIS 871, pp. 207–219, 2018.
https://doi.org/10.1007/978-3-319-93800-4_17

where $u(t)$ is the advertising effort, i.e. a control variable. Here m is a saturation level of the sale rate, γ is a response constant characterizing efficiency of advertising, and δ is a decay constant that determines the rate at which consumers are lost due to product obsolescence.

The classical monopoly model having numerous applications was suggested by Sethi in 1983 [4]. Here the dynamics of the market share $x(t)$ is given by the non-linear differential equation

$$\dot{x}(t) = \rho u(t)\sqrt{1 - x(t)} - \delta x(t), x(t_0) = x_0. \tag{2}$$

Equation (2) was generalized on oligopoly by Sorger [5]. The further development of the Sorger model was made in [6]. The discrete models of optimal advertising were considered in [7–10].

In the present work, we consider the discrete oligopoly model. In this model competitive firms fight for a market share using the advertising.

In addition to the main players in the market, there are usually some small sellers and producers. They do not take part in model as players. They can have different purposes. We will consider their acts as the uncontrollable factors. Below we regard their acts as the uncertainty.

2   Games in Discrete-Time Under Uncertainty

In this section we will follow [11] and will consider a multistage positional two-player game in discrete-time under uncertainty

$$\langle \{1,2\}, \Sigma, \{\mathfrak{U}_i\}_{i=1,2}, \mathcal{Z}, \{\mathcal{J}_i(U, Z, t_0, x_0)\}_{i=1,2}\rangle. \tag{3}$$

Here player 1 and player 2 are denoted by $\{1,2\}$, Σ is an operated system with increasing time $t = t_0, t_0 + 1, \ldots, T - 1$, where T is a positive integer. The dynamic of Σ is described by difference vector equation

$$x(t+1) = F(x(t), u_1(t), u_2(t), z(t)), \tag{4}$$

where $x \in R^n$ is the phase vector, the pair $(t, x) \in T \times R^n$ is the position of the game at the time t, $(t_0, x_0) \in \{0, 1, \ldots, T - 1, T\} \times R^n$ is the starting position. Denote by $u_i \in R^n$ the control action of the player i and by $z \in R^n$ an uncertain factor.

The strategy $U_i(t)$ of the player i at the time $t = 0, 1, \ldots, T - 1$ depends on the time t and the position x. $U_i(t)$ is positional because it depends on the realizing position (t, x) of the game.

The set of strategies for the player i in the game (3) with initial position (t_0, x_0) will be denote by

$$\mathfrak{U}_i = \{U_i = (U_i(t_0), U_i(t_0 + 1), \ldots, U_i(T - 1) \,|\, U_i(t)$$
$$\div u_i(t, x), t = t_0, \ldots, T - 1, x \in R^n\}.$$

Also we will use the set $\mathfrak{U} = \mathfrak{U}_1 \times \mathfrak{U}_2$ of situations $U = (U_1, U_2)$.

The uncertainty $Z(t)$ at the time $t = 0, 1, \ldots, T-1$ will be identified with the n-vector function $z(t, x, u_1, u_2)$. The uncertainty $Z(t)$ at the time t forms on the basis of knowledge about game position (t, x) and the realization of the strategy of the both players at this point. In other words it means that the uncertainty forms in the class of counter-strategies. The set of uncertainties in the game (3) with initial position (t_0, x_0) well be denoted by

$$\mathcal{Z} = \{Z = (Z(t_0), Z(t_0 + 1), \ldots, Z(T-1) \mid Z(t) \div z(t, x, u_1, u_2)\}.$$

The party of the game (3) develops by the following way. Let (t_0, x_0) be the initial position. The players chose it and use their specific strategies $U_i^* \in \mathfrak{U}_i$ $(i = 1, 2), U_i^* \div (u_i^*(t_0), \ldots, u_i^*(T-1))$. In the game the specific uncertainty is realized on base of these strategies

$$Z^* \in \mathcal{Z}, \quad Z^* = (Z^*(t_0), \ldots, Z^*(T-1)).$$

Then system (4) takes the form

$$x(t + 1) = F(t, u_1^*(t), u_2^*(t), z^*(t)), \quad t = t_0, \ldots, T-1.$$

and we obtain the following sequences: the sequence of values of the phase vectors $\{x(T)\}_{t=t_0}^{T}$, the two sequences of realizations of the strategies U_i^* $(i = 1, 2)$ that are chosen by the players $\{u_i[t]\}_{t=t_0}^{T-1}$ $(i = 1, 2)$, and the sequence of realization of uncertainties $\{z[t]\}_{k=t_0}^{T}$.

By means of these sequences we introduce the payoff function for i-th player

$$\mathcal{J}_i(U^*, Z^*, t_0, x_0) = \sum_{t=0}^{T} G(x(t), u_1[t], u_2[t], z[t]). \tag{5}$$

The value of this function is the payoff for this player.

The goal of i-th player is to choose his strategy $U_i \in \mathfrak{U}_i$ in such a way that the payoff of i-th player will be the largest. Moreover, the player does not have to unite in the coalition with the other player. In this situation the both players taking into account the realization of any uncertainty $Z \in \mathcal{Z}$.

The formation of the strong guaranteed equilibrium occurs in two stages [11].

I stage: The interior minimum
For each i $(i = 1, 2)$ we fix the situation U^* in the game (3) and solve the problem

$$\min_{Z \in \mathcal{Z}} \mathcal{J}_i(U^*, Z, t_0, x_0) = \mathcal{J}_i(U^*, Z^{(i)}, t_0, x_0).$$

II stage: The external Nash equilibrium
Now we construct "the game of guarantees" which follows from (3). Here each player $i(i = 1, 2)$ is taking into account the realization of the worst uncertainty $Z^{(i)}$ for him.

$$\langle \{1, 2\}, \{\Sigma(Z = Z^{(i)})\}_{i=1,2}, \{\mathfrak{U}_i\}_{i=1,2}, \{\mathcal{J}_i(U_1, U_2, Z^{(i)}, t_0, x_0)\}_{i=1,2} \rangle. \tag{6}$$

In the game (6) we find the Nash equilibrium situation $U^e = (U_1^e, U_2^e)$ from the equalities

$$\min_{U_1 \in \mathfrak{U}_1} \mathcal{J}_1(U_1, U_2^e, Z^{(i)}, t_0, x_0) = \mathcal{J}_1(U^e, Z^{(i)}, t_0, x_0) = \mathcal{J}_1^e[t_0, x_0],$$

$$\min_{U_2 \in \mathfrak{U}_2} \mathcal{J}_2(U_1^e, U_2, Z^{(i)}, t_0, x_0) = \mathcal{J}_2(U^e, Z^{(i)}, t_0, x_0) = \mathcal{J}_2^e[t_0, x_0].$$

The pair $U^e, \mathcal{J}^e[t_0, x_0] = (\mathcal{J}_1^e[t_0, x_0], \mathcal{J}_2^e[t_0, x_0])$ will be called the strongly guaranteed Nash equilibrium for the game (3). U^e is called by the strongly guaranteed strategy profile and $\mathcal{J}_i^e[t_0, x_0]$ is the strongly guaranteed payoff for i-th player ($i = 1, 2$).

3   Model

Let us consider the problem of finding the optimal cost of the advertising effort for a competitive market. There are two competitive firms that occupy only a part of the market. The remaining part of the market is occupied by small companies. We can not predict their strategy and the stochastic features of their behaviour are unknown. We will interpret this uncertainty as the availability of the third dummy player on the market. This player does not have its own profit functional. In this case we can describe the sales dynamics by two equations from the system of non-linear difference equations in Prasad-Sethi model for three competitive firms.

$$\begin{cases} x_1(t+1) = (1-\delta)x_1(t) + \rho_1 u_1(t)\sqrt{1-x_1(t)} - \frac{\rho_2}{2}u_2(t)\sqrt{1-x_2(t)} \\ \qquad\quad - \frac{\rho_3}{2}z(t)\sqrt{x_1(t) + x_2(t)} + \frac{\delta}{3}, \\ x_2(t+1) = (1-\delta)x_2(t) + \rho_2 u_2(t)\sqrt{1-x_2(t)} - \frac{\rho_1}{2}u_1(t)\sqrt{1-x_1(t)} \\ \qquad\quad - \frac{\rho_3}{2}z(t)\sqrt{x_1(t) + x_2(t)} + \frac{\delta}{3}. \end{cases} \quad (7)$$

Here $t = 1, \ldots, T$ is the time when decisions were made (steps of dynamic process); T is the number of the time of decision acceptance (planing horizon); $\delta \in [0; 1]$ is a market share decay expenditure rate at the time t; ρ_1, ρ_2 are response constants; $x_1(t), x_2(t)$ are market shares at the time t; $u_1(t), u_2(t)$ are advertising expenditure rates at the time t; the control $z(t)$ for the third dummy player is the uncertainty in this model.

The quality of companies functionals is determined by the following functionals

$$J_1 = \frac{m_1 x_1(t)}{(1+r)^t} + \sum_{k=0}^{T-1} \frac{m_1 x_1(k) - \frac{c}{2}u_1^2(k) + \frac{1}{2}z^2(k)}{(1+r)^k}, \quad (8)$$

$$J_2 = \frac{m_2 x_2(t)}{(1+r)^t} + \sum_{k=0}^{T-1} \frac{m_2 x_2(k) - \frac{c}{2}u_2^2(k) + \frac{1}{2}z^2(k)}{(1+r)^k}, \quad (9)$$

where m_1, m_2 are revenues potentials (a margin per unit product), r is the discount rate, c is the coefficient characterizing the cost of advertising.

It is assumed that each of the competitive firms optimizes their advertising costs. We will follow the principle of Hermeyer and we will assume that when the player is making the decision he has to orientate on the maximum opposition to the uncertainty.

For the solution of the problem we will use the notion of the strong guaranteed equilibrium which described in the previous section.

4   The Construction of the Strongly Guaranteed Nash Equilibrium

First of all we transcribe the problem in the matrix form.

Introduce vectors:

$$X(t) = \begin{pmatrix} x_1(t) \\ x_2(t) \end{pmatrix}, \quad U(t) = \begin{pmatrix} u_1(t) \\ u_2(t) \end{pmatrix}, \quad D = \frac{1}{3} \begin{pmatrix} \delta \\ \delta \end{pmatrix}, \quad Z(t) = \begin{pmatrix} z(t) \\ z(t) \end{pmatrix}.$$

Then the initial system of equations takes this form:

$$X(t+1) = (1-\delta)X(t) + \begin{pmatrix} \rho_1 \sqrt{1-x_1(t)} & -\frac{\rho_2 \sqrt{1-x_2(t)}}{2} \\ -\frac{\rho_1 \sqrt{1-x_1(t)}}{2} & \rho_2 \sqrt{1-x_2(t)} \end{pmatrix} U(t)$$
$$- \frac{\rho_3}{2} \sqrt{x_1(t) + x_2(t)} Z(t) + D.$$

We represent the matrix which acts on the vector U in the form:

$$\begin{pmatrix} \rho_1 \sqrt{1-x_1(t)} & -\frac{\rho_2 \sqrt{1-x_2(t)}}{2} \\ -\frac{\rho_1 \sqrt{1-x_1(t)}}{2} & \rho_2 \sqrt{1-x_2(t)} \end{pmatrix}$$
$$= \begin{pmatrix} 1 & -\frac{1}{2} \\ -\frac{1}{2} & 1 \end{pmatrix} \begin{pmatrix} \rho_1 \sqrt{1-x_1(t)} & 0 \\ 0 & \rho_2 \sqrt{1-x_2(t)} \end{pmatrix}$$
$$= \begin{pmatrix} 1 & -\frac{1}{2} \\ -\frac{1}{2} & 1 \end{pmatrix} \begin{pmatrix} \rho_1 & 0 \\ 0 & \rho_2 \end{pmatrix} \begin{pmatrix} \sqrt{1-x_1(t)} & 0 \\ 0 & \sqrt{1-x_2(t)} \end{pmatrix}.$$

Introduce matrices: $K = \begin{pmatrix} 1 & -\frac{1}{2} \\ -\frac{1}{2} & 1 \end{pmatrix}, \rho = \begin{pmatrix} \rho_1 & 0 \\ 0 & \rho_2 \end{pmatrix}.$

Aside from that we denote $\begin{pmatrix} \sqrt{1-x_1(t)} & 0 \\ 0 & \sqrt{1-x_2(t)} \end{pmatrix} = \sqrt{I - x(t)}$, where

$$x(t) = \begin{pmatrix} x_1 & 0 \\ 0 & x_2 \end{pmatrix}.$$

If we introduce the diagonal matrix

$$I - x(t) = \begin{pmatrix} 1 - x_1(t) & 0 \\ 0 & 1 - x_2(t) \end{pmatrix},$$

it will conform with the matrix function root from the diagonal matrix.

Thus the initial system in the matrix form takes the following form:

$$X(t+1) = (1-\delta)X(t) + K\rho\sqrt{I - x(t)}U(t) - \frac{\rho_3}{2}\sqrt{x_1(t) + x_2(t)}Z(t) + D. \quad (10)$$

Also we write functionals (8), (9) in the vector form

$$J = \begin{pmatrix} J_1 \\ J_2 \end{pmatrix} = \frac{1}{(1+r)^T} \begin{pmatrix} m_1 & 0 \\ 0 & m_2 \end{pmatrix} X(T)$$

$$+ \sum_{k=0}^{T-1} \left[\frac{1}{(1+r)^k} \begin{pmatrix} m_1 & 0 \\ 0 & m_2 \end{pmatrix} X(k) - \frac{c}{2(1+r)^k} \begin{pmatrix} u_1^2(k) \\ u_2^2(k) \end{pmatrix} + \frac{1}{2(1+r)^k} \begin{pmatrix} z^2(k) \\ z^2(k) \end{pmatrix} \right].$$

Introduce matrix

$$m = \begin{pmatrix} m_1 & 0 \\ 0 & m_2 \end{pmatrix} \text{ and vector } U^2(k) = \begin{pmatrix} u_1^2(k) \\ u_2^2(k) \end{pmatrix} \text{ (note that here and below } U^2,$$

Z^2 are only the notations without any vectors multiplication).

Than

$$J = \frac{mX(T)}{(1+r)^T} + \sum_{k=0}^{T-1} \left[\frac{mX(k)}{(1+r)^k} - \frac{cU^2(k)}{2(1+r)^k} + \frac{Z^2(k)}{2(1+r)^k} \right].$$

We introduce the Bellman vector function on the step k:

$$V^{(T)} = \begin{pmatrix} V_1^{(T)} \\ V_2^{(T)} \end{pmatrix} = \frac{mX(T)}{(1+r)^T}.$$

For uniformity we write $V^{(T)}$ in the form:

$$V^{(T)} = \frac{m}{(1+r)^T}\left(\alpha(T)X(T) + \beta(T)\right),$$

where $\alpha(T) = I_2 = \begin{pmatrix} 1 & 0 \\ 0 & 1 \end{pmatrix}$, $\beta(T) = 0 = \begin{pmatrix} 0 \\ 0 \end{pmatrix}$.

We recursively denote the auxiliary vector function $W^{(k)}$ and the vector function $V^{(k)}$:

$$W^{(k)} = \begin{pmatrix} W_1^{(k)} \\ W_2^{(k)} \end{pmatrix} = \frac{mX(k)}{(1+r)^k} - \frac{cU^2(k)}{2(1+r)^k} + \frac{Z^2(k)}{2(1+r)^k} + V^{(k+1)}, k = T, \ldots, 1, 0,$$

where

$$V^{(k)} = \begin{pmatrix} \max_{u_1(k)} \min_{z(k)} W_1^{(k)} \\ \max_{u_2(k)} \min_{z(k)} W_2^{(k)} \end{pmatrix}.$$

In order to further differentiate each vector coordinate on its argument we will use the projections operators on the first or the second coordinate:

$$pr_1 W^{(k)} = W_1^{(k)}, \quad pr_2 W^{(k)} = W_2^{(k)}.$$

It is similar for other vectors.

Theorem 1. *The sequence of Bellman functions and the sequence of optimal controls and the sequence of the uncertainty are obtained by the following recurrent formulas*

$$V^{(k)} = \frac{m}{(1+r)^T} \big[\alpha_k X(k) + \beta_k\big], \tag{11}$$

$$U^*(k) = \begin{pmatrix} u_1^*(k) \\ u_2^*(k) \end{pmatrix} = \frac{m\rho\sqrt{I - x(k)}}{c(1+r)^{T-k}} \begin{pmatrix} \alpha(k+1)_{11} - \frac{\alpha(k+1)_{12}}{2} \\ -\frac{\alpha(k+1)_{21}}{2} + \alpha(k+1)_{22} \end{pmatrix}, \tag{12}$$

$$Z^*(k) = \begin{pmatrix} z_1^*(k) \\ z_2^*(k) \end{pmatrix} = \frac{m\rho_3\alpha(k+1)}{2(1+r)^{T-k}} \sqrt{x_1(k) + x_2(k)} \begin{pmatrix} 1 \\ 1 \end{pmatrix}, \tag{13}$$

where

$$\alpha(k) = (1+r)^{T-k}I + (1-\delta)\alpha(k+1) - \frac{\rho_3^2}{4}\alpha(k+1)m\alpha(k+1)K_1 \tag{14}$$

$$+ \frac{m\rho^2}{2c(1+r)^{T-k}} \begin{pmatrix} (\alpha_{11}(k+1) - \frac{\alpha_{12}(k+1)}{2})^2 & 0 \\ 0 & (-\frac{\alpha_{21}(k+1)}{2} + \alpha_{22}(k+1))^2 \end{pmatrix}$$

$$+ \frac{m\rho_3^2}{8(1+r)^{T-k}} \begin{pmatrix} (\alpha_{11}(k+1) + \alpha_{12}(k+1))^2 & 0 \\ 0 & (\alpha_{21}(k+1) + \alpha_{22}(k+1))^2 \end{pmatrix} K_1$$

$$- \frac{\alpha(k+1)Km\rho^2}{c(1+r)^{T-k}} \begin{pmatrix} \alpha_{11}(k+1) - \frac{\alpha_{12}(k+1)}{2} & 0 \\ 0 & -\frac{\alpha_{21}(k+1)}{2} + \alpha_{22}(k+1) \end{pmatrix},$$

$$\beta(k) = \beta(k+1) + \frac{\alpha(k+1)\delta}{3}\begin{pmatrix} 1 \\ 1 \end{pmatrix} \tag{15}$$

$$- \frac{m\rho^2}{2c(1+r)^{T-k}} \begin{pmatrix} (\alpha_{11}(k+1) - \frac{\alpha_{12}(k+1)}{2})^2 & 0 \\ 0 & (-\frac{\alpha_{21}(k+1)}{2} + \alpha_{22}(k+1))^2 \end{pmatrix}\begin{pmatrix} 1 \\ 1 \end{pmatrix}$$

$$+ \frac{\alpha(k+1)Km\rho^2}{c(1+r)^{T-k}} \begin{pmatrix} \alpha_{11}(k+1) - \frac{\alpha_{12}(k+1)}{2} & 0 \\ 0 & -\frac{\alpha_{21}(k+1)}{2} + \alpha_{22}(k+1) \end{pmatrix}\begin{pmatrix} 1 \\ 1 \end{pmatrix}.$$

Proof. Let us define the Bellman vector function on the step $k = T - 1$.

First of all we search $\min W_1^{(T-1)} = pr_1 W^{(T-1)}$ $\min W_2^{(T-1)} = pr_2 W^{(T-1)}$ on the variable $z(T-1)$.

We will find the partial derivative from the vector function $W^{(T-1)}$ on the variable

$$\left(W^{(T-1)}\right)'_{z(T-1)} = \frac{Z(T-1)}{(1+r)^{T-1}} + \left(V^{(T)}\right)'_{z(T-1)}$$

$$= \frac{Z(T-1)}{(1+r)^{T-1}} + \frac{m}{(1+r)^T} \left(X\right)'_{z(T-1)}$$

$$= \frac{Z(T-1)}{(1+r)^{T-1}} - \frac{m\rho_3}{2(1+r)^T} \sqrt{x_1(T-1) + x_2(T-1)} \begin{pmatrix} 1 \\ 1 \end{pmatrix}.$$

Here we are using that $\left(Z^2\right)' = \frac{1}{2}Z$, it is obviously.

We find the critical points $z_1^*(T-1)$ for $W_1^{(T-1)}$ and $z_2^*(T-1)$ for $W_2^{(T-1)}$ from the vector equation $(W^{(T-1)})'_{z(T-1)} = 0$. They give the vector

$$Z^*(T-1) = \begin{pmatrix} z_1^*(T-1) \\ z_2^*(T-1) \end{pmatrix} = \frac{\rho_3 m}{2(1+r)} \sqrt{x_1(T-1) + x_2(T-1)} \begin{pmatrix} 1 \\ 1 \end{pmatrix}.$$

Thus, it is obviously that

$$Z^*(T-1) = \frac{\rho_3}{2(1+r)} \sqrt{x_1(T-1) + x_2(T-1)} \begin{pmatrix} m_1 \\ m_2 \end{pmatrix}.$$

Since $\left(W_{1,2}^{(T-1)}\right)''_{z(T-1)} > 0$ so the point $z_1^*(T-1)$ for the function W_1^{T-1} and the point $z_2^*(T-1)$ for the function W_2^{T-1} are really the minima w.r.t. variable $z(T-1)$.

Substitute $Z^*(T-1)$ in the formulas for $W^{(T-1)}$

$$W^{(T-1)}\big|_{Z(T-1)=Z^*(T-1)} = \frac{mX(T-1)}{(1+r)^{T-1}}$$

$$-\frac{cU^2(T-1)}{2(1+r)^{T-1}} + \frac{\rho_3(x_1(T-1)+x_2(T-1))}{8(1+r)^{T+1}} \begin{pmatrix} m_1^2 \\ m_2^2 \end{pmatrix} + V^{(T)}\big|_{Z(T-1)=Z^*(T-1)}$$

$$= \frac{\rho_3^2 m^2}{8(1+r)^{T+1}} \begin{pmatrix} 1 & 1 \\ 1 & 1 \end{pmatrix} X(T-1) + V^{(T)}(Z^*(T-1)).$$

Now we find $\max_{u_1(T-1)} W_1^{(T-1)}$. For this we have to find the partial derivative for the vector function $W^{(T-1)}$ under the variable $u_1(T-1)$:

$$(W^{(T-1)})'_{u_1(T-1)} = -\frac{c}{(1+r)^{T-1}} \begin{pmatrix} u_1(T-1) \\ 0 \end{pmatrix} + (V^{(T)})'_{u_1(T-1)}(Z^*(T-1)).$$

So

$$(V^{(T)})'_{u_1(T-1)} = \frac{m}{(1+r)^T}(X(T))'_{u_1(T-1)} = \frac{m}{(1+r)^T}\left[K\rho\sqrt{1-x(T-1)}\begin{pmatrix} 1 \\ 0 \end{pmatrix}\right],$$

then

$$(W^{(T-1)})'_{u_1(T-1)}$$

$$= -\frac{c}{(1+r)^{T-1}} \begin{pmatrix} u_1(T-1) \\ 0 \end{pmatrix} + \frac{m}{(1+r)^T}\left[K\rho\sqrt{1-x(T-1)}\begin{pmatrix} 1 \\ 0 \end{pmatrix}\right].$$

It easy to see that

$$\frac{m}{(1+r)^T}\left[K\rho\sqrt{1-x(T-1)}\begin{pmatrix} 1 \\ 0 \end{pmatrix}\right]$$

$$= -\frac{c}{(1+r)^{T-1}} \begin{pmatrix} u_1(T-1) \\ 0 \end{pmatrix} + \frac{1}{(1+r)^T}\begin{pmatrix} m_1\rho_1\sqrt{1-x_1(T-1)} \\ * \end{pmatrix}.$$

In the critical points for $W_1(T-1)$ it should be

$$pr_1(W^{(T-1)})'_{u_1(T-1)} = 0.$$

So we get

$$u_1^*(T-1) = \frac{1}{c(1+r)} m_1 \rho_1 \sqrt{1 - x_1(T-1)}.$$

Similarly we find

$$u_2^*(T-1) = \frac{1}{c(1+r)} m_2 \rho_2 \sqrt{1 - x_2(T-1)}$$

and so

$$U^*(T-1) = \frac{m\rho}{c(1+r)} \sqrt{I - x(T)} \begin{pmatrix} 1 \\ 1 \end{pmatrix}.$$

We need $Z^{*2}(T-1)$ and $U^{*2}(T-1)$. Calculations show that

$$Z^{*2}(T-1) = \frac{\rho_3^2 m^2}{4(1+r)^2} \begin{pmatrix} 1 & 1 \\ 1 & 1 \end{pmatrix} X(T-1)$$

and

$$U^{*2}(T-1) = \frac{m^2\rho^2}{c^2(1+r)^2} \left[\begin{pmatrix} 1 \\ 1 \end{pmatrix} - X(T-1) \right].$$

Now we can find

$$V^{(T-1)} = W^{(T-1)} \bigg|_{\substack{Z(T-1)=Z^*(T-1) \\ U(T-1)=U^*(T-1)}}.$$

When we substitute the critical points we obtain

$$V^{(T-1)} = \frac{m}{(1+r)^T} \left(\left[[(1+r) + (1-\delta)]I + \frac{(I-K)m\rho^2}{c(1+r)} - \frac{m\rho_3^2}{8(1+r)} \begin{pmatrix} 1 & 1 \\ 1 & 1 \end{pmatrix} \right] X(T-1) \right.$$
$$\left. + \left[\frac{(K-I)m\rho^2}{c(1+r)} \begin{pmatrix} 1 \\ 1 \end{pmatrix} + D \right] \right)$$

So we obtain the representation

$$V^{(T-1)} = \frac{m}{(1+r)^T} [\alpha(T-1)X(T-1) + \beta(T-1)],$$

where

$$\alpha(T-1) = \left[(1+r) + (1-\delta) \right] I + \frac{(I-K)m\rho^2}{c(1+r)} - \frac{m\rho_3^2}{8(1+r)} \begin{pmatrix} 1 & 1 \\ 1 & 1 \end{pmatrix},$$

$$\beta(T-1) = \frac{(K-I)m\rho^2}{c(1+r)} \begin{pmatrix} 1 \\ 1 \end{pmatrix} + D.$$

Now we obtain the analogous representation by induction for $V^{(k)}$ assuming that

$$V^{(k+1)} = \frac{m}{(1+r)^T}\left[\alpha(k+1)X(k+1) + \beta(k+1)\right].$$

To find $\min\limits_{z(k)} W_1^{(k)}$, $\min\limits_{z(k)} W_2^{(k)}$ we find the partial derivative of the vector function $W^{(k)}$ on $z(k)$:

$$
\begin{aligned}
(W^{(k)})'_{z(k)} &= \frac{Z(k)}{(1+r)^k} + (V^{(k+1)})'_{z(k)} \\
&= \frac{Z(k)}{(1+r)^k} + \frac{m\alpha(k+1)}{(1+r)^T}(X(k+1))'_{z(k)} \\
&= \frac{Z(k)}{(1+r)^k} - \frac{m\rho_3\alpha(k+1)}{2(1+r)^T}\sqrt{x_1(k)+x_2(k)}\begin{pmatrix}1\\1\end{pmatrix}.
\end{aligned}
$$

The critical point $z_1^*(k)$ for $W_1^{(k)}$ and the critical point $z_2^*(k)$ for $W_2^{(k)}$ are found from the vector equation $(W^{(k)})'_{z(k)} = 0$. They give the vector

$$Z^*(k) = \begin{pmatrix}z_1^*(k)\\z_2^*(k)\end{pmatrix} = \frac{m\rho_3\alpha(k+1)}{2(1+r)^{T-k}}\sqrt{x_1(k)+x_2(k)}\begin{pmatrix}1\\1\end{pmatrix}.$$

Since $(W_{1,2}^{(k)})''_{z(k)} > 0$ then at the points $z_1^*(k)$ and $z_2^*(k)$ the functions $W_1^{(k)}$ and $W_2^{(k)}$ actually have a minimum in the variable z.

Now we find $\max\limits_{u_1(k)} W_1^{(k)}$ and $\max\limits_{u_2(k)} W_2^{(k)}$. Since

$$(W^{(k)})'_{u_1(k)} = -\frac{c}{(1+r)^k}\begin{pmatrix}u_1(k)\\0\end{pmatrix} + \frac{m\alpha(k+1)}{(1+r)^T}K\rho\sqrt{I-x(k)}\begin{pmatrix}1\\0\end{pmatrix},$$

so the stationary point $W_1^{(k)}$ which is found from the equation $pr_1(W^{(k)})'_{u_1(k)} = 0$ has the following form

$$u_1^*(k) = \frac{m_1\rho_1\sqrt{1-x_1(k)}}{c(1+r)^{T-k}}\left(\alpha_{11}(k+1) - \frac{1}{2}\alpha_{12}(k+1)\right).$$

We similarly obtain

$$u_2^*(k) = \frac{m_2\rho_2\sqrt{1-x_2(k)}}{c(1+r)^{T-k}}\left(-\frac{1}{2}\alpha_{21}(k+1) + \alpha_{22}(k+1)\right).$$

Thus

$$U^*(k) = \begin{pmatrix}u_1^*(k)\\u_2^*(k)\end{pmatrix} = \frac{m\rho\sqrt{I-x(k)}}{c(1+r)^{T-k}}\begin{pmatrix}\alpha(k+1)_{11} - \frac{\alpha(k+1)_{12}}{2}\\-\frac{\alpha(k+1)_{21}}{2} + \alpha(k+1)_{22}\end{pmatrix}.$$

To facilitation the further calculations we will use next formulas

$$\sqrt{x_1(k) + x_2(k)}\, Z^*(k) = \tfrac{\rho_3 m \alpha(k+1)}{2(1+r)^{T-k}} K_1 X(k),$$

$$(Z^*(k))^2$$
$$= \tfrac{m^2 \rho_3^2}{4(1+r)^{2T-2k}} \begin{pmatrix} (\alpha_{11}(k+1) + \alpha_{12}(k+1))^2 & 0 \\ 0 & (\alpha_{21}(k+1) + \alpha_{22}(k+1))^2 \end{pmatrix}$$
$$\times K_1 X(k),$$

$$\sqrt{1 - x(k)}\, U^*(k)$$
$$= \tfrac{(1+r)^{k-T}}{c} M \rho \begin{pmatrix} \alpha_{11}(k+1) - \tfrac{\alpha_{12}(k+1)}{2} & 0 \\ 0 & -\tfrac{\alpha_{21}(k+1)}{2} + \alpha_{22}(k+1) \end{pmatrix}$$
$$\times \left[\begin{pmatrix} 1 \\ 1 \end{pmatrix} - X(k) \right],$$

$$(U^*(k))^2$$
$$= \tfrac{m^2 \rho^2}{c^2(1+r)^{2T-2k}} \begin{pmatrix} (\alpha_{11}(k+1) - \tfrac{\alpha_{12}(k+1)}{2})^2 & 0 \\ 0 & (-\tfrac{\alpha_{21}(k+1)}{2} + \alpha_{22}(k+1))^2 \end{pmatrix}$$
$$\times \left[\begin{pmatrix} 1 \\ 1 \end{pmatrix} - X(k) \right].$$

Here $K_1 = \begin{pmatrix} 1 & 1 \\ 1 & 1 \end{pmatrix}$.

Now we can find $V^{(k)}$. For this we substitute $Z^*(k)$, $U^*(k)$ in $W^{(k)}$. Simple calculations give the following result

$$V^{(k)} = \tfrac{m}{(1+r)^T} \left[\alpha(k) X(k) + \beta(k) \right],$$

$$\alpha(k) = (1+r)^{T-k} I + (1-\delta)\alpha(k+1) - \tfrac{\rho_3^2}{4} \alpha(k+1) m \alpha(k+1) K_1$$
$$+ \tfrac{m\rho^2}{2c(1+r)^{T-k}} \begin{pmatrix} (\alpha_{11}(k+1) - \tfrac{\alpha_{12}(k+1)}{2})^2 & 0 \\ 0 & (-\tfrac{\alpha_{21}(k+1)}{2} + \alpha_{22}(k+1))^2 \end{pmatrix}$$
$$+ \tfrac{m\rho_3^2}{8(1+r)^{T-k}} \begin{pmatrix} (\alpha_{11}(k+1) + \alpha_{12}(k+1))^2 & 0 \\ 0 & (\alpha_{21}(k+1) + \alpha_{22}(k+1))^2 \end{pmatrix} K_1 +$$
$$- \tfrac{\alpha(k+1)K m\rho^2}{c(1+r)^{T-k}} \begin{pmatrix} \alpha_{11}(k+1) - \tfrac{\alpha_{12}(k+1)}{2} & 0 \\ 0 & -\tfrac{\alpha_{21}(k+1)}{2} + \alpha_{22}(k+1) \end{pmatrix},$$

$$\beta(k) = \beta(k+1) + \tfrac{\alpha(k+1)\delta}{3} \begin{pmatrix} 1 \\ 1 \end{pmatrix}$$
$$- \tfrac{m\rho^2}{2c(1+r)^{T-k}} \begin{pmatrix} (\alpha_{11}(k+1) - \tfrac{\alpha_{12}(k+1)}{2})^2 & 0 \\ 0 & (-\tfrac{\alpha_{21}(k+1)}{2} + \alpha_{22}(k+1))^2 \end{pmatrix} \begin{pmatrix} 1 \\ 1 \end{pmatrix}$$
$$+ \tfrac{\alpha(k+1)K m\rho^2}{c(1+r)^{T-k}} \begin{pmatrix} \alpha_{11}(k+1) - \tfrac{\alpha_{12}(k+1)}{2} & 0 \\ 0 & -\tfrac{\alpha_{21}(k+1)}{2} + \alpha_{22}(k+1) \end{pmatrix} \begin{pmatrix} 1 \\ 1 \end{pmatrix}.$$

We obtain the explicit algorithm for solving this optimization problem:

1. Find matrices $\alpha(k)$ and vectors $\beta(k)$, $k = T, \ldots, 1, 0$ by recurrent formulae (14), (15).
2. Find the minimizing sequence $Z^*(k)$ by formula (13) and optimal control $U^*(k)$ by formula (12).
3. For $Z(k) = Z^*(k)$ and $U(k) = U^*(k)$ we find $X(k)$, $k = 1, \ldots, T$ by formula (10).
4. Construct the sequence $V^{(k)}$, $k = T, \ldots, 1, 0$ by formula (11).

The value $V^{(0)}$ gives the maximal value for payoff functionals.

5   Conclusion

In this paper, we have developed the discrete model of the advertising control at the competitive market. The optimal control problem for our model can be solved explicitly. The proposed algorithm can be applied to construct the optimal advertising strategy in real markets. In the future, we will study the effect of advertising in supply chains [13–16] under uncertainty.

Acknowledgments. This research was supported by Act 211 Government of the Russian Federation, contract no. 02.A03.21.0011 and by Grant of the Foundation for perspective scientific researches of Chelyabinsk State University (2018).

References

1. Erickson, G.M.: Empirical analysis of closed-loop duopoly advertising strategies. Manag. Sci. **38**, 1732–1749 (1992)
2. Fruchter, G., Kalish, S.: Closed-loop advertising strategies in a duopoly. Manag. Sci. **43**(1), 54–63 (1997)
3. Vidale, M.L., Wolf, H.B.: An operation research study of sales response to advertising. Oper. Res. **5**(3), 370–381 (1957)
4. Sethi, S.P.: Deterministic and stochastic optimization of a dynamic advertising model. Optimal Control Appl. Methods **4**(2), 179–184 (1983)
5. Sorger, G.: Competitive dynamic advertising: a modification of the case game. J. Econ. Dynamics Control **13**(1), 55–80 (1989)
6. Sethi, S.P., Prasad, A., He, X.: Optimal advertising and pricing in a new-product and advertising competition. J. Econ. Manag. Strategy **4**(3), 351–360 (2008)
7. Chiarella, C., Szidarovszky, F.: Discrete dynamic oligopolies with intertemporal demand interactions. Math. Pannon. **19**(1), 107–115 (2008)
8. Gracheva, S.S., Pershin, M.A.: Optimal advertising expenditures of a monopolist with an only good in discrete linear model. Control Econ. Syst. Electron. J. **3**, 26–44 (2013). (in Russian)
9. Gracheva, S.S.: Optimization of advertising strategy of a company in case of nonlinear function of demand. Vestnik of Samara State University. Econ. Manag. **2**(113), 180–185 (2014). (in Russian)

10. Adukov, V.M., Adukova, N.V., Kudryavtsev, K.N.: On a discrete model of optimal advertising. J. Comput. Eng. Math. **2**(3), 13–24 (2015)
11. Zhukovskiy, V.I., Kudryavtsev, K.N.: Balancing conflicts and applications, 304. URSS, Moscow (2012). (in Russian)
12. Zhukovskiy, V.I., Kudryavtsev, K.N.: Balancing conflicts under uncertainty. II. Analog of maximin. Math. Game Theory Appl. **5**(20), 3–45 (2013). (in Russian)
13. Aust, G.: Vertical cooperative advertising and pricing decisions in a manufacturer-retailer supply chain: a game-theoretic approach. Vertical Cooperative Advertising in Supply Chain Management. CMS, pp. 65–99. Springer, Cham (2015). https://doi.org/10.1007/978-3-319-11626-6_4
14. Karray, S., Amin, S.H.: Cooperative advertising in a supply chain with retail competition. Int. J. Prod. Res. **53**(1), 88–105 (2015)
15. Lau, A.H.L., Lau, H.S.: Some two-echelon supply-chain games: improving from deterministic-symmetric-information to stochastic-asymmetric-information models. Eur. J. Oper. Res. **161**(1), 203–223 (2005)
16. Esmaeili, M., Aryanezhad, M.B., Zeephongsekul, P.: A game theory approach in seller-buyer supply chain. Eur. J. Oper. Res. **195**(2), 442–448 (2009)

Public-Private Partnership Models
with Tax Incentives: Numerical Analysis
of Solutions

Sergey Lavlinskii[1,2,3]([✉]), Artem A. Panin[1,2], and Aleksandr V. Plyasunov[1,2]

[1] Sobolev Institute of Mathematics, Novosibirsk, Russia
{lavlin,apljas}@math.nsc.ru, aapanin1988@gmail.com
[2] Novosibirsk State University, Novosibirsk, Russia
[3] Zabaikalsky State University, Chita, Russia

Abstract. An analysis is presented of a public-private partnership model in the natural resources sector of Russia, whereby the government provides tax incentives and supports the investor in infrastructure development and, to some extent, in the implementation of mandatory environmental measures. The analysis builds on the Stackelberg model and an original iterative solution algorithm based on probabilistic local search. A full-size model test site is constructed to demonstrate the capabilities of the approach. The actual data and dimensions of the model test site capture the specificity of the modeled object and make possible a practical study of the properties of the Stackelberg equilibrium. Based on the modeling results, an assessment is made of the impact of various factors on the effectiveness of the subsoil development program and the basic principles are formulated for government decision-making in this area.

Keywords: Stackelberg game
Bilevel mathematical programming problems
Mineral resources development program
Probabilistic local search algorithm

1 Introduction

In modern Russia, the agenda of sustainable development of natural resource based regions with poor production infrastructure assumes specific features, necessitating the use of additional tools to influence the investor and launch public-private partnerships (PPPs), which are designed to reduce risks and distribute responsibility between the subsoil usage stakeholders – the government and private businesses [1–3]. One of these partnership models was examined in [4,5] under the assumption that the government provides support to the investor in underdeveloped regions not only for infrastructure development but also, in part, for the implementation of mandatory environmental protection measures. This specifically Russian model was applied in the infrastructure development

© Springer International Publishing AG, part of Springer Nature 2018
A. Eremeev et al. (Eds.): OPTA 2018, CCIS 871, pp. 220–234, 2018.
https://doi.org/10.1007/978-3-319-93800-4_18

projects financed from the Investment Fund of Russia. However, this initiative was not successful, primarily because all attempts of the Russian government to encourage partnerships with private businesses were not grounded in economically sound administrative decisions [6–8].

Simultaneously with the PPP initiatives, the government designed a legislative framework for substantial profit tax incentives for large natural resource development projects in underdeveloped regions. The new tax regulations open a possibility for combining the previously applied PPP models with the newly introduced tax incentives. So far, no large project of this kind has been implemented in Russia, but the previous negative experience of the Investment Fund necessitates an analysis of the combined model based on the PPP and taxation laws.

This issue is the focus of this paper, which aims at analyzing the efficiency of the PPP schemes that use the full range of opportunities provided by the law and are based on the Stackelberg game-theoretic model. This approach, dictated by the hierarchy of interaction between the government and the private investor in the natural resources sector, allows one to reach a compromise of interests and ensure long-term efficiency not only for the private investor but also for the government, which sets for itself the task of strategic management of the natural resource sector.

This study continues our research in [4,5], which focused on a PPP model whereby the government confined its support to investments in infrastructure development and environmental protection. Here, we investigate a generalized version of the model, which includes the possibility of using tax incentives within the PPP. In further research, this problem statement can be developed to consider transaction costs. These costs are known to be high in Russia and have a tangible effect on the tax incentive mechanism. Models of this kind may be of great use in spatial planning and in addressing strategic administrative tasks in natural resource based regions.

2   Mathematical Model

A formal description of the PPP model can be presented as follows. We use the following notations:

T is a planning horizon; I is a set of investment projects;
J is a set of infrastructure development projects; K is a set of environmental projects; M is a set of tax incentive levels.
Investment project i in year t:
CFP_i^t is the cashflow (the difference between the incomes and expenses of all kinds, taking into account a transaction costs);
EPP_i^t is the environmental damage from the implementation of project;

DBP_i^t is the government revenue from the implementation of project;
ZPP_i^t is the population income (wages) from the implementation of project;
TP_{im}^t is a tax incentive of level m for project (the part of profit tax determined by the incentive level).
Infrastructure development project j in year t:
ZI_j^t is the costs of implementation of project;
EPI_j^t is the environmental damage from the implementation of project;
VDI_j^t is the government revenue from local economic development as a result of the implementation of project;
ZPI_j^t is the population income (wages) from the implementation of project.
Environmental project k in year t:
ZE_k^t is the costs of implementation of project;
ZPE_k^t is the local population direct income (wages);
EDE_k^t is the valuation of additional incomes of the local population.

The matrices μ and ν define the relationship between the projects, where μ_{ij} is a coherence indicator for the infrastructure and investment projects, $i \in I$, $j \in J$, and ν_{ij} is a coherence indicator for the environmental and investment projects, $i \in I$, $k \in K$:

$$\mu_{ij} = \begin{cases} 1, \text{ if the implementation of investment project } i \\ \quad \text{requires the implementation of infrastructure development project } j, \\ 0 \text{ otherwise;} \end{cases}$$

$$\nu_{ik} = \begin{cases} 1, \text{ if the implementation of investment project } i \\ \quad \text{requires the implementation of environmental project } k, \\ 0 \text{ otherwise.} \end{cases}$$

The discounts of the government and the investor:
DG is the discount of the government; DI is the discount of the investor;
The budget constraints:
b_t^G is the government budget in year t; b_t^O is the investor budget in year t.
We use the following variables:

$$x_j = \begin{cases} 1, \text{ if the government launches infrastructure development project } j, \\ 0 \text{ otherwise;} \end{cases}$$

$$\bar{y}_k = \begin{cases} 1, \text{ if the government is prepared to launch environmental project } k \\ \quad \text{(the government has included it into the budget expenses),} \\ 0 \quad \text{otherwise;} \end{cases}$$

$$y_k = \begin{cases} 1, \text{ if the government launches environmental project } k \\ \quad \text{as agreed with the investor,} \\ 0 \text{ otherwise;} \end{cases}$$

$$\bar{\varphi}_{im} = \begin{cases} 1, \text{ if the government is prepared to provide the investor with a tax} \\ \quad \text{incentive of level } m \text{ for investment project } i, \\ 0 \text{ otherwise;} \end{cases}$$

$$\varphi_{im} = \begin{cases} 1, \text{ if the government provides the investor with a tax incentive} \\ \quad \text{ of level } m \text{ for investment project } i, \\ 0 \text{ otherwise;} \end{cases}$$

$$z_i = \begin{cases} 1, \text{ if the investor launches investment project } i, \\ 0 \text{ otherwise;} \end{cases}$$

$$u_k = \begin{cases} 1, \text{ if the investor launches environmental project } k, \\ 0 \text{ otherwise.} \end{cases}$$

The government problem $\mathcal{PS}$:

$$\sum_{t \in T} \Big(\sum_{i \in I} (DBP_i^t + ZPP_i^t - EPP_i^t) z_i - \sum_{i \in I} \sum_{m \in M} TP_{im}^t \varphi_{im}$$

$$+ \sum_{j \in J} (VDI_j^t + ZPI_j^t - EPI_j^t - ZI_j^t) x_j + \sum_{k \in K} (EDE_k^t + ZPE_k^t - ZE_k^t) y_k$$

$$+ \sum_{k \in K} (EDE_k^t + ZPE_k^t) u_k \Big) / (1 + DG)^t \to \max_{x, \bar{y}, \bar{\varphi}, \varphi, y, z, u} \tag{1}$$

subject to:

$$\sum_{j \in J} ZI_j^t x_j + \sum_{k \in K} ZE_k^t \bar{y}_k \le b_t^G; t \in T; \tag{2}$$

$$\sum_{m \in M} \bar{\varphi}_{im} \le 1; i \in I; \tag{3}$$

$$(y, z, u, \varphi) \in \mathcal{F}^*(x, \bar{y}, \bar{\varphi}); \tag{4}$$

$$\bar{\varphi}_{im}, x_j, \bar{y}_k, \in \{0, 1\}; i \in I, j \in J, k \in K, m \in M. \tag{5}$$

The objective function of the government represents the net discounted income received by the government and the local population (1). Constraints (2) guarantee that the government expenses on infrastructure and environmental protection stay within the budget. Constraints (3) forbid the government to provide several tax incentives within one project. Constraint (4) means that the investor acts in an optimal way, which implies solving the low-level problem (the investor problem). The set $\mathcal{F}^*(x, \bar{y}, \bar{\varphi})$ is a set of optimal solutions of the low-level parametric problem.

The investor problem $\mathcal{PI}(x, \bar{y}, \bar{\varphi})$:

$$\sum_{t \in T} \Big(\sum_{i \in I} (CFP_i^t z_i + \sum_{m \in M} TP_{im}^t \varphi_{im}) - \sum_{k \in K} ZE_k^t u_k \Big) / (1 + DI)^t \to \max_{z, u, y, \varphi} \tag{6}$$

subject to:

$$\sum_{k \in K} ZE_k^t u_k - \sum_{i \in I} (CFP_i^t z_i - \sum_{m \in M} TP_{im}^t \varphi_{im}) \le b_t^O; t \in T; \tag{7}$$

$$\sum_{t \in T} \left(\sum_{i \in I} (ZPP_i^t - EPP_i^t) z_i + \sum_{j \in J} (ZPI_j^t - EPI_j^t) x_j \right.$$
$$\left. + \sum_{k \in K} (EDE_k^t + ZPE_k^t)(y_k + u_k) \right) / (1 + DI)^t \geq 0; \qquad (8)$$

$$\sum_{t \in T} \left(\sum_{i \in I} CFP_i^t z_i + \sum_{m \in M} TP_{im}^t \varphi_{im} - \sum_{k \in K} ZE_k^t u_k \right) / (1 + DI)^t \geq 0; \quad (9)$$

$$x_j \geq \mu_{ij} z_i; i \in I, j \in J; \qquad (10)$$

$$y_k + u_k \leq 1; k \in K; \qquad (11)$$

$$y_k + u_k \geq \nu_{ik} z_i; i \in I, k \in K; \qquad (12)$$

$$y_k \leq \bar{y}_k; k \in K; \qquad (13)$$

$$\varphi_{im} \leq \bar{\varphi}_{im}; i \in I; m \in M; \qquad (14)$$

$$\sum_{m \in M} \varphi_{im} \leq z_i; i \in I; \qquad (15)$$

$$y_k, z_i, u_k, \varphi_{im} \in \{0, 1\}; i \in I, k \in K, m \in M. \qquad (16)$$

The income of the private investor is determined by objective function (6). From (7) it follows that the investor's costs in each year do not exceed the budget, considering the income received from the investment projects and the tax incentives. Satisfying constraint (8) means reaching a compromise between the interests of the population, the government, and the private investor. Constraint (9) ensures that the investor makes an average-industry profit rate. Constraints (10)–(12) ensure technological coherence of the projects and prevent the situation when the investor and the government implement the same environmental projects simultaneously. Constraints (13) guarantee that the government implements only those environmental projects that are included in the budget.

The choice of the government's objective function reflects the specifics of modern Russia. The rent estimate of a given natural resource deposit is defined as the NPV (net present value) of the deposit development project with an industry-average discount, minus the NPV of the corresponding environmental project. This is exactly how the objective function, in the part related to the government, is designed: the first term reflects the budget revenues and the third one, the costs of the environmental projects financed by the government. This circumstance allows us to say that, solving this problem, we, to some extent, solve the problem of maximizing the government's share of the rent. Thus, solving the problem, we find a compromise between the interests of all the parties involved in the PPP project: the government, which receives the maximum possible share of the rent; the private investor, who maximizes their discounted income; and the local population, who benefit from new jobs and from the wages they earn within the projects and lose from the adverse impacts on the environment.

To solve the PPP planning problem, we applied an approximate hybrid algorithm based on the ideas of local descent [9–17] and the CPLEX package.

The latter is applied for solving both the one-level problem, where the government decides for the investor, and the investor problem. The local descent is used to search for a good approximate solution for the government.

Since the problem being studied has two levels and an arbitrary feasible solution $(x, \bar{y}, \varphi, \bar{\varphi}, y, z, u)$ contains the optimal solution (y, z, u, φ) of the parametric investor problem with the parameters x, $\bar{y}$ and $\bar{\varphi}$, we call the solution $(x, \bar{y}, \bar{\varphi})$ an almost feasible solution if it satisfies constraints (2), (3), and (5) and the investor problem with the parameters $(x, \bar{y}, \bar{\varphi})$ is solvable.

Algorithm parameters:

$mIter$ is the maximum number of iterations in the algorithm for finding the initial solution (Step 2);

$cfBound$ is the coefficient of constraint relaxation by the value of objective function (1) in solving the auxiliary problem in Step 2.3.

Hybrid algorithm:

Step 1. Calculate the upper bound $Bound$ by solving the government problem with the low-level constraints (i.e., the problem with objective function (1) and constraints (2), (3), (5) and (7)–(16)).

Step 2. Find a feasible solution $(x^0, \bar{y}^0, \bar{\varphi}^0)$ (which will later be used as an initial solution in the local search algorithm):

Step 2.1. $iter := 1$.

Step 2.2: If $iter \leq mIter$, then solve the investor problem with the government's variables and constraints (i.e., the problem with objective function (6) and constraints (2), (3), (5), and (7)–(16)) and the additional constraint that the government's objective function (1) is no less than $(Bound - 1)/iter$. Otherwise, proceed to Step 3.

Step 2.3: If the problem in the previous step is solvable and $(x, \bar{y}, \varphi, \bar{\varphi}, y, z, u)$ is an optimal solution, then calculate the value f of objective function (1) by solving the problem $\mathcal{PI}(x, \bar{y}, \bar{\varphi})$. If $f < (Bound - 1)/(iter * cfBound)$ or the problem in the previous step has no solution, then assume that $iter := iter + 1$ and proceed to Step 2.2; otherwise, assume $x^0 := x$, $\bar{y}^0 := \bar{y}$, $\bar{\varphi}^0 := \bar{\varphi}$, and $f^0 := f$ and proceed to Step 3.

Step 3: If we could not find a feasible solution in Step 2, then we use a zero solution as a feasible one; i.e., we assume that $x^0 := 0$, $\bar{y}^0 := 0$ and $\bar{\varphi}^0 := 0$, and calculate the value f^0 of objective function (1) by solving the problem $\mathcal{PI}(x^0, \bar{y}^0, \bar{\varphi}^0)$. Then we apply the local improvement algorithm:

Step 3.1: Take $(x, \bar{y}, \bar{\varphi}) := (x^0, \bar{y}^0, \bar{\varphi}^0)$ as a starting solution and $f := f^0$ as a record value.

Step 3.2: Find the best neighbor $(x^*, \bar{y}^*, \bar{\varphi}^*)$ in the neighborhood of the solution $(x, \bar{y}, \bar{\varphi})$.

Step 3.3: If the value of the objective function $f(x^*, \bar{y}^*, \bar{\varphi}^*) > f$, then assume that $x := x^*$, $\bar{y} := \bar{y}^*$, $\bar{\varphi} := \bar{\varphi}^*$ and $f := f(x^*, \bar{y}^*, \bar{\varphi}^*)$ and proceed to Step 3.2; otherwise, stop the algorithm.

We used as a neighborhood in the local improvement algorithm the following randomized neighborhood with precisely one neighbor. The randomized neighborhood of the solution $(x, \bar{y}, \bar{\varphi})$ has precisely one solution $(x', \bar{y}', \bar{\varphi}')$, which was obtained as follows. Each component of the vector $x^{'}$ is a random value which with probability $1 - 1/|J|$ equals the corresponding component of the vector x and with probability $1/|J|$, of vector $1 - x$. It is analogous for the vectors $\bar{y}^{'}$ and $\bar{y}$, but the probabilities are $1 - 1/|K|$ and $1/|K|$, respectively. For the tax incentives, the probabilities are $1 - 1/(|I||M|)$ and $1/(|I||M|)$, respectively.

The local improvement algorithm with the randomized neighborhood is stopped after 5000 iterations, the parameters $mIter$ and $cfBound$ were 30 and 3, respectively.

3   Analysis of the Properties of Equilibrium Solutions

To demonstrate the methodology of application of the described tools, we designed a special model test site, whose prototype was a set of 50 polymetallic ore deposits in the Zabaikalskii krai (Transbaikalia). For this model test site, we composed a set of 10 infrastructure projects, some of which are being implemented (railroads and powerlines) while others make up for the infrastructure that is currently missing but is necessary for the deposit development projects (powerlines and highways). For each of the deposits, there are five levels of tax incentives and a set of compensating environmental activities integrated into the relevant environmental project. The calculations were carried out on a personal computer (i7 3621QM processor, 4 GB RAM), and the counting time was about 30 min.

In this way, the model test site captures the specificity of the object being modeled and provides an information base for studying the properties of the Stackelberg equilibrium. The methodology is based on analyzing the sensitivity of the solutions of the corresponding bilevel Boolean programming problem to changes in the basic parameters of the model. This issue is of fundamental importance, primarily because for many of the model parameters, we know only the operational ranges of their values.

Under certain assumptions about the government's information capabilities, the initial bilevel statement of the planning problem can be substantially simplified and reduced to a one-level mathematical programming problem. This is possible if the government has access to sufficiently detailed information on the deposit development projects, forecasts of market prices, and the budget constraints of the investor. This means that the decision-making on subsoil use is supported by the relevant governmental institutions that assess subsoil development projects from the perspective of the government and society.

Then, the original bilevel model transforms into a Boolean programming problem with the variables $y_k, z_i, u_k, x_j, \varphi_{im} \in \{0, 1\}$, the government's objective function (1), and constraints (2), (5), (7)–(12), (15) and (16). This one-level problem statement applies more to an economy with an informed government, which dominates in the natural resources sector; however, this statement proves to be convenient for understanding what we get from the transition from the one- to bilevel statement.

The figures show the results of the calculations that analyzed the sensitivity of the solution to changes in the key model parameters, i.e., the discounts of the investor and the government. In the calculations, we compared four models: the basic model without tax incentive in the two- (A) and one-level (B) statements and the two- and one-level models with tax incentives (C and D).

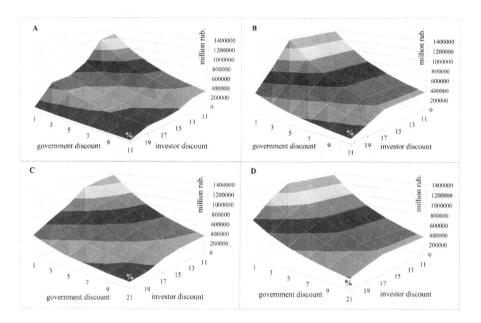

Fig. 1. The government objective function and the partner discounts

Figures 1, 2, 3, 4 and 5 show the results of applying the third lever of government support to the investor, i.e., tax incentives in the working ranges of the partners' discounts. From the point of view of the government's objective function, the introduction of tax incentives into the bilevel problem statement removes the cavity on the surface in Model A, ensuring greater stability of the partnership results in Model C with the increase in the investor discount (Fig. 1). In the one-level problem statement, the effect of tax incentives manifests itself most saliently at high investor discounts, resulting in substantially greater values of the functional.

It follows from Fig. 2 that, other things being equal, the bilevel statement (A and C) yields an increase in the investor's objective function, compared with the one-level statement (B and D). The effect of the tax incentives is most tangible for the investor in the bilevel statement. Unlike in the one-level models B and D, in which the government, informed about the investor's production capacity, simply nullifies its partner's NPV at high investor discounts, the leader-follower relationship pattern provides the investor with a positive income at high discounts, coupled with an additional increase in the income from preferential taxation.

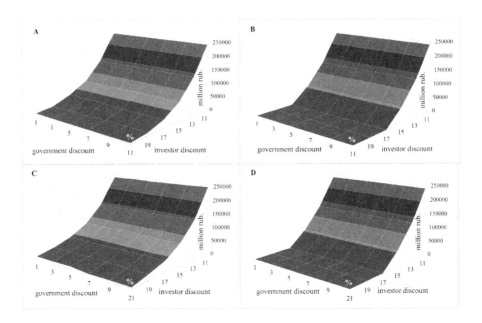

Fig. 2. The investor objective function and the partner discounts

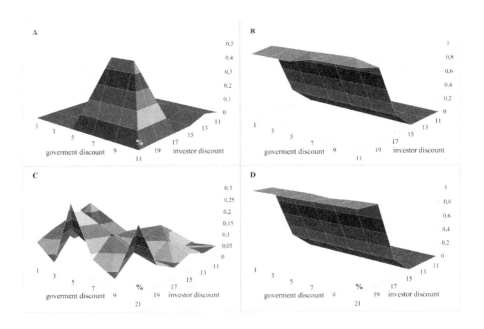

Fig. 3. The partner discounts and the government share in the environmental costs

How does the government combine the two levers of investor support, i.e., environmental initiatives and tax incentives? In the one-level problem statement, the choice of the support lever depends little on the government discount. The government supports the investor, starting from average investor discounts, and rapidly increases its support of environmental projects to reach a maximum at high discounts (Figs. 3B and D). The government uses tax incentives only at the largest investor discounts, selecting from a third to a half of the deposit development projects (Fig. 4D) and using virtually the maximum level of tax incentives (Fig. 5D).

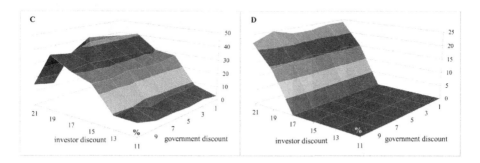

Fig. 4. The partner discounts and the number of tax incentives

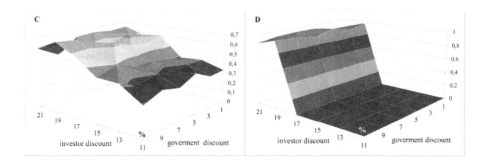

Fig. 5. The partner discounts and the level of tax incentives

In the bilevel models, the government uses the two investor support levers in a more sophisticated way across the entire range of the partner discounts. The use of tax incentives expands the support program in environmental construction (Figs. 3A and C), with the number and level of tax incentives increasing with the increase in the investor discount.

The Figs. 6, 7 and 8 show the results of the calculations that analyzed the sensitivity of the solution to changes in the other key model parameters, i.e.,

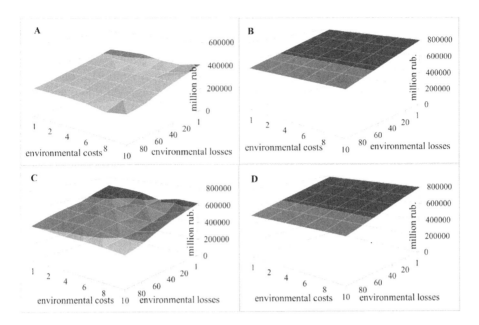

Fig. 6. The environmental parameters of the model and the objective function of the government

environmental costs and environmental losses. To this end, we fix the government and investor discounts at the average level ($DG = 5\%$ and $DI = 15\%$) and investigate the dependence of the main results of the cooperation on the environmental parameters of the model.

It turns out that at these discounts, the government in the one-level models B and D provides no tax incentives and no support for environmental projects. The reason is that the informed government implements a full-scale infrastructure development program, knowing that the investor is able, on their own without any tax incentives, to develop all the deposits, taking over all the environmental costs, and yet remain within the area of positive NPV (9).

The functionals of the government and the investor in the one-level models look accordingly (Figs. 6B, D, 7B, and D): the value of the government's objective function does not depend on environmental costs and decreases linearly with growing environmental losses. The value of the investor's objective function does not depend on environmental losses and decreases linearly with increasing environmental costs.

In the bilevel model A, the government cuts the infrastructure program at all values of the environmental parameters, especially at high environmental costs and losses (Fig. 8A). In model C, the use of tax incentives allows the government to increase the number of infrastructure development projects (Fig. 8C). The growing infrastructure support manifests itself not only in the noticeably higher functional of the government (Figs. 6A and C) but also in the higher value of the

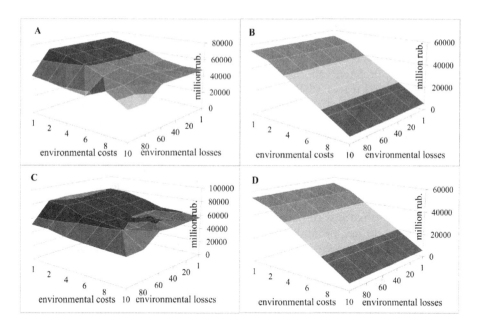

Fig. 7. The environmental parameters of the model and the objective function of the investor

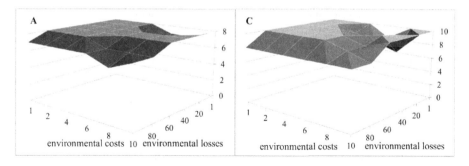

Fig. 8. The environmental parameters of the model and the number of infrastructure development projects launched by the government

investor's objective function, which increased due to the tax incentives (Fig. 7A and C).

The government coordinates tax incentives and the environmental support program, depending on the problem statement. In the one-level models B and D, the investor gets neither environmental support nor tax incentives over the entire range of the environmental parameters. In the bilevel models, the government implements some of the environmental projects and provides medium-level tax incentives for approximately a half of the deposit development projects.

4   Results and Discussion

The legal framework, established by the government, within which it can provide substantial tax preferences for large-scale natural resource development projects allows a combined use of PPPs and tax incentives. The modeling results show the practical relevance of the above-described models and define a hierarchy of factors that affect the efficiency of partnership relations between the government and the private investor in the development of natural resource deposits in poorly developed regions.

The bilevel problem statement adequately reflects modern Russian realities and captures the process of finding a compromise between the interests of the government and the investor. Analysis of the project documentation for several PPPs in the natural resources sector shows that expert assessments focus on proving the project profitability for the investor. This situation is consistent with the initial assumptions of the Stackelberg model, in which the government solves the upper level problem, knowing nothing about the intricacies of technology and market conditions, and the rent-seeking investor strives for the maximum profitability in the cooperation with the government. Adding tax incentives to the original PPP reflects the current situation and expands the partnership potential; we saw this effect in Figs. 1 and 2, where tax preferences raise the value of the objective function for both the government and the investor at high discounts.

The one-level model assumption that the government has complete information about the investor's technologies and opportunities seems inadequate in modern Russia. However, the solution of the one-level problem sets the upper bound on the government's functional in the bilevel model. This circumstance allows us to consider the one-level model as a limiting case of the bilevel model and to use the solution of the one-level problem as an ideal baseline describing the behavior of a rational, fully informed government.

That is why the above example for fixed discount values, whereby the government in the one-level models provides neither tax incentives nor environmental support, suggests that in real life, government support in forms other than infrastructure development projects may sometimes be superfluous. In our example, the superfluity of government support to the rent-seeking investor is evident in Figs. 7A and C, which show a stably high value of the investor functional over the entire range of environmental parameters. This situation suggests that the investor reaches a profitability substantially higher than the discount and, therefore, achieves an excessive safety margin due to the investor's superfluous share of rental incomes.

We see that within the above-described PPP models, the government shows complex behavioral patterns not only in choosing infrastructure and environmental projects for support but also in selecting the objects of preferential taxation. This behavior is rational, but it requires a well-adjusted approach to determining the specific amounts of government support.

The modeling results show the impact of various factors on the efficiency of the sub-soil development programs generated in the different models.

Thus, calculations showed that the most significant factor is the discount rate for the investor. To increase the social efficiency of the current partnership schemes with the rent-seeking investor and to ensure a fair distribution of the natural resource rent, one needs to reduce the investor discount in every possible way.

Since the solution of the one-level problem with an informed government determines the upper bound on the value of the functional in the bilevel model, which most adequately describes the current style of governance in the natural resources sector, the government decision-making should be guided by the following rules. Firstly, the government should not use small values of its own discount. Secondly, the government should seek a detailed understanding of the natural resource objects intended for development and the relevant development projects. This is the only way, considering the support provided to the investor, the government can build relationships as the owner of natural resources, who receives a possibly greater share of the natural resource rent as the value created by nature.

Relying on this methodological groundwork, one can address the core issue of natural resource based regions, i.e., a comprehensive scenario for the development of local natural resources, including infrastructure development plans and investment proposal packages containing rules for granting tax incentives and implementing various stages in spatial development. Here, it is crucial to consider the interests of society, as well as private businesses. The search for options to harmonize these interests is not an easy task, and the tools proposed in this paper are aimed at solving it.

Acknowledgements. This work was financially supported by the Russian Science Foundation (project No. 16-18-00073, problem statement), the Russian Foundation for Humanities (project No. 16-02-00049, hybrid algorithm), the Russian Foundation for Basic Research (project No. 16-06-00046, numerical analysis).

References

1. Reznichenko, N.V.: Public-private partnership models. Bulletin of St. Petersburg University, Series 8 Management **4**, 58–83 (2010). (in Russian)
2. Quiggin, J.: Risk, PPPs and the public sector comparator. Aust. Account. Rev. **14**(33), 51–61 (2004)
3. Bennett, J., Iossa, E.: Delegation of contracting in the private provision of public services. Rev. Industr. Organ. **29**(1), 75–92 (2006)
4. Lavlinskii, S., Panin, A., Pliasunov, A.: A two-level planning model for public-private partnership. Autom. Remote Control **11**, 89–103 (2015). https://doi.org/10.1134/S0005117915110077
5. Lavlinskii, S., Panin, A., Pliasunov, A.: Comparison of models of planning the public-private partnership. J. Appl. Industr. Math. **10**(3), 1–17 (2016). https://doi.org/10.1134/S1990478916030017
6. Lavlinskii, S.M.: Public-private partnership in a natural resource region: ecological problems, models, and prospects. Stud. Russ. Econ. Dev. **21**(1), 71–79 (2010). https://doi.org/10.1134/S1075700710010089

7. Glazyrina, I.P., Kalgina, I.S., Lavlinskii, S.M.: Problems in the development of the mineral and raw material base of Russia's far east and prospects for the modernization of the region's economy in the framework of Russian-Chinese cooperation. Reg. Res. Russ. **3**(4), 21–29 (2013). https://doi.org/10.1134/S2079970514010055

8. Glazyrina, I.P., Lavlinskii, S.M., Kalgina, I.S.: Public-private partnership in the mineral resources complex of Zabaikalskii krai: problems and prospects. Geogr. Nat. Resour. **35**(4), 359–364 (2014). https://doi.org/10.1134/S1875372814040088

9. Dempe, S.J.: Foundations of Bilevel Programming. Kluwer Academic Publishers, Dordrecht (2002)

10. Talbi, El.-G. (ed.): Metaheuristics for Bi-level Optimization. Studies in Computational Intelligence. Springer, Heidelberg (2013). https://doi.org/10.1007/978-3-642-37838-6

11. Raidl, G.R., Baumhauer, T., Hu, B.: Speeding up logic-based benders' decomposition by a metaheuristic for a bi-level capacitated vehicle routing problem. In: Blesa, M.J., Blum, C., Vob, S. (eds.) 9th International Workshop on Hybrid Metaheuristics, 11–13 June, Hamburg, Germany, pp. 183–197 (2014)

12. Alekseeva, E., Kochetov, Y.U., Dempe, S.: Local search approach for the competitive facility location problem in mobile networks. Int. J. Artif. Intell. **16**(1), 130–143 (2018)

13. Davydov, I., Kochetov, Yu., Talbi, El-G.: A matheuristic for the discrete bilevel problem with multiple objectives at the lower level. Int. Trans. Oper. Res. **24**(5), 959–981 (2017)

14. Davydov, I., Kochetov, Y.U., Plyasunov, A.: On the complexity of the (r|p)-centroid problem in the plane. TOP **22**(2), 614–623 (2014)

15. Plyasunov, A.V., Panin, A.A.: The pricing problem, Part I: exact and approximate algorithms. J. Appl. Industr. Math. **7**(2), 1–14 (2013)

16. Plyasunov, A.V., Panin, A.A.: The pricing problem, Part II: computational complexity. J. Appl. Industr. Math. **7**(3), 1–13 (2013)

17. Kononov, A.V., Kochetov, Yu.A., Plyasunov, A.V.: Competitive facility location models. Comput. Math. Math. Phys. **49**(6), 994–1009 (2009)

Fuzzy Core Allocations in a Mixed Economy of Arrow-Debreu Type

Valery A. Vasil'ev[1,2(✉)]

[1] Sobolev Institute of Mathematics, Pr. Acad. Koptyuga 4,
630090 Novosibirsk, Russia
vasilev@math.nsc.ru

[2] Novosibirsk State University, Pirogova Str. 1, 630090 Novosibirsk, Russia

Abstract. An important feature of the mixed economic system under consideration, besides the presence of a mixed production sector, is that two different regulation mechanisms function jointly: central planning and flexible market prices. Thus, this model is characterized by the presence of dual markets. In the first market, prices are stable and the allocation of commodities is determined by rationing schemes and governmental orders. In the second market, prices are flexible and are formed by the standard mechanism of equating demand and supply. We assume that the excess of any commodity purchased in the first market may be resold by any economic agent at flexible market prices. Whereas a lot of papers are devoted to existence and efficiency of mixed market equilibria, this paper investigates extremal properties of equilibrium allocations in a mixed economy of Arrow-Debreu type. A notion of fuzzy domination in a mixed environment is given, and coincidence of the fuzzy core and equilibrium allocations in certain specifications of economy in question is shown to hold.

Keywords: Rationing · Governmental order · Equilibrium
Fuzzy core allocation

1 Introduction

In this paper, a notion of fuzzy domination in a mixed environment is given, and coalitional stability of equilibrium allocations in a mixed economy of Arrow-Debreu type is established. Moreover, conditions are proposed that ensure coincidence of the fuzzy core and the set of equilibrium allocations for some important specifications of the general model of mixed economy suggested in [11].

An essential feature of the mixed economic system under consideration is that two different regulation mechanisms function jointly: central planning and flexible market prices. Thus, unlike the classical models, these models are characterized by the presence of dual markets. In the first market, prices are stable

This research was supported by the program of fundamental scientific researches of the SB RAS I.5.1., project 0314-2016-0018, and by RFBR grant 16-06-00101.

© Springer International Publishing AG, part of Springer Nature 2018
A. Eremeev et al. (Eds.): OPTA 2018, CCIS 871, pp. 235–248, 2018.
https://doi.org/10.1007/978-3-319-93800-4_19

and the allocation of commodities to a great extent is determined by rationing schemes and governmental orders. In the second market, prices are flexible and are formed by the standard mechanism of equating demand and supply. We assume that the excess of any commodity purchased in the first market may be resold by any agent at the flexible market prices.[1]

The problem of properly defining of a fuzzy core allocation in a mixed economic system arises from both the presence of fixed prices for rationed commodities and the multiplicity of types of coalition stability of equilibrium allocations which correspond to different types of flexible prices in the second market. A universal way to overcome the difficulties indicated is based on the use of an appropriate linear approximation of nonlinear income functions and leads to the formation of several types of cores which characterize all possible variants of coalition stability. The result is that testing of the famous Edgeworth's conjecture reduces to analyzing asymptotic behavior of each core separately (detailed analysis of the fuzzy domination for the mixed pure exchange model see in [8]).

Note, that we introduce all types of coalitional stability of equilibrium allocations, but restrict ourselves to special class of mixed economies (satisfying so-called "willingness to buy at the second market" condition) in studying the problem of decentralization of the fuzzy core allocations in a mixed economy of Arrow-Debreu type. Mixed economies from the special class mentioned admit a global linearization of the budget constraints. This fact allows to limit considerations to a common type of coalition stability for all equilibrium allocations, and, consequently, to get a proper analog of the classic core equivalence result in terms of fuzzy domination (for classic original see, e.g., [1,2]). Some results on the efficiency and general existence problem for equilibria in a mixed economy may be found in [3,11]. As to the cooperative characterization of equilibrium allocations, several mixed exchange models with both finite and infinite number of economic agents have been studied in [8–10]. In this paper we present some extensions of the results obtained for the finite mixed pure exchange models to the case of finite mixed economies of Arrow-Debreu type with rationing and governmental order in both consumption and production sectors.

The rest of the paper is organized as follows. In the second section, we introduce basic notations and fundamental definitions related to the equilibrium allocations. In the third section, we introduce several types of fuzzy dominations depending on the structure of equilibrium prices, and propose some conditions providing rather simple equilibrium presentation of the fuzzy core allocations for the mixed economy of Arrow-Debreu type.

2   The Model

This section contains basic notation and fundamental definitions, related to the concept of equilibrium in a mixed economy of Arrow-Debreu type. We provide a brief interpretation for some of these notions, which are taken from [11].

---

[1] More details on the mixed economic system under consideration may be found in description of the first model of this type, introduced by V.L. Makarov (cf. [4,5]).

2.1   Basic Notation

A mixed economy of Arrow-Debreu type is characterized by substantial inter-actions between the fix-price and flexible-price markets. It is described by the following basic data:

$$\mathcal{E} = \left\langle L, (X_i', X_i'', u_i, \beta^i, \theta^i, \omega^i)_{i \in N_1}, (Y_j, \vartheta^j)_{j \in N_2}, (s_{ij}', s_{ij}'')_{(i,j) \in N_1 \times N_2}, q, P \right\rangle,$$

where $N_1 = \{1, \ldots, n_1\}$ is the set of consumers, $N_2 = \{n_1 + 1, \ldots, n_2\}$ the set of producers (firms), $L = \{1, \ldots, l\}$ the set of commodities, $X_i' \subseteq \mathbf{R}^l$ the consumption set of $i \in N_1$ in the first market, $X_i'' \subseteq \mathbf{R}^l$ the consumption set of $i \in N_1$ in the second market, u_i the utility function of agent $i \in N_1$ on $X_i' \times X_i''$, $\omega^i \in R_+^l$ the initial endowment of $i \in N_1$, $\omega^o := \sum_{i \in N_1} \omega^i$ the aggregate initial endowment, $\beta^i \in R_+^l$ the maximal amount of rationed goods, available to $i \in N_1$ in the first market, implying $x^{i\prime} \leq \beta^i$ for any feasible consumption plan $x^{i\prime} \in X_i'$, $\beta^o := \sum_{i \in N_1} \beta^i$ the aggregate amount of rationed goods, available to consumers, $\theta^i \in \mathbf{R}^l$ the governmental order assigned to agent $i \in N_1$, $\theta^o := \sum_{i \in N_1} \theta^i$ the aggregate governmental order assigned to consumers, $Y_j \subseteq \mathbf{R}^l$ the production set of firm $j \in N_2$, $\vartheta^j \in \mathbf{R}^l$ the governmental order (plan) assigned to producer $j \in N_2$, $\vartheta^o := \sum_{j \in N_2} \vartheta^j$ the aggregate governmental order (plan) assigned to producers, s_{ij}' the share of the fix-price profit $q \cdot \vartheta^j$ of producer $j \in N_2$ transferred to consumer $i \in N_1$, s_{ij}'' the share of the flexible-price profit $p \cdot (y^j - \vartheta^j)$ of producer $j \in N_2$, transferred to consumer $i \in N_1$, $q \in \mathbf{R}_+^l$ the vector of fixed prices for consumers and producers in the first market, $P = \mathbf{R}_+^l$ is the set of flexible prices.[2]

As usual, the shares s_{ij}' and s_{ij}'' are assumed to satisfy the standard con-straints: s_{ij}', $s_{ij}'' \geq 0$ for $i \in N_1$, $j \in N_2$, and $\sum_{i \in N_1} s_{ij}' = \sum_{i \in N_1} s_{ij}'' = 1$ for each $j \in N_2$. To simplify the presentation, throughout in the paper we presuppose $\beta^o \neq 0$ and q to be strictly positive vector. It is also assumed everywhere below, like in [11], that the rationing schemes and governmental orders in consumption and production sectors are compatible: $\beta^o = \theta^o + \vartheta^o$. So, we presuppose that the amount $\sum_{N_1} \beta^i$ of the rationed commodities maximally available at fixed prices q has to be provided by the aggregate governmental orders to consumers and producers. The government thereby assumes that consumers will exhaust the rationing quota at the given fixed prices on the first market either for individual consumption or for reselling.

As in the standard models of equilibrium analysis, consumers maximize util-ities given their budgets, and producers maximize their profits under the flexible prices. The government collects the amounts θ^i, $i \in N_1$, and ϑ^j, $j \in N_2$, of the commodities and offers the amounts β^i, $i \in N_1$, to the households at the current fixed prices. Households may resell any excess amount on the second market at the prevailing flexible prices.

Some of the above concepts, which are not relevant for the standard Arrow-Debreu model [6] need some explanation. Namely, the fact that utility is defined

---

[2] As usual, $x \cdot y$ stands for the scalar product $\sum_{k=1}^{l} x_k y_k$ of the vectors $x = (x_1, \ldots, x_l)$ and $y = (y_1, \ldots, y_l)$ from $\mathbf{R}^l$.

on $X_i' \times X_i''$ indicates that it might matter to households $i \in N_1$, whether the consumption commodities come from the first or the secondary markets, even if a surplus of a particular commodity bought on the first market is resold on the second. Thus, this assumption has something to do with the perception the individuals have of the various markets. Specifically, it is a common knowledge to the consumers in the countries with economy under transition that due to the market disorder at the first stage of transition to free market system there are, as a rule, a lot of falsificated commodities at both, first and secondary markets. Falsification level may vary considerably from one market to another, yielding various trust levels granted to the markets under consideration. Of course, on the way towards the perfect competition these levels tend to be equal, but at the moment the phenomenon of nonequivalent evaluation of the different markets (in terms of utility functions) still prevails.

These basic notations will now be applied to introduce the concepts, which are fundamental for a model of a mixed economy.

2.2   Main Definitions

The following definitions, like in pure exchange models, refer to the familiar concepts of feasible states, budget sets, and equilibrium allocations. They are with minor, but nevertheless essential modifications of the same nature as in the standard Arrow-Debreu model [6], and are again taken from [11].

To define the budget set $B_i(p)$ of agent $i \in N_1$, let the flexible price vector $p \in \mathbf{R}_+^l$ be given. The basic income $\nu_i(p)$ of $i \in N_1$ consists of the following terms:

$$\alpha_i(p) := q \cdot \theta^i + p \cdot (\omega^i - \theta^i), \quad \delta_i' := \sum_{j \in N_2} s_{ij}' q \cdot \vartheta^j, \quad \delta_i''(p) := \sum_{j \in N_2} s_{ij}'' \pi_j(p).$$

The profit $\pi_j(p)$ of firm $j \in N_2$ earned on the secondary market is defined by $\pi_j(p) := \sup\{p \cdot (y^j - \vartheta^j) \mid y^j \in Y_j\}$. One has to take into account the governmental order ϑ^j, which has to be supplied by firm j at fixed prices q. Hence, the term $p \cdot \vartheta^j$ in the above formula represents the opportunity cost of the governmental order.

Put $\nu_i(p) := \alpha_i(p) + \delta_i' + \delta_i''(p)$ and let $w_i(p, x'^i) := \nu_i(p) + (p-q)^+ \cdot (\beta^i - x'^i)$ be the total income of agent $i \in N_1$ (the additional term $(p-q)^+ \cdot (\beta^i - x'^i)$ results from reselling the surplus quantities $(\beta^i - x'^i)$ of the rationed commodities on the secondary market). Here and below, for any $a \in \mathbf{R}^L$ we denote by a^+ vector from $\mathbf{R}^L$ with components $a_k^+ = a_k$ if $a^k > 0$, and $a_k^+ = 0$ if $a^k \leq 0$ ("positive variation" of a).

Remark 1. For the case of $\beta_h^i > x_h'^i$ for an agent $i \in N_1$ and a particular commodity $h \in L$ with $p_h > q_h$, it is thus assumed that the consumer nevertheless exploits the quota β_h^i entirely, in order to resell the excess amount on the flexible price market.

Definition 1. *Given the above concepts, the budget set of agent $i \in N_1$ is defined as follows:*

$$B_i(p) := \{x^i = (x'^i, x''^i) \in X_i(\beta) \mid q \cdot x'^i + p \cdot x''^i \le w_i(p, x'^i)\},$$

with $X_i(\beta) := \{x^i \in X_i \mid x'^i \le \beta^i\}$ and $X_i := X_i' \times X_i''.$

The individual demand and supply correspondences, $D_i(p)$, $i \in N_1$, and $S_j(p)$, $j \in N_2$, are defined in the standard way

$$D_i(p) := \{x^i \in B_i(p) \mid \mathcal{P}_i^\beta(x^i) \cap B_i(p) = \emptyset\},$$
$$S_j(p) := \{y^j \in Y_j \mid \pi_j(p) = p \cdot (y^j - \vartheta^j)\}$$

(with $\mathcal{P}_i^\beta(x^i) := \{\tilde{x}^i \in X_i(\beta) \mid u_i(\tilde{x}^i) > u_i(x^i)\}$). Then the excess demand correspondence $E(p)$ is given by $E(p) := \sum_{i \in N_1} D_{oi}(p) - \sum_{j \in N_2} S_j(p) - \sum_{i \in N_1} \omega^i$, with $D_{oi}(p) := \{x^{oi} = x'^i + x''^i \mid (x'^i, x''^i) \in D_i(p)\}$.

Next we consider the set of feasible allocations of $\mathcal{E}$. Put $X := \prod_{i \in N_1} X_i$, $X(\beta) := \prod_{i \in N_1} X_i(\beta)$, $Y := \prod_{j \in N_2} Y_j$, $Z := X \times Y$, and $Z(\beta) := X(\beta) \times Y$.

Definition 2. *The set $Z_\beta(\mathcal{E})$ of feasible allocations of $\mathcal{E}$ is given by*

$$Z_\beta(\mathcal{E}) := \{z = (x, y) \in Z(\beta) \mid \sum_{i \in N_1} x^{oi} = \sum_{j \in N_2} y^j + \sum_{i \in N_1} \omega^i\}.$$

The concept of an equilibrium allocation is then introduced in the usual way:

Definition 3. *A feasible allocation $(\bar{x}, \bar{y}) = ((\bar{x}^i)_{i \in N_1}, (\bar{y}^j)_{j \in N_2}) \in Z_\beta(\mathcal{E})$ is an equilibrium allocation of the mixed economy $\mathcal{E}$, if there exists a price vector $\bar{p} \in \mathbf{R}_+^l$ such that $\bar{x}^i \in D_i(\bar{p})$ for $i \in N_1$, and $\bar{y}^j \in S_j(\bar{p})$ for $j \in N_2$. Again, as usual, $\bar{p} \in \mathbf{R}_+^l$ is an equilibrium price vector, and $(\bar{p}, \bar{z})$ with $\bar{z} = (\bar{x}, \bar{y})$ is an equilibrium state of the mixed economy $\mathcal{E}$. The set of equilibrium allocations of $\mathcal{E}$ is denoted by $\mathcal{W} = \mathcal{W}(\mathcal{E})$.*

We introduce further the concept of fuzzy blocking (domination) and corresponding notion of the fuzzy core for our mixed economy of Arrow-Debreu type. It is based, mostly, on two principles: (1) privatization, and (2) linearization. The former one deals with the assignment to the consumers not only the profit shares, but the shares of production sets themselves (privatization, or distribution of the production activity between the consumers, like in [6]). As to the latter principle, it concerns some special tools allowing to eliminate nonlinearity of income functions (at least, up to some extent).

Observe, that the payments $\alpha_i' = q \cdot \theta^i$ and $\delta_i' = \sum_{j \in N_2} s_{ij}' q \cdot \vartheta^j$ of the central agency for the delivery θ^i and $\vartheta'^i = \sum_{j \in N_2} s_{ij}' \vartheta^j$ of the obligatory supplies are equal to their values against the fixed state prices q. Hence, on the state market both the selling and purchasing prices are equal to q. So, the total payment for the delivery of obligatory supply, $\sum_{N_1} (\alpha_i' + \delta_i')$, provides consumption of the maximum total amount, $\sum_{N_1} \beta^i$, of rationed goods against fixed state prices q. Note, that an extra consumption takes place against the market prices.

The principal difficulty of cooperative characterization of the set $\mathcal{W} := \mathcal{W}(\mathcal{E})$ of equilibrium allocations of mixed economy $\mathcal{E}$ relates to nonlinearity of the total income functions $w_i(p, x'^i)$ with respect to the flexible prices p. To overcome this difficulty we make use of the fact that by definition functions w_i are piecewise linear w.r.t. p, and partition $\mathcal{W}$ into several components according to the types of possible equilibrium prices p. Subsequent analysis of coalitional stability of the equilibrium allocations is then carried out separately for each of these components.

We now present the formal definitions adapted to consideration of coalitional stability of the equilibrium allocations related to different types of equilibrium prices. Given $K \subseteq L$, put

$$P_K := \{p \in \mathbf{R}_+^l \mid p_k \geq q_k,\ k \in K;\quad p_j \leq q_j,\ j \in J\},$$

where J denotes the complement of K, i.e., $J := L \backslash K$. The components mentioned above are subsets of the set $\mathcal{W}$, which correspond to a particular type of the equilibrium prices classified with the help of the convex polyhedrons P_K:

$$\mathcal{W}_K = \mathcal{W}_K(\mathcal{E}) := \{z \in \mathcal{W} \mid \exists p \in P_K : (p, z) \text{ is an equilibrium state of } \mathcal{E}\}.$$

Allocations in $\mathcal{W}_K$ are called K-*equilibrium allocations of the economy* $\mathcal{E}$.

3   Core Allocations in the Mixed Economy $\mathcal{E}$

In this section, we introduce a concept of fuzzy core allocations in terms of appropriate fuzzy domination relations in $\mathcal{E}$, and provide rather natural conditions that make it possible to get equilibrium presentation of the fuzzy core allocations in the mixed economy of Arrow-Debreu type.

3.1   Fuzzy K-Core in the Mixed Economy $\mathcal{E}$

We turn now to present definition of fuzzy domination in the mixed economy $\mathcal{E}$. Fix $K \subseteq L$ and, for each vector $a \in \mathbf{R}^L$, denote by a_K its projection onto $\mathbf{R}^K \times \{0\} \subseteq \mathbf{R}^L$:

$$(a_K)_k := \begin{cases} a_k, & k \in K, \\ 0, & k \in J = L \backslash K. \end{cases}$$

In case $b = (b^1, \ldots, b^{n_1}) \in (\mathbf{R}^L)^{N_1}$ we write $b_K := (b_K^1, \ldots, b_K^{n_1})$. Below we use also the following abridged notations. For every $i \in N_1$, we denote

$$\vartheta'^i := \sum_{j \in N_2} s'_{ij} \vartheta^j, \quad \vartheta''^i := \sum_{j \in N_2} s''_{ij} \vartheta^j, \quad \widehat{\vartheta}^i := \vartheta'^i - \vartheta''^i,$$

$$\gamma'^i := \theta^i + \vartheta'^i - \beta^i, \quad \gamma''^i := \theta^i + \vartheta''^i - \beta^i.$$

To introduce modified initial endowments $\widehat{\omega}'^i$, $\widehat{\omega}''^i$, and privatized production sets $\widetilde{Y}_i$ and $\widehat{Y}_i$ we put

$$\widehat{\omega}'^i := \omega^i - \gamma'^i, \quad \widehat{\omega}''^i := \omega^i - \gamma''^i, \quad \widehat{Y}_i := \widehat{\vartheta}^i + \widetilde{Y}_i, \quad i \in N_1,$$

where $\widetilde{Y}_i := \sum_{j \in N_2} s''_{ij} Y_j$, $i \in N_1$.

Remark 2. It follows from the definition of $\widehat{Y}_i$ that in case $\vartheta''^i = \vartheta'''^i$, each $i \in N_1$, we have: $\widehat{Y}_i = \widetilde{Y}_i$ for every $i \in N_1$. Specifically, $\widehat{Y}_i = \widetilde{Y}_i$, whenever $s'_{ij} = s''_{ij}$ for every $i \in N_1$, $j \in N_2$.

As it was noted in Introduction, the proof of the Edgeworth conjecture for the model $\mathcal{E}$ may be given on the basis of a characterization of the sets $\mathcal{W}_K$ in terms of appropriate fuzzy dominations. To introduce a formal definition, recall [1], that a *fuzzy coalition* is a nonzero element $\tau = (\tau_1, \ldots, \tau_{n_1})$ of the unit hypercube $I^{n_1} = [0,1]^{n_1}$. Here, the components of τ measure the degree of participation of the economic agents in an ordinary coalition $N(\tau) = \operatorname{supp}\tau$ with

$$N(\tau) := \{i \in N_1 \mid \tau_i > 0\},$$

referred to as the *support* of τ. We denote $\mathcal{T} := I^{n_1} \backslash \{0\}$ and, for arbitrary $\tau = (\tau_1, \ldots, \tau_{n_1}) \in \mathcal{T}$ and $b = (b^1, \ldots, b^{n_1}) \in (\mathbf{R}^l)^{N_1}$, write $b(\tau) := \sum_{N_1} \tau_i b^i$.

To ease the notations, we make use of the following shortenings

$$\widehat{Y} := \prod_{i \in N_1} \widehat{Y}_i \quad \text{and} \quad \widehat{Z}(\beta) := X(\beta) \times \widehat{Y}.$$

Further, for any $z = (x, y) \in \widehat{Z}(\beta)$ we use the notations $x' = (x'^1, \ldots, x'^{n_1})$, $x'' = (x''^1, \ldots, x''^{n_1})$, $y = (y^1, \ldots, y^{n_1})$, and $x^0 = x' + x''$. Like in case of rationing, $\beta = (\beta^1, \ldots, \beta^{n_1})$, and governmental order in consumption, $\theta = (\theta^1, \ldots, \theta^{n_1})$, we put $\omega := (\omega^1, \ldots, \omega^{n_1})$, $\vartheta' := (\vartheta'^1, \ldots, \vartheta'^{n_1})$, and $\widehat{\omega}' := \omega - \theta - \vartheta' + \beta$.

Definition 4. *A fuzzy coalition τ K-dominates a feasible allocation $\bar{z} \in Z_\beta(\mathcal{E})$ if there exists an allocation $z = (x^i, y^i)_{i \in N_1} \in \prod_{i \in N_1} X_i(\beta) \times \widehat{Y}_i$ such that*

$KF1.$ $u_i(x^i) > u_i(\bar{x}^i)$, $i \in N(\tau)$,
$KF2.$ $x^0_K(\tau) \leq \widehat{\omega}'_K(\tau) + y_K(\tau)$,
$KF3.$ $q_{K \cup I} \cdot x^0(\tau) + q_{J \backslash I} \cdot x'(\tau) \leq q_{K \cup I} \cdot (\omega + y)(\tau) + q_{J \backslash I} \cdot (\theta + \vartheta')(\tau)$, each $I \subseteq J$.

In particular, for $K = L$, a fuzzy coalition $\tau \in \mathcal{T}$ L-dominates a feasible allocation $\bar{z} \in Z_\beta(\mathcal{E})$ if there exists an allocation $z = (x^i, y^i)_{i \in N_1} \in \prod_{i \in N_1} X_i(\beta) \times \widehat{Y}_i$ such that

$LF1.$ $u_i(x^i) > u_i(\bar{x}^i)$, $i \in N(\tau)$,
$LF2.$ $x^0(\tau) \leq \widehat{\omega}'(\tau) + y(\tau)$,
$LF3.$ $q \cdot x^0(\tau) \leq q \cdot \omega(\tau) + q \cdot y(\tau)$.

The requirement that the coalition allocation $(x^i, y^i)_{i \in N(\tau)}$ be balanced at the prices q, represented by condition $LF3$, reflects a specific feature of the economy $\mathcal{E}$, the fixed prices in the first market.

Remark 3. Observe, that condition $LF3$ is essential only when nonequivalent exchange (at fixed prices q) of the initial endowment ω^i for $\widehat{\omega}'^i$ takes place: if $q \cdot (\theta^i + \vartheta'^i) = q \cdot \beta^i$ for all $i \in N_1$, then, evidently, $LF3$ is a direct consequence of $LF2$. In general case, however, each of $2^{|J|}$ inequalities appearing in $KF3$, matters and, as with the case of L-domination, reflects the requirement that $(x^i, y^i)_{i \in N(\tau)}$ be balanced with respect to the corresponding possible "extreme" realizations p^* of the flexible market prices from P_K, which have the form

$$p^*_{K \cup I} = q_{K \cup I}, \quad p^*_{J \setminus I} = 0, \quad I \subseteq J.$$

Here it is taken into account that, at zero flexible prices for commodities in $J \setminus I$, the entire income of an agent $i \in N_1$ from selling the corresponding initial endowment $\omega^i_{J \setminus I}$ consists only of the amount $q_{J \setminus I} \cdot (\theta^i + \vartheta'^i)$ guaranteed by the governmental order.

Definition 5. *The fuzzy K-core of the economy $\mathcal{E}$ is the set $C_{K,F}(\mathcal{E})$ of all allocations $\bar{z} \in Z_\beta(\mathcal{E})$ that cannot be K-dominated by any fuzzy coalition τ:*

$$C_{K,F} = C_{K,F}(\mathcal{E}) := \left\{ \bar{z} \in Z_\beta(\mathcal{E}) \mid \text{there exists no } \tau \in \mathcal{T} \text{ that } K\text{-dominates } \bar{z} \right\}.$$

So, roughly speaking, any allocation from the fuzzy core $C_{K,F}$ is *coalitional stable* in the sense there is no fuzzy coalition that K-dominates it.

3.2 Fuzzy Core Equivalence for L-Equilibrium Allocations

By applying approach developed in [8–10] for the mixed economic systems without production, we can obtain rather mild conditions providing the inclusions

$$\mathcal{W}_K \subseteq C_{K,F}, \quad K \subseteq L \tag{1}$$

(see, e.g., Assumptions $(A1), (A2)$, below). To verify these inclusions, which are valid under rather mild assumptions, we can use a special representation of the budget sets $B_i(p)$ defined by the type of flexible prices p. For the inclusions reversed to (1) and their refinements to hold (that imply equivalence of the cooperative and equilibrium mechanisms in $\mathcal{E}$) several additional assumptions must be made. We write

$$X' := \prod_{N_1} X'_i, \quad X'' := \prod_{N_1} X''_i,$$
$$X'(\mathcal{E}) := \Pr_{X'} Z_\beta(\mathcal{E}), \quad X''(\mathcal{E}) := \Pr_{X''} Z_\beta(\mathcal{E}), \quad X_\beta(\mathcal{E}) := \Pr_X Z_\beta(\mathcal{E}),$$

and offer these assumptions (below, $a \gg 0$ means $a_k > 0$, each component of a):

(A1) $X'_i = X''_i = R^l_+$, for all $i \in N_1$,
(A2) Y_j are convex sets, containing zero vector, for all $j \in N_2$,
(A3) $\forall x \in X_\beta(\mathcal{E}) \left[\sum_{i \in N_1} x^{0i} \gg 0 \right]$,
(A4) u_i are continuous and concave, for all $i \in N_1$,
(A5) u_i are strictly increasing in x''^i, for all $i \in N_1$.

To prevent too lengthy and tedious considerations, we demonstrate the main ideas of the approach proposed for the special class of mixed economies of Arrow-Debreu type, only. Namely, below we deal with the models satisfying so-called "willingness to by at the second market" condition (providing that free market is "weakly preferred" to the governmental one). This condition, formulated in [11], is as follows:

A6) $\forall x^i \in X_i(\beta) \forall \Delta \in R_+^l \left[\Delta \le x'^i \Rightarrow u_i(x'^i - \Delta, x''^i + \Delta) \ge u_i(x'^i, x''^i) \right]$, for all $i \in N_1$.

It can easily be verified that under strict positivity of the rations assumptions $(A5)$ and $(A6)$ imply the following property of the economy $\mathcal{E}$ (see [11]).

Proposition 1 (Vasil'ev-Wiesmeth). *Let mixed economy $\mathcal{E}$ satisfies assumptions $(A5)$, $(A6)$, and, besides, its rations β^i are strictly positive for each $i \in N_1$. Then the set of equilibrium allocations $\mathcal{W}$ reduces to the set of L-equilibrium allocations, $\mathcal{W}_L$:*

$$\mathcal{W}(\mathcal{E}) = \mathcal{W}_L(\mathcal{E}).$$

Remark 4. Assumption $(A6)$ is obviously satisfied in case of the neutrality of preferences of economic agents with respect to the different markets, when the utility of consumption bundle $x^i = (x'^i, x''^i)$ depends only on the total consumption $x^{oi} = x'^i + x''^i$, or, in more details, when $u_i(x^i) = u_i(\hat{x}^i)$ for any $x^i, \hat{x}^i \in X_i$ such that $x'^i + x''^i = \hat{x}'^i + \hat{x}''^i$.

To test that the fuzzy core coincides with equilibrium allocations under the assumptions $(A1)-(A6)$, we need in dual characteristics of the convex cone T_L, defined by (below, $t^0 := t' + t''$ for any $t = (t', t'')$)

$$T_L := \left\{ t = (t', t'') \in R^{2l} \mid t'' \le 0, \ q \cdot t^0 \le 0 \right\}.$$

We present one of these characteristics by description of the polar cone $(T_L)^0$.

Proposition 2. *The polar cone $T_L^0 = (T_L)^0 := \left\{ h \in R^{2l} \mid h \cdot t \le 0, \ t \in T_L \right\}$ is given by the formula*

$$T_L^0 = \left\{ (\lambda q, p) \in R^{2l} \mid \lambda \ge 0, \ \lambda q \le p \right\}. \tag{2}$$

Proof. We have to prove that a concrete description of the polar cone T_L^0 may be given by formula (2) (remind that in the first line of Proposition 2 just the formal definition of T_L^0 is presented). Put $H_L := \left\{ (\lambda q, p) \in R^{2l} \mid \lambda \ge 0, \ \lambda q \le p \right\}$. Since the cone H_L is convex and closed, by the bipolar theorem (see, e.g., [7]), we have $H_L^{00} = H_L$. Therefore, to prove the equality $H_L = T_L^0$, it suffices to show that

$$H_L^0 = T_L. \tag{3}$$

To establish (3), we make use of the fact that the extreme rays of the cone H_L are generated by the vectors $\left\{ (0, e^k) \right\}_{k \in L} \cup \left\{ (q, q) \right\}$. To prove this fact, observe that $q \gg 0$ and, for each element $(\lambda q, p) \in H_L$, we have:

$$(\lambda q, p) = \lambda(q, q) + (0, s), \tag{4}$$

where $s = p - \lambda q$, and, in addition, by definition of H_L, the inequality $s \geq 0$ holds. Therefore, one can rewrite (4) as follows: $(\lambda q, p) = \lambda(q, q) + \sum_{k \in L} s_k(0, e^k)$, which proves our claim, since λ and s_k, $k \in L$, are nonnegative.

Based on the above, (2) is implied by definition of the sets T_L and H_L and the evident relation $(t', t'') \cdot (q, q) = q \cdot t^0$, $\quad t \in R^{2l}$. $\square$

Introduce now an almost ordinary economy $\mathcal{E}^0$ of Arrow-Debreu type (a reduced version of the mixed economy $\mathcal{E}$ with $q = 0$), which will be useful for description of the L-core allocations in the economy, satisfying assumption $(A6)$. $\mathcal{E}^0$ is characterized by the parameters

$$\mathcal{E}^0 = \left\langle L, (X_i(\beta), u_i, \widehat{\omega}''^i)_{i \in N_1}, (Y_j)_{j \in N_2}, (s''_{ij})_{(i,j) \in N_1 \times N_2}, P \right\rangle, \tag{5}$$

with initial endowments $\widehat{\omega}''^i$ resulting from reallocation of the bundles ω^i through rationing $\beta = (\beta^1, \ldots, \beta^{n_1})$, and central orders in consumption $\theta = (\theta^1, \ldots, \theta^{n_1})$ and production $\vartheta = (\vartheta^{n_1+1}, \ldots, \vartheta^{n_2})$ by the formula $\widehat{\omega}''^i = \omega^i - \theta^i - \vartheta''^i + \beta^i$.

Remark 5. It is easy to verify that the budget sets of the reduced mixed economy $\mathcal{E}^0$ with zero fixed prices take the form standard for Arrow-Debreu models (see, e.g., [6]): $B_i^0(p) := \left\{ x^i \in X_i(\beta) \mid p \cdot x^{0i} \leq p \cdot \widehat{\omega}''^i + \widetilde{\pi}_i(p) \right\}$, $\quad i \in N_1$, where, as above, $\widehat{\omega}''^i = \omega^i - \gamma''^i$, $\gamma''^i = \theta^i + \vartheta''^i - \beta^i$, and

$$\widetilde{\pi}_i(p) := \sup \left\{ p \cdot \widetilde{y}^i \mid \widetilde{y}^i \in \widetilde{Y}_i \right\}, \quad \widetilde{Y}_i = \sum_{j \in N_2} s''_{ij} Y_j, \quad i \in N_1.$$

Minor differences consist of the "compound" character of the consumption sets $X_i(\beta) = \left\{ (x'^i, x''^i) \in X_i \mid x'^i \leq \beta^i \right\}$, $i \in N_1$, and the utility functions u_i dependent on $2l$ variables. It is easily seen, also, that the fuzzy L-domination in $\mathcal{E}^0$ is analogous to the traditional one (see, e.g., [1]): a fuzzy coalition $\tau \in \mathcal{T}$ L-dominates an allocation $\bar{z} \in Z_\beta(\mathcal{E}^0)$ if there exists an allocation $z = (x, \widetilde{y})$ from $\widetilde{Z}(\beta) := X(\beta) \times \widetilde{Y}$ (with $X(\beta) := \prod_{N_1} X_i(\beta)$ and $\widetilde{Y} := \prod_{N_1} \widetilde{Y}_i$) such that

$LF1^0.$ $u_i(x^i) > u_i(\bar{x}^i)$, $\quad i \in N(\tau)$,
$LF2^0.$ $x^0(\tau) \leq \widehat{\omega}''(\tau) + \widetilde{y}(\tau)$.

By applying formulae $\widehat{Y}_i = \widehat{\vartheta}^i + \widetilde{Y}_i$, $i \in N_1$, and equalities $\widehat{\omega}''^i = \widehat{\omega}'^i + \widehat{\vartheta}^i$, $i \in N_1$, one can easily prove that this domination is stronger than the fuzzy L-domination in original economy $\mathcal{E}$. Therefore, the following inclusions hold

$$\mathcal{W}_L^0 \subseteq C_{L,F}^0 \subseteq C_{L,F}, \tag{6}$$

where, as above, we denote by $\mathcal{W}_L^0$ and $C_{L,F}^0$ the set of L-equilibrium allocations and the fuzzy L-core of the model $\mathcal{E}^0$, respectively. Since the fixed prices in this model are equal to zero, $\mathcal{W}_L^0$ coincides with $\mathcal{W}^0 := \mathcal{W}(\mathcal{E}^0)$, the set of Walras allocations of $\mathcal{E}^0$, and the fuzzy L-core coincides with the ordinary fuzzy core of $\mathcal{E}^0$ (some details, concerning fuzzy domination in the classic Arrow-Debreu model can be found in [1]). Thus, from inclusions (6) it follows

$$\mathcal{W}(\mathcal{E}^0) \subseteq C_{L,F}(\mathcal{E}). \tag{7}$$

Turn, finally, to the description of the L-core allocations in terms of market mechanism of a mixed economy of Arrow-Debreu type. The following theorem provides such description for the economies satisfying "willingness to buy at the second market" condition.

Theorem 1. *Let a mixed economy $\mathcal{E}$ satisfies assumptions $(A1)-(A6)$, and, besides, its rations β^i are strictly positive for each $i \in N_1$. Then the set of fuzzy L-core allocations of $\mathcal{E}$ admits the following presentation:*

$$C_{L,F}(\mathcal{E}) = \mathcal{W}(\mathcal{E}) \cup \mathcal{W}(\mathcal{E}^0),$$

with economic system $\mathcal{E}^0$ defined by the formula (5).

To prove this theorem we need in some auxiliary constructions. First, denote by Γ_L the linear operator from R^{2l} into R^{2l} acting by the rule

$$\Gamma_L(t) := (0, t^0), \quad t = (t', t'') \in R^{2l},$$

where, as above, $t^0 = t' + t''$. Second, for any $i \in N_1$ put

$$\omega_{L,i} := \left(\gamma'^i, \widehat{\omega}'^i\right), \quad \widehat{Y}_{0i} := \{0\} \times \widehat{Y}_i.$$

Define, finally, the main auxiliary objects, which proved to be useful in the proof of Theorem 1. For that end, for any feasible allocation $\bar{z} = (\bar{x}, \bar{y})$ of the mixed economy $\mathcal{E}$ introduce the following sets:

$$\mathcal{M}_{L,i}(\bar{z}) := \Gamma_L \left(\mathcal{P}_i^\beta(\bar{x}^i)\right) - \widehat{Y}_{0i} - \{\omega_{L,i}\}, \quad i \in N_1,$$

$$\mathcal{M}_L(\bar{z}) := \left\{ x(\tau) \,\middle|\, x \in \prod_{i \in N_1} \mathcal{M}_{L,i}(\bar{z}), \, \tau \in \mathcal{T} \right\}.$$

Observe, that from the very definition of the fuzzy L-domination we get (in terms of the set $\mathcal{M}_L(\bar{z})$) a rather simple criterion for $\bar{z} \in Z_\beta(\mathcal{E})$ to belong to the fuzzy L-core of the mixed economy $\mathcal{E}$.

Lemma 1. *For any $\bar{z} \in Z_\beta(\mathcal{E})$ it holds: $\bar{z} \in C_{L,F} \Leftrightarrow \mathcal{M}_L(\bar{z}) \cap T_L = \emptyset$.*

One more useful property of the set $\mathcal{M}_L(\bar{z})$ is as follows.

Lemma 2. *If $\mathcal{E}$ satisfies Assumptions $(A1), (A2)$, and $(A4)$, then, for any allocation $\bar{z} \in Z_\beta(\mathcal{E})$, the set $\mathcal{M}_L(\bar{z})$ is convex.*

Proof. By applying a standard convexity argumentation. □

We are now in position to prove the main result of the paper.

Proof (of Theorem 1). To check inclusion $\mathcal{W}(\mathcal{E}^0) \cup \mathcal{W}(\mathcal{E}) \subseteq C_{L,F}(\mathcal{E})$, note first that due to the insertion (7) considered above we need in the proof of the inclusion $\mathcal{W}(\mathcal{E}) \subseteq C_{L,F}(\mathcal{E})$, only. So, take an arbitrary equilibrium allocation $\bar{z} = ((\bar{x}^i)_{i \in N_1}, (\bar{y}^j)_{j \in N_2})$ with equilibrium prices $\bar{p}$ satisfying inequality $\bar{p} \geq q$.

Assume that there exists a fuzzy coalition τ that L-dominates $\bar{z}$ via an allocation $z = (x^i, y^i)_{i \in N_1} \in \prod_{i \in N_1} X_i(\beta) \times \widehat{Y}_i$. Put $\tilde{y}^i = \sum_{j \in N_2} s''_{ij} \bar{y}^j$ and $\bar{y}^i = \widehat{\vartheta}^i + \tilde{y}^i$, each $i \in N_1$. Observe that from the definition of equilibrium allocation it follows: $\pi_i(\bar{p}) = \bar{p} \cdot (\tilde{y}^i - \vartheta''^i) \geq \bar{p} \cdot (y^i - \vartheta''^i)$, each $i \in N_1$. Hence, due to the inclusions $\bar{x}^i \in D_i(\bar{p})$, $i \in N_1$, inequalities $u_i(x^i) > u_i(\bar{x}^i)$, $i \in N(\tau)$, imply (from the very definition of the budget sets $B_i(\bar{p})$):

$$\bar{p} \cdot (x^{0i} - y^i - \widehat{\omega}'^i) - q \cdot \gamma'^i \geq \bar{p} \cdot (x^{0i} - \bar{y}^i - \widehat{\omega}'^i) - q \cdot \gamma'^i > 0, \ i \in N(\tau), (8)$$

where, as before, $\gamma'^i = \theta^i + \vartheta'^i - \beta^i$, $i \in N_1$. Put $s = \bar{p} - q$ and taking into account equalities $\widehat{\omega}'^i = \omega^i - \gamma'^i$, rewrite inequalities $\bar{p} \cdot (x^{0i} - y^i - \widehat{\omega}'^i) - q \cdot \gamma'^i > 0$, arising from (8), in more convenient form

$$q \cdot (x^{0i} - y^i - \omega^i) + s \cdot (x^{0i} - y^i - \widehat{\omega}'^i) > 0, \quad i \in N(\tau). \tag{9}$$

Summing up the inequalities (9) that are multiplied by the corresponding components of τ, we arrive at the inequality

$$q \cdot (x^0(\tau) - y(\tau) - \omega(\tau)) + s \cdot (x^0(\tau) - y(\tau) - \widehat{\omega}'(\tau)) > 0. \tag{10}$$

But from the definition of fuzzy L-domination it follows that

$$q \cdot (x^0(\tau) - y(\tau) - \omega(\tau)) \leq 0, \quad x^0(\tau) - y(\tau) - \widehat{\omega}'(\tau) \leq 0,$$

which contradicts (10) as far as vectors q and s are nonnegative. This contradiction proves inclusion $\mathcal{W}(\mathcal{E}) \subseteq C_{L,F}(\mathcal{E})$.

To prove the reverse inclusion $C_{L,F}(\mathcal{E}) \subseteq \mathcal{W}(\mathcal{E}) \cup \mathcal{W}(\mathcal{E}')$ consider an arbitrary allocation $\bar{z} = (\bar{x}, \bar{y})$ that belongs to $C_{L,F} = C_{L,F}(\mathcal{E})$. Applying Lemma 1, we have $\mathcal{M}_L(\bar{z}) \cap T_L = \emptyset$. Further, Assumption $(A5)$ and Lemma 2 imply that the set $\mathcal{M}_L(\bar{z})$ is nonempty and convex. Hence, from the Minkowski separation theorem it follows that there exists a nonzero functional $\widehat{p} = (p', p'')$ separating $\mathcal{M}_L(\bar{z})$ and T_L:

$$\sup\left\{\widehat{p} \cdot t \mid t \in T_L\right\} \leq \inf\left\{\widehat{p} \cdot t \mid t \in \mathcal{M}_L(\bar{z})\right\}. \tag{11}$$

Remind, that T_L is a cone. Since T_L has its vertex at the origin, we conclude that $\sup\{\widehat{p} \cdot t \mid t \in T_L\} = 0$. Consequently, $\widehat{p}$ belongs to the polar T_L^0 of the cone T_L. By Proposition 2, the separating functional $\widehat{p}$ has the form $(\lambda q, \bar{p})$ with $\lambda \geq 0$, where $\bar{p}$ satisfies the inequality $\bar{p} \geq \lambda q$. Let, for definiteness, $\lambda > 0$ (case $\lambda = 0$ will be analyzed below, separately). Then, without loss of generality, we may and shall assume that $\lambda = 1$ and, respectively, $\widehat{p} = (q, \bar{p})$.

By (11) and obvious inclusions $\mathcal{M}_{L,i}(\bar{z}) \subseteq \mathcal{M}_L(\bar{z})$, which are valid for all $i \in N_1$, we have

$$-q \cdot \gamma'^i + \bar{p} \cdot (x^{0i} - y^i - \widehat{\omega}'^i) \geq 0, \quad i \in N_1, \tag{12}$$

whenever $x^i \in \mathcal{P}_i^\beta(\bar{x}^i)$, and $y^i \in \widehat{Y}_i$, $i \in N_1$. Therefore, due to the strict monotonicity of u_i in x''^i, by choosing $y^i = \bar{y}^i$ with $\bar{y}^i := \widehat{\vartheta}^i + \sum_{j \in N_2} s''_{ij} \bar{y}^j \in \widehat{Y}_i$, and passing to the limit $x_n^i \to \bar{x}^i$ with $x_n^i \in \mathcal{P}_i^\beta(\bar{x}^i)$, we obtain

$$F_i^L(\bar{p}, \bar{x}^i, \bar{y}^i) \geq 0, \quad i \in N_1, \tag{13}$$

where $F_i^L(\bar{p}, \bar{x}^i, \bar{y}^i)$ stands for the left-hand side of (12) with $x^i = \bar{x}^i$ and $y^i = \bar{y}^i$. Summing up the inequalities (13) and making use the equalities

$$\sum_{i \in N_1} \gamma'^i = \sum_{i \in N_1} \gamma''^i = \sum_{i \in N_1} \theta^i + \sum_{j \in N_2} \vartheta^j - \sum_{i \in N_1} \beta^i = 0,$$

$$\sum_{i \in N_1} \bar{y}^i = \sum_{i \in N_1} \sum_{j \in N_2} s_{ij}'' \bar{y}^j = \sum_{j \in N_2} \bar{y}^j, \ \sum_{i \in N_1} \bar{x}^{0i} = \sum_{i \in N_1} \omega^i + \sum_{i \in N_1} \bar{y}^i,$$

we have

$$-q \cdot \sum_{N_1} \gamma'^i + \bar{p} \cdot \sum_{N_1} \left(\bar{x}^{0i} - \widehat{\omega}'^i - \bar{y}^i \right) = \bar{p} \cdot \left(\sum_{N_1} \bar{x}^{0i} - \sum_{N_1} \omega^i - \sum_{N_2} \bar{y}^j \right) = 0.$$

This implies that each inequality in (13) holds, in fact, as equality

$$-q \cdot \gamma'^i + \bar{p} \cdot \left(\bar{x}^{0i} - \bar{y}^i - \widehat{\omega}'^i \right) = 0, \quad i \in N_1. \tag{14}$$

Consequently, due to the fact that $\bar{p} \cdot (\bar{y}^i - \vartheta'^i) = \sum_{j \in N_2} s_{ij}'' \bar{p} \cdot (\bar{y}^j - \vartheta^j) \leq \pi_i(\bar{p})$, we get: $\bar{x}^i \in B_i(\bar{p})$ for all $i \in N_1$. Moreover, passing to the limit $x_n^i \to \bar{x}^i$ with $x_n^i \in \mathcal{P}_i^\beta(\bar{x}^i)$ in (12), and applying equalities (14), we obtain obvious relations: $F_i^L(\bar{p}, \bar{x}^i, y^i) \geq 0 = F_i^L(\bar{p}, \bar{x}^i, \bar{y}^i)$ for any $y^i \in \widehat{Y}_i$ and $i \in N_1$. These relations immediately imply the inequalities: $\bar{p} \cdot \bar{y}^i \geq \bar{p} \cdot y^i$, each $y^i \in \widehat{Y}_i$, $i \in N_1$. Hence, $\bar{p} \cdot \sum_{j \in N_2} s_{ij}'' \bar{y}^j \geq \bar{p} \cdot \sum_{j \in N_2} s_{ij}'' y^j$ for any $i \in N_1$ and $y^j \in Y_j$, $j \in N_2$. Therefore, the production vectors $\bar{y}^j$ belong to the corresponding supply sets $S_j(\bar{p})$:

$$\bar{p} \cdot \bar{y}^j \geq \bar{p} \cdot y^j \quad \text{for any} \ y^j \in Y_j, \ j \in N_2.$$

Thus, to complete the proof of inclusion $\bar{z} \in \mathcal{W}(\mathcal{E})$ we have only to check the relations: $\mathcal{P}_i^\beta(\bar{x}^i) \cap B_i(\bar{p}) = \emptyset$, $i \in N_1$. Since $\bar{p} \geq q$, it follows from $q \gg 0$ that $\bar{p}_k > 0$ for all $k \in L$. Put $\delta_i(\bar{p}, \bar{y}^i) := q \cdot \gamma'^i + \bar{p} \cdot (\widehat{\omega}'^i + \bar{y}^i)$ and, taking into account inclusion $\bar{p} \in P_L$, rewrite the budget constraints of the agents as

$$\bar{p} \cdot x^{0i} \leq \delta_i(\bar{p}, \bar{y}^i), \quad i \in N_1.$$

By Assumption $(A1)$ we have $x^{0i} \geq 0$, each $i \in N_1$. Hence, $\delta_i(\bar{p}, \bar{y}^i) \geq 0$. Note that equality $\delta_i(\bar{p}, \bar{y}^i) = 0$ implies $B_i(\bar{p}) = \{(0,0)\}$. Therefore, it suffices to verify the relations $\mathcal{P}_i^\beta(\bar{x}^i) \cap B_i(\bar{p}) = \emptyset$ in case $\delta_i(\bar{p}, \bar{y}^i) > 0$. Let $x^i \in \mathcal{P}_i^\beta(\bar{x}^i)$ and $\bar{p} \cdot x^{0i} = \delta_i(\bar{p}, \bar{y}^i) > 0$. Then $x^i \neq (0,0)$ and, without loss of generality, we may assume that there exists an $\tilde{x}^i \neq x^i$ such that $\tilde{x}^i \leq x^i$ and $\tilde{x}^i \in X_i(\beta)$. Consequently, due to strict positivity of $\bar{p}$ and q, we can choose a sequence $\{x_n^i\}_{n=1}^\infty \subseteq X_i(\beta)$ such that $\lim x_n^i = x^i$ and

$$\bar{p} \cdot x_n^{0i} < \delta_i(\bar{p}, \bar{y}^i), \quad n \geq 1. \tag{15}$$

By continuity of the functions u_i, we obtain $x_n^i \in \mathcal{P}_i^\beta(\bar{x}^{\,i})$ for n sufficiently large. But this, together with (15), is inconsistent with (12). Thus, for $\lambda > 0$, the inclusion $\bar{z} \in \mathcal{W}(\mathcal{E})$ is established.

Consider the case $\lambda = 0$. Note, that due to the assumption $(A3)$ one can rather easily verify that the condition $\widehat{p} = (0, \bar{p}) \neq 0$ yields positivity of $\bar{p}$. Next, arguing as in the proof of relations $\mathcal{P}_i^\beta(\bar{x}^{\,i}) \cap B_i(\bar{p}) = \emptyset$ for $\lambda > 0$, we arrive at $\bar{p} \cdot \bar{x}^{\,0i} = \bar{p} \cdot \left(\widehat{\omega}^{\,\prime i} + \bar{y}^i\right)$, $i \in N_1$, with $\bar{p} \cdot x^{0i} > \bar{p} \cdot \left(\widehat{\omega}^{\,\prime i} + y^i\right)$ for all $x^i \in \mathcal{P}_i^\beta(\bar{x}^{\,i})$, $y^i \in \widehat{Y}_i$. But this means, due to the equality $\widehat{\omega}^{\,\prime\prime i} = \widehat{\omega}^{\,\prime i} + \widehat{\vartheta}^i$, that $\bar{z}$ belongs to $\mathcal{W}(\mathcal{E}^o)$. This completes the proof of Theorem 1. $\square$

Acknowledgement. The author would like to thank the Program of Fundamental Scientific Researches of the SB RAS I.5.1. (project 0314-2016-0018), and Russian Foundation for Basic Research (grant No. 16-06-00101) for partial financial support.

References

1. Aubin, J.-P.: Mathematical Methods of Game and Economic Theory. North-Holland, Amsterdam, New York, Oxford (1979)
2. Ekeland, I.: Éléments d'Économie Mathématique. Hermann, Paris (1979)
3. van der Laan, G., Vasil'ev, V.A., Venniker, R.J.G.: On the transition from the mixed economy to the market. Sib. Adv. Math. **10**, 1–33 (2000)
4. Makarov, V.L., Vasil'ev, V.A., et al.: On some problems and results of the modern mathematical economics. Optimizatsija **30**, 5–87 (1982). (in Russian)
5. Makarov, V.L., Vasil'ev, V.A., et al.: Equilibrium, rationing and stability. Matekon **25**, 4–95 (1989)
6. Nikaido, H.: Convex Structures and Economic Theory. Academic Press, New York (1968)
7. Rockafellar, R.T.: Convex Analysis. Princeton University Press, Princeton (1970)
8. Vasil'ev, V.A.: On Edgeworth equilibria for some types of nonclassic markets. Sib. Adv. Math. **6**, 96–150 (1996)
9. Vasil'ev, V.A.: On the coincidence of the cores and coordinated allocations in mixed economic systems. Doklady Akademii Nauk **352**, 446–450 (1997). (in Russian)
10. Vasil'ev, V.A.: Core equivalence in a mixed economy. In: Herings, J.J.P., van der Laan, G., Talman, A. (eds.) The Theory of Markets, pp. 59–82. North-Holland, Amsterdam, Oxford, New York, Tokyo (1999)
11. Vasil'ev, V.A., Wiesmeth, H.: Equilibrium in a mixed economy of Arrow-Debreu type. J. Math. Econ. **44**(2), 132–147 (2008)

Applied Optimization Problems and Metaheuristics

Convergence Analysis of Swarm Intelligence Metaheuristic Methods

Tatjana Davidović$^{(\boxtimes)}$ and Tatjana Jakšić Krüger

Mathematical Institute of Serbian Academy of Sciences and Arts, Belgrade, Serbia
{tanjad,tatjana}@mi.sanu.ac.rs

Abstract. Intensive applications and success of metaheuristics in practice have initiated research on their theoretical analysis. Due to the unknown quality of reported solution(s) and the inherently stochastic nature of metaheuristics, the theoretical analysis of their asymptotic convergence towards a global optimum is mainly conducted by means of probability theory. In this paper, we show that principles developed for the theoretical analysis of Bee Colony Optimization metaheuristic hold for swarm intelligence based metaheuristics: they need to implement learning mechanisms in order to properly adapt the probability rule for modification of a candidate solution. We propose selection schemes that a swarm intelligence based metaheuristic needs to incorporate in order to assure the so-called *model convergence*.

Keywords: Optimization problems · Solution quality
Nature-inspired methods · Asymptotic properties · Stochastic processes

1 Introduction

Let us consider an optimization problem that requires minimization of a real-valued function f on a feasible space $\mathcal{X}$. More precisely, $f : S \to \mathbb{R}$ with a domain $S \subseteq \mathbb{R}^n$. S is also called a *set of solutions* (or *solution space*) for the considered optimization problem, while each $x \in S$ represents a *solution*. A solution $x = (x^1, x^2, \ldots, x^n)$ is an array in the n-dimensional space and it consists of components x^i, $i = 1, \ldots, n$. $\mathcal{X} \subseteq S$ is called a *set of feasible solutions* and it contains only the solutions that satisfy constraints defined within the considered optimization problem, i.e., $x \in \mathcal{X}$ represents a *feasible solution*. An *optimal solution* (or *optimum*) of the considered optimization problem is $x^* \in \mathcal{X}$ such that $f(x^*) \leq f(x)$ for all $x \in \mathcal{X}$. All other solutions $x \in \mathcal{X}$ are called *sub-optimal*. The optimal solution may not exist and then the considered optimization problem is unfeasible. If the optimal solution exists it may not be unique, and therefore, the solving of the considered optimization problem means finding one or more (or sometimes even all) of its optimal solutions.

Finding the optimal solutions is usually a very hard task and it involves the application of exact methods that are time and/or space consuming. The efficient, problem-specific methods designed to find sub-optimal solutions very fast

© Springer International Publishing AG, part of Springer Nature 2018
A. Eremeev et al. (Eds.): OPTA 2018, CCIS 871, pp. 251–266, 2018.
https://doi.org/10.1007/978-3-319-93800-4_20

are called heuristics. Metaheuristic methods have been developed to eliminate limitations of exact and heuristic methods, i.e., to satisfy requirements for less computational resources and to obtain a better quality of sub-optimal solutions at the same time. Today we distinguish various types of metaheuristics [25]. The population-based methods explore the idea that combining or modifying existing solutions can produce new and (hopefully) better ones. Swarm Intelligence (SI), especially its engineering stream, is a discipline of Artificial Intelligence (AI) that studies actions of individuals in various decentralized systems [1]. It explores the behavior of natural entities (consisting of many individuals) in order to build artificial systems for solving problems of practical relevance [4]. Therefore, among population-based we distinguish SI-based metaheuristics such as Ant Colony Optimization (ACO) [6], Artificial Bee Colony (ABC) [18], Bee Colony Optimization (BCO) [3], and Particle Swarm Optimization (PSO) [26]. An exhaustive list of SI-based metaheuristics methods may be found in [21].

Although practically very useful, metaheuristic algorithms suffer from a theoretical disadvantage: it is hard to evaluate the quality of the reported solution. The obtained solution may be even optimal, however, it is almost impossible to prove that. In the literature, metaheuristics are primarily investigated experimentally, usually associated with their concrete engagement and implementations. In addition, some theoretical research related to the convergence analysis of metaheuristic methods has already been conducted [2,10,20,22,28].

The importance of metaheuristics' theoretical background has inspired our work on proving convergence properties of the SI-based metaheuristic methods. The difficulty of theoretical analysis of stochastic search methods, in general, may be found in complex, highly nonlinear and stochastic correlations between their constituting parts [28]. Furthermore, theoretical analysis of metaheuristics commonly implies mathematical verification of the asymptotic convergence of the reported solution towards an optimal one, under some predefined conditions. Assuming that a considered optimization problem is solvable, investigating convergence properties of a metaheuristic algorithm is related to the question: is the optimal solution reachable if the algorithm is given enough time and resources [5]. Inspired by [10,12], two types of convergence for BCO, the so-called *best-so-far* convergence and sophisticated *model* convergence, have been utilized in [15,16]. Therefore, our intention now is to apply the gained insights to all SI-based metaheuristics. The main contributions of this paper are fourfold. First, we systematically review the existing notations and definitions related to the model convergence of metaheuristic methods. Then we provide an extension of the generic procedure (proposed in [10]) in such a way that it reflects the main steps of SI metaheuristic methods. Third, we recommend learning rules that assure model convergence of SI methods. Finally, we provide a systematic proof of model convergence of SI methods towards a global optimum when the recommended learning rules are applied.

In Sect. 2 we review some of the known results related to the SI methods and their convergence analysis. After a short survey of general notation and properties of stochastic sequences in Sect. 3, in Sect. 4 we present conditions that are sufficient to obtain convergence of SI-based metaheuristics to desired optimal solution. The last section contains concluding remarks.

2   Swarm Intelligence Metaheuristic Methods

From the biological perspective, swarm behavior (such as fish schools, flocks of birds, herds of land animals, insects' communities, etc.) is founded on existential needs of individuals to collaborate without any central control. In such a way they increase the probability to survive because predators mostly attack isolated individuals. This type of behavior is first and foremost characterized by autonomy, distributed functioning and self-organization. With this main idea in mind, SI investigates cooperation of individuals in biological systems and implements them to solve various practical problems [1].

Typical examples of SI metaheuristic algorithms are: ACO, PSO and various bees algorithms. Practicality and usefulness of these algorithms verified experimentally have motivated their theoretical research. Related to PSO, theoretical verification of convergence may be found in [7,17,26,27,29], for ACO in [8,9,12,19,24,30] and for BCO in [15,16]. In the above mentioned papers conditions of convergence are given for the specific method or a specific implementation of the corresponding SI metaheuristic methods. In the case of BCO, the authors of [15,16] have presented sufficient conditions for convergence of a constructive version of the BCO algorithm (BCOc). Theoretical analysis of the improvement-based version of the BCO algorithm (BCOi) is given in [14]. Therefore, our goal is to recognize conditions of asymptotic convergence under a general framework that describes all known SI metaheuristics w.r.t. to both types of generating solutions (constructive and the improvement one).

Constructive SI methods are building solutions by adding components to an empty solution or to already generated partial solutions. The examples of constructive methods are ACO and early versions of BCO, called BCOc. On the contrary, the improvement-based ones are modifying the existing complete solutions in an attempt to improve their quality. Typical examples of improvement-based SI methods are PSO, and BCOi, while ABC represents combination of these two approaches.

2.1   Instance- and Model-Based Algorithms

In [30] the authors proposed a framework that should improve the performance of majority of metaheuristic methods from a theoretical aspect. This framework is based on analyzing parameters of a metaheuristic method. Borrowing the notation from the machine learning field, the authors of [30] recognize two types of metaheuristic methods: *instance-based* and *model-based*. To generate new candidate solutions an instance-based algorithm utilizes only a current solution or a set of current solutions. On the contrary, model-based algorithms utilize a parameterized probabilistic scheme (called model) to generate candidate solutions.

A metaheuristic method is said to satisfy the *model-based search properties* if it iteratively explores the following two steps [30]:

- Generates (constructs or transforms) candidate solutions using some parameterized probabilistic model.
- Modifies the model (i.e., instantiate *update rule*) using candidate solutions such that the search is directed towards more promising regions.

Such a metaheuristic adopts the *model-based parameter scheme*, thus establishing a basis for the *model convergence* [10]. It requires learning properties which may be implemented in a form of an *update rule* for the method's parameters and/or structure [30,31]. The update rule represents the utilization of information extracted during the search in order to update the model.

2.2   Generic Procedure

The authors of [10,11,30] have presented a general framework called *generic algorithm* (i.e., *generic procedure*) to encompass most (or all) known metaheuristics for combinatorial optimization problems. The generic procedure allows a rather flexible description of different segments (modules) of a metaheuristic. It can be viewed as an iterative algorithm that utilizes two different structures: (1) m_t - a state of memory, and (2) L_t - a list of N *sample points*, i.e., solutions ($\boldsymbol{x}_s \in \mathcal{X}, 1 \le s \le N, N \in \mathbb{N}$) in iteration t. The stopping criterion is defined as the maximum number of iterations. The main steps of the generic procedure are as follows:

1. $t \leftarrow 1$;
2. Initialization of memory m_t;
3. Until stopping criterion is satisfied:
   (a) Determine the list L_t as a function $g(m_t, \xi_t)$ of memory state m_t and a random influence ξ_t;
   (b) Determine a value of objective function $f(\boldsymbol{x}_s)$ for all $\boldsymbol{x}_s \in L_t$ and generate a list L_t^+ of pairs $(\boldsymbol{x}_s, f(\boldsymbol{x}_s))$;
   (c) Determine new memory state m_{t+1} as a function $h(m_t, L_t^+, \xi_t')$ of current memory state m_t, current list L_t^+ and random influence ξ_t';
   (d) $t \leftarrow t + 1$.

The state of memory m_t may be further defined by two components: the so-called *sample-generating part* (m_t^s) and the *reporting part* (m_t^r). The sample-generating part holds all necessary information to generate L_t in iteration t. All other relevant information are stored in m_t^r, such as solution of the highest quality found so far, namely the *best-so-far* solution, $\boldsymbol{x}^{bsf}$. The procedure to determine $\boldsymbol{x}^{bsf}$ depends on the best solution found in iteration t, i.e., on the *iteration best* solution ($\hat{\boldsymbol{x}}_t$). The solution $\hat{\boldsymbol{x}}_t$ is determined as $f(\hat{\boldsymbol{x}}_t) = \min_{1 \le s \le N} f(\boldsymbol{x}_s)$. We should emphasize that for the purpose of convergence analysis $\boldsymbol{x}^{\overline{bsf}}$ represents a current approximation of an optimal solution [10]. Function $g(m_t, \xi_t)$ determines a probability distribution of new solutions to enter the list L_t, while function

$h(m_t, L_t^+, \xi_t')$ defines rules to determine the state of memory m_{t+1} in the next iteration. Moreover, $g(m_t, \xi_t)$ refers to a set of possible transformations (i.e., possible moves) that generate new solutions depending on the current model. Function $h(m_t, L_t^+, \xi_t')$ is responsible for modifying (updating) the model by integrating learning properties ([10], p. 168). It is important to note that in this formalism, information about a problem instance may be used as an argument of functions g and h. Consequently, the update rules differentiate w.r.t. the type of optimization problem being solved and the type of heuristic rules that either construct or modify solutions during the search. The provided generic procedure accommodates well all notable SI metaheuristic algorithms as it was shown for some versions of ACO, PSO [10] and BCO [14].

Here, we present a modification of generic procedure that contains more details about the steps of typical SI method. Finding a solution of an optimization problem requires all of its components to be determined. Constructive methods are selecting values for components and building a solution step by step, while the improvement-based metaheuristics are transforming the current values of solution components in order to improve the quality of the considered solution. The SI methods that satisfy model-based search properties involve some learning steps that influence the solution generation process. More precisely, the determination of solution components values is influenced by the quality of previously generated solutions. The main focus of learning is the modification of selection scheme for values of components (expressed by selection probability values $p_{i,j}$). As we are considering the combinatorial optimization problems, the set of possible components values is finite (or at most countable). Consequently, selection probability $p_{i,j}$ measures the chances that component i will take value j. If the SI method is model-based the values for selection probabilities will change from iteration to iteration, i.e., we will have different values for $p_{i,j}(t)$, $t = 1, 2, \ldots$

SI metaheuristics are population-based methods, and therefore their m_t consists of N current solutions, best-so-far solution, current selection probability values, and some other data specific for each particular method (we do not go into details here and consider only the first three items). The pseudo-code of model-based SI metaheuristic method is presented by the Algorithm 1.

Each SI metaheuristic method has its specificities, however, they can be described by some general steps. At the beginning, the reading of problem data and setting of parameter values is performed. Then, some initialization is required, and we described it in the following way: best-so-far solution (the final solution to be reported to the user at the end of execution) is initialized to an empty solution and the corresponding objective function value is set to a large enough constant. These steps are not necessary yet they ensure the compactness of our pseudo-code. The values for selection probabilities $p_{i,j}(1)$ are set to given initial values, denoted by $p_{i,j}(0)$, $\nu(i)$ denotes the number of possible values for the component i. The values $p_{i,j}(0)$ may be different in various SI methods, although usually all selection probabilities have the same initial value. Some of the methods involve restarts in their executions and then the values of $p_{i,j}(t)$ are set to $p_{i,j}(0)$ again. We also assume that the initial state of

```
t ← 1
Read(Problem input data, method parameters)
x^{bsf} ← ∅, f(x^{bsf}) ← ∞ {initialization of m_t}
for i ← 1, n do
    for j ← 1, ν(i) do
        p_{i,j}(t) ← p_{i,j}(0)
    end for
end for
{main loop}
repeat
    for s ← 1, N do
        x_s ← CreateSolution(m_t, p_{i,j}(t), rand)
        if f(x_s) < f(x^{bsf}) then
            x^{bsf} ← x_s
        end if
    end for
    for i ← 1, n do
        for j ← 1, ν(i) do
            p_{i,j}(t + 1) ← Update(p_{i,j}(t))
        end for
    end for
    t ← t + 1
until stopping criterion is satisfied
Return(x^{bsf}, f(x^{bsf}))
```

Algorithm 1: Pseudo-code for SI method

m_t (actually the content of m_1) does not include any solution from the population. This enables to capture the characteristics of both constructive and improvement-based methods. As our REPEAT loop is executed for $t = 1$ first, each particular method can generate initial population according to its own set of rules (realized by procedure CREATESOLUTION). The same (or similar) procedure can be used in all remaining iterations to generate the new population of solutions by the considered SI method and possibly update x^{bsf}. The final step of each iteration consists of updating values for selection probabilities to be used in the next iteration.

3   Convergence Analysis

Convergence of a stochastic sequence addresses the question whether or not a series of random variables $(X_1, X_2, \ldots)$ converges to a new random variable X^*. In case of metaheuristic algorithms we observe a sequence of solutions produced at the end of each iteration. Having in mind that SI metaheuristics are population-based methods, we need to determine a single solution $x_t = (x_t^1, x_t^2, \ldots, x_t^n) \in \mathcal{X}$ that is reported at the end of iteration t. Here, $x_t^i \in \mathbb{R}$ represents the i-th component of the solution x_t. Usually, $x_t = \hat{x}_t$.

Because the algorithm incorporates a global knowledge exchange among the iterations, we are able to obtain best-so-far solution x_t^{bsf} in any iteration t. Here, we use a strict policy for the update of x_t^{bsf}. Namely, in the initialization phase of the search the value of variable x^{bsf} is set arbitrary until the condition $f(\hat{x}_t) < f(x^{bsf})$ is satisfied and then x^{bsf} copies a value from $\hat{x}_t$. Consequently, we become interested in the sequence of solutions x_t^{bsf} generated by the meta-heuristic method, i.e., we observe the sequence $(x_1^{bsf}, x_2^{bsf}, \dots, x_t^{bsf}, \dots)$.

The best-so-far convergence analyzes conditions under which the sequence $(x_t^{bsf})_{t=1}^{\infty}$ converges to an optimal solution x^*, more precisely, it evaluates probability that an optimal solution will be found at least once during the search. Accordingly, one should find the conditions that guarantee a sequence of objective function values $f(x_t^{bsf})$ to converge ("w. pr. 1" or "in probability") to $f^* = \min\{f(x) : x \in \mathcal{X}\}$. As shown in [10,23,24] the only requirement is that, in any iteration, (p^*) – the probability to find an optimal solution is strictly greater than zero. However, the concept of best-so-far convergence is too "generous" as it appears that even a random search, known as quite inefficient algorithm, converges to a global optimum [23]. A superior behavior may be expected only in the cases when the set of current solutions (also referred to as model) tends to be modified into the set of optimal and high quality solutions, as $t \to \infty$. This type of modification is named the *model convergence* [10] and it tends to evaluate the probability that the algorithm reaches a state in which it generates only optimal solution(s). Model convergence is hard to prove as it requires adequate balance between exploration and exploitation of the search obtained by fine-tuning of the algorithm's parameters.

4   Model Convergence of SI Optimization Methods

In this section we present conditions that are sufficient for any SI based meta-heuristic algorithm to find an optimal solution. We start with necessary notation and remind on known theoretical results in the literature. The basic tools of probability theory are utilized, such as limit theorems for stochastic sequences, in particular the second Borel-Cantelli lemma which we do not repeat here.

To express the model-based search properties, the SI method needs to be well organized (structured) algorithm which utilizes the information about the performance from the previous stages of its execution. Consequently, the global knowledge exchange between iterations is the main assumption in our analysis. These requirements are fulfilled if CREATESOLUTION procedure of the SI generic pseudo-code (Algorithm 1) includes some learning properties.

4.1   Preliminary Conditions

To explain the course of our analysis, the following events should be defined (borrowing the notation from [10,13]).

Definition 1. *For any feasible solution x of the considered problem and an iteration counter $t \geq 1$ assume that:*

- $C(t)$ denotes the event that $x_t = x$, i.e., the examined solution was generated in iteration t. $C^c(t)$ is used to represent the complementary event.
- $B(t)$ marks the event that x was not visited during the first t iterations i.e., $B(t) = C^c(1) \cap C^c(2) \cap \cdots \cap C^c(t)$. The complementary event is $B^c(t)$.
- $B = \bigcap_{t=1}^{\infty} B(t)$ represents the event that the algorithm cannot generate the examined solution x, i.e., $x \neq x_t$ for all $t = 1, 2, \ldots$
- $r(t) = \Pr(B^c(t)|B(t-1)) = \Pr(C(t)|B(t-1))$ means the probability that x is generated in the iteration t, although it has not been obtained in any of the previous iterations. $\Diamond$

According to Definition 1, $B(1)$ denotes that x was not produced by the algorithm in the first iteration, $B(2)$ describes the situation that x was not visited in the first and second iteration, and so on. Consequently, $\{B(t)\}_{t=1}^{\infty}$ represents a non-increasing sequence of events[1], more precisely:

$$B(1) \supseteq B(2) \supseteq \cdots \supseteq B(t) \supseteq B(t+1) \supseteq \cdots .$$

Based on Definition 1 and definition of convergence in probability from [10, 14, 16], $\Pr(C(t)) \to 1$ as $t \to \infty$ denotes that the sequence x_t converges in probability to the set $\mathcal{X}^*$ (that contains only optimal solutions). Moreover, based on the definition of events B and $B(t)$ we can conclude that

$$\Pr(B(t)) \to P(B) = \Pr\left(\left\{\bigcap_{t=1}^{+\infty} B(t)\right\}\right) \text{ as } t \to +\infty.$$

To establish connection between events introduced in Definition 1 we present here the theorem about the convergence of Generalized Hill Climbing algorithm (GHC) proven in [13]. The pseudo-code for GHC is presented by the Algorithm 2 in Appendix.

Theorem 1. *A GHC algorithm converges in probability to $\mathcal{X}^*$ if and only if the following two conditions are satisfied:*

(i) $\sum_{t=1}^{+\infty} r(t) = +\infty,$

(ii) $\Pr(\{C^c(t)|B^c(t-1)\}) \to 0$ as $t \to \infty.$

Then, an equivalent form for (i) can be shown, i.e.,

Lemma 1. $\Pr(B) = 0 \Leftrightarrow \sum_{t=1}^{+\infty} r(t) = +\infty.$

Lemma 1 was obtained for the single-solution metaheuristic, however, it is straight-forward to extend it to the population-based methods by observing the sequence of best-so-far solutions.

---

[1] In [10] the author mistakenly reported that the sequence $\{B(t)\}_{t=1}^{\infty}$ is non-decreasing.

In order to identify the model convergence properties for majority of the SI metaheuristic methods, we need to define four modification schemes for the values of selection probabilities. Based on our previous experience, besides constructive and improvement-based SI metaheuristics we need to distinguish between two types of optimization problems. The first type includes optimization problems for which the size of the solution is smaller then the size of the problem. More precisely, these are the so called selection problems (like p-median or p-center location problems). In these kind of problems we are usually given a number of possibilities to choose a subset of their values that will constitute a solution of the considered problem. In this case, modification scheme for selection probability values needs to consider only components' affiliation within x^{bsf}. The second type of optimization problems are the ones whose solutions represent orderings and/or groupings of all given elements. For these problems the length of the solution vector is equal to the size of the problem. Typical examples of the second type problems are Traveling Salesman Problem (TSP), Vehicle Routing Problem (VRP) and various scheduling problems. For these problems the probability update rules should be modified in such a way to take care of ordering or group affiliation of solution components.

4.2 Modification Schemes in the Cases When Subset of Data Constitutes a Solution

For the constructive SI method the selection probability for component i taking value j in the iteration $t + 1$ should be modified in the following way:

$$p_{i,j}(t+1) = \begin{cases} 1 - \lambda_t \cdot (1 - p_{i,j}(t)) & \text{if } j \in x^{bsf}; \\ \lambda_t \cdot p_{i,j}(t) & \text{if } j \notin x^{bsf}; \\ p_{i,j}(0) & \text{if } j \text{ was not chosen before.} \end{cases} \tag{1}$$

where $0 < \lambda_t \leq 1$ represents the time dependent *learning rate*. As we already mentioned, $p_{i,j}(0)$ represents the initial value for selection probability. The idea is to learn the influence of the component's value to the quality of generated solutions. This means that if value j belongs to the best-so-far solution, the probability that j will be included as a value for component i of some solution constructed in the next iteration increases. If the value j is not in the best-so-far solution but was selected before (in some of the previous iterations) the probability of its selection in the next iteration decreases. For all values that were not considered in any of previous iterations the selection probability value remains unchanged (its value is still equal to the initial one).

In addition, we define the probability of generating an optimal solution for this type of problems by the constructive SI method as the following indicator function of the pair (i, j):

$$p_{i,j}^* = \begin{cases} 1 \text{ if } j \in x^*; \\ 0 \text{ otherwise.} \end{cases}$$

In the case of the improvement-based metaheuristic method the selection probability that component i should take value j in the iteration $t+1$ should be modified as follows (assuming $0 < \lambda_t \leq 1$):

$$p_{i,j}(t+1) = \begin{cases} 1-\lambda_t \cdot (1-p_{i,j}(t)) & \text{if } j \in \boldsymbol{x}^{bsf} \text{ and } i \notin \boldsymbol{x}^{bsf}; \\ \lambda_t \cdot p_{i,j}(t) & \text{otherwise}; \end{cases} \qquad (2)$$

where $i \notin \boldsymbol{x}^{bsf}$ is an abbreviation for $x_s^i \notin \boldsymbol{x}^{bsf}$ and it actually refers to the previous value of the component i in any solution $\boldsymbol{x}_s$ considered for transformation. If value j was a part of the best-so-far solution and the current value of component i was not, the probability that j will substitute the current value of component i in the next iteration is increased. In all other cases we decrease the probability of selecting value j in the next iteration.

The corresponding indicator function is defined in the following way:

$$p_{i,j}^* = \begin{cases} 1 \text{ if } i \notin \boldsymbol{x}^*, j \in \boldsymbol{x}^*; \\ 0 \text{ otherwise}. \end{cases}$$

4.3 Modification Schemes in the Cases When All Data Constitute a Solution

First, we consider the problems (such as TSP) requiring to properly order solution's components. The selection probability modification schemes in the case of constructive and improvement-based SI methods have the same form:

$$p_{i,j}(t+1) = \begin{cases} 1-\lambda_t \cdot (1-p_{i,j}(t)) & \text{if } (i,j) \in \boldsymbol{x}^{bsf}; \\ \lambda_t \cdot p_{i,j}(t) & \text{otherwise}; \end{cases} \qquad (3)$$

where $0 < \lambda_t \leq 1$. Here, we calculate the probability that component i should be assigned value j either by adding value in a constructive method or by transforming its previous value in an improvement-based method. The notation $(i,j) \in \boldsymbol{x}^{bsf}$ describes the case that components $i-1$ and i are getting the same combination of values as in the best-so-far solution. The corresponding indicator function can be defined as follows:

$$p_{(i,j)}^* = \begin{cases} 1 \text{ if } (i,j) \in \boldsymbol{x}^*; \\ 0 \text{ otherwise}. \end{cases}$$

The same rule is applied to the VRP type problems, while for scheduling like problems pair (i,j) denotes that the task i should be allocated to the group j. Consequently, the modification (selection) scheme (3) is applicable in this case as well.

4.4 Sufficient Conditions for Model Convergence of SI Methods

Let us assume that the considered SI algorithm is applying one of previously defined schemes for modifying selection probabilities. In this section we provide the sufficient conditions for convergence in probability of the SI obtained solution toward an optimal one. In order to assure model convergence of the SI algorithm, two conditions should be satisfied: (i) all feasible solutions need to be reachable from any initial solution and (ii) upon an optimal solution is found its generation has to be favored.

The first condition represents the best-so-far convergence and it is satisfied if, with probability one, there exists an iteration $t > t_0$ in which any considered solution $\boldsymbol{x}$ (optimal included) is found. This is consistent with the condition (i) from the Theorem 1, i.e., the probability that an optimal solution will never be generated tends to zero when $t \to \infty$. The second condition is related to model convergence. It requires to prove that, after generating an optimal solution, by applying the corresponding update rule defined by one of the Eqs. (1), (2) or (3), the generation of optimal solutions will be supported for the considered SI algorithm. This actually means that $p_{i,j}(t)$ converges to $p_{i,j}^*$.

Theorem 2. *The conditions*

$$1 \geq \lambda_t \geq \frac{\log t}{\log(t+1)} \ \textit{for all } t \geq t_0, \ (t_0 \geq 2), \tag{4}$$

and

$$\sum_{t=1}^{+\infty}(1-\lambda_t) = +\infty, \tag{5}$$

are sufficient for the corresponding SI method to converge in probability toward an optimal solution $\boldsymbol{x}^*$ from $\mathcal{X}^*$.

Proof: (i) (best-so-far convergence) We actually prove that $\Pr(B) = 0$, i.e., the equivalent condition from Lemma 1. Let $\boldsymbol{x}$ be a given feasible solution. Then $C(t)$ means that $\boldsymbol{x}$ is found for the first time in iteration t. As

$$B = C^c(1) \cap C^c(2) \cap \cdots \Rightarrow \boldsymbol{x} \text{ is never found}$$

then it holds

$$\Pr(B) = \Pr(\{C^c(1) \cap C^c(2) \cap \cdots\}) \leq \Pr(\{\boldsymbol{x} \text{ is never found}\})$$

$$= \prod_{t=1}^{+\infty} \Pr(\{\boldsymbol{x} \text{ is not found in iteration } t | \boldsymbol{x} \text{ is not found in iteration } k < t\}). \tag{6}$$

If we refer to solution components and selection probability update rules (1), (2) and (3), in the worst case for all pairs of components (i, j) not being established in iterations $1, \ldots, t$, it holds:

$$p_{i,j}(t) = \left[\prod_{k=1}^{t-1} \lambda_k\right] \cdot p_{i,j}(0),$$

(justifiable easily by induction). Applying the first condition of the Theorem 2, it holds

$$\left[\prod_{k=1}^{t-1} \lambda_k\right] \cdot p_{i,j}(0) \geq \left[\prod_{k=1}^{t_0-1} \lambda_k\right] \cdot \left[\prod_{j=t_0}^{t-1} \frac{\log j}{\log(j+1)}\right] \cdot p_{i,j}(0)$$

$$= \left[\prod_{k=1}^{t_0-1} \lambda_k\right] \cdot \frac{\log t_0}{\log t} \cdot p_{i,j}(0) = \frac{const}{\log t}.$$

In such a way, for any pair of components (i,j), we obtained a lower bound of the worst case selection scenario. Consecutively, even in the worst case, for the probability to find the solution x by the considered SI method it holds:

$$\prod_{(i,j)\in x} p_{i,j}(t) \geq \left(\frac{const}{\log t}\right)^n,$$

n being the number of components in the solution x.
Assuming that the solution x was not found by the SI method, an upper bound on the right hand side of the relation (6) is obtained:

$$\prod_{t=t_0}^{+\infty} \left[1 - \left(\frac{const}{\log t}\right)^n\right].$$

Applying logarithm and the convexity of the exponential function indicated by $(1-a) \leq e^{-a}, \forall a \in (0,1)$ (i.e., $\log(1-a) \leq -a$), from the previous term we obtain:

$$\sum_{t=t_0}^{+\infty} \log\left(1 - \left(\frac{const}{\log t}\right)^n\right) \leq -\sum_{t=t_0}^{+\infty} \left(\frac{const}{\log t}\right)^n = -\infty.$$

Now we deduce that

$$\prod_{t=t_0}^{+\infty} \left[1 - \left(\frac{const}{\log t}\right)^n\right] = 0,$$

i.e., $\Pr(B) \leq 0$ in (6). As $\Pr(B) \geq 0$ always holds, it is obvious that $\Pr(B) = 0$.

(ii) (Model convergence) Let us assume that x^* is generated in the iteration m for the first time. Then $x_t^{bsf} = x^*$ for all $t \geq m$. Moreover, we can prove that in all iterations $t > m$ the selection probability for pairs (i,j), not included in x^*, decreases, i.e., converges to zero as $t \to \infty$.
Let $(i,j) \notin x^*$. According to (1), (2) and (3), the selection probability of pair (i,j) in iteration $m+r$, $r = 1, 2, ...$ will be modified as follows:

$$p_{i,j}(m+r) = \left[\prod_{k=m+1}^{m+r} \lambda_k\right] \cdot p_{i,j}(m).$$

The previous claim can be verified easily by induction. In addition, the condition (5) gives us

$$\sum_{t=1}^{+\infty}(1 - \lambda_t) = +\infty, \quad \text{which is equivalent to,} \quad \prod_{k=1}^{+\infty}\lambda_k = 0.$$

Consequently, for the probability that $(i, j) \notin \boldsymbol{x}^*$ will be used again after the optimal solution $\boldsymbol{x}^*$ was generated, it holds:

$$\lim_{t\to+\infty} p_{i,j}(t) = \lim_{t\to+\infty}\left[\prod_{k=m+1}^{t-1}\lambda_k\right] \cdot p_{i,j}(m) = 0.$$

That is exactly what needed to be proved for model convergence part of the theorem.   ∎

5   Conclusion and Future Work

We provide the sufficient conditions for the model convergence of the SI-based metaheuristics: (1) all feasible solutions must be reachable from any point in the solution space; (2) once an optimal solution is found, its generation is favored. To fulfill these requirements, the SI algorithm needs to incorporate selection schemes that exploit the knowledge from the previous search. We have provided three modification schemes w.r.t. the characteristics of the SI method and the considered optimization problem. Although the established conditions guarantee the asymptotic convergence of the SI generated solution toward one of the optimal ones, the question of practical usability of this result still remains open. Namely, in practice we cannot perform an infinite number of iterations, and therefore, we need also the evaluation of convergence speed. Consequently, future work should include runtime analysis of an expected time (iteration index) to obtain the optimal solution, the so-called *first hitting time*.

Acknowledgments. This research was supported by the Serbian Ministry of Education, Science and Technological Development, Grant Nos. 174033, 044006 and F159.

Appendix

To describe the GHC algorithm, we remind on several definitions as provided in [13]. An objective function $f : S \to [0, +\infty)$ is defined on a finite set S of all possible solutions. Two important components of GHC are: (a) *neighborhood function* $\eta : S \to 2^S$, where $\eta(\boldsymbol{x}) \subseteq S$, for all $\boldsymbol{x} \in S$, and (b) *hill climbing random variables* $R_k : S \times S \to \mathbb{R}$, where $k = 1, 2, \dots$ indicates an iteration counter of the outer loop controlled by STOP_OUTER. The pseudo-code for GHC algorithm is given as Algorithm 2. At each iteration i of GHC's inner loop, a candidate solution $\boldsymbol{x}$ is generated uniformly at random among all neighbours of the solution $\boldsymbol{x}_i \in S$ i.e., according to probability mass (density) function $h_{\boldsymbol{x}_i}(\boldsymbol{x}) = 1/|\eta(\boldsymbol{x})|$.

The hill climbing random variables R_k are utilized to accept or reject neighbour solution. STOP_INNER is the stopping criterion for inner loop utilized to inspect if a current solution is a local optimum and if the counter k should be incremented. According to [13], it is important to assume that the verification if the current solution is a local optimum can be conducted in polynomial time. This is possible if there are a polynomial number of neighboring solutions of the current solution. As a consequence, the generation of a local optimum implies that a new random variable will be used within the next iteration.

Define function η and a set of random variables R_k;
$k \leftarrow 1$; /* initialize outer loop counter */
$i \leftarrow 0$; /* initialize inner loop counter */
Select an initial solution $\boldsymbol{x}_0 \in S$;
while not STOP_OUTER **do**
  while not STOP_INNER **do**
    Generate $\boldsymbol{x} \in \eta(\boldsymbol{x}_i)$ according to $h_{\boldsymbol{x}_i}(\boldsymbol{x})$;
    $\Delta(\boldsymbol{x}_i, \boldsymbol{x}) = f(\boldsymbol{x}) - f(\boldsymbol{x}_i)$;
    if $R_k \geq \Delta(\boldsymbol{x}_i, \boldsymbol{x})$ **then**
      Accept solution $\boldsymbol{x}$;
    else
      Reject solution $\boldsymbol{x}$;
    end if
    $i \leftarrow i + 1$;
  end while
  $k \leftarrow k + 1$;
  Update R_k;
end while

Algorithm 2: Pseudo-code of the GHC algorithm.

It is important to note that the search space S must be *reachable*, that is, all solutions are accessible regardless of a starting point $\boldsymbol{x}_0$ [13].

References

1. Bonabeau, E., Dorigo, M., Theraulaz, G.: Swarm Intelligence: From Natural to Artificial Systems. Oxford University Press, New York (1999)
2. Chen, J., Ni, J., Hua, M.: Convergence analysis of a class of computational intelligence approaches. Math. Probl. Eng. **2013**, 1–10 (2013)
3. Davidović, T., Teodorović, D., Šelmić, M.: Bee Colony Optimization- Part I: the algorithm overview. Yugoslav J. Oper. Res. **25**(1), 33–56 (2015)
4. Dorigo, M., Birattari, M.: Swarm intelligence. Scholarpedia **2**(9), 1462 (2007). http://www.scholarpedia.org/Swarm_intelligence
5. Dorigo, M., Blum, C.: Ant Colony Optimization theory: a survey. Theor. Comput. Sci. **344**, 243–278 (2005)

6. Dorigo, M., Stützle, T.: Ant Colony Optimization: overview and recent advances. In: Gendreau, M., Potvin, J.Y. (eds.) Handbook of Metaheuristics. ISOR, vol. 146, pp. 227–263. Springer, Boston (2010). https://doi.org/10.1007/978-1-4419-1665-5_8

7. Garcia-Gonzalo, E., Fernandez-Martinez, J.L.: A brief historical review of Particle Swarm Optimization (PSO). J. Bioinform. Intell. Contr. **1**(1), 3–16 (2012)

8. Gutjahr, W.J.: A graph-based ant system and its convergence. Future Gener. Comput. Syst. **16**(8), 873–888 (2000)

9. Gutjahr, W.J.: ACO algorithms with guaranteed convergence to the optimal solution. Inf. Process. Lett. **82**(3), 145–153 (2002)

10. Gutjahr, W.J.: Convergence analysis of metaheuristics. In: Maniezzo, V., Stützle, T., Voß, S. (eds.) Matheuristics: Hybridizing Metaheuristics and Mathematical Programming. AOIS, vol. 10, pp. 159–187. Springer, Boston (2009). https://doi.org/10.1007/978-1-4419-1306-7_6

11. Gutjahr, W.J.: Stochastic search in metaheuristics. In: Gendreau, M., Potvin, J.Y. (eds.) Handbook of Metaheuristics. ISOR, vol. 146, pp. 573–597. Springer, Boston (2010). https://doi.org/10.1007/978-1-4419-1665-5_19

12. Gutjahr, W.J.: Ant colony optimization: recent developments in theoretical analysis. In: Auger, A., Doerr, B. (eds.) Theory of Randomized Search Heuristics: Foundations and Recent Developments, pp. 225–254. World Scientific (2011). https://www.worldscientific.com/worldscibooks/10.1142/7438

13. Jacobson, S.H., Yücesan, E.: Global optimization performance measures for generalized hill climbing algorithms. J. Global Optim. **29**(2), 173–190 (2004)

14. Jakšic Krüger, T.: Development, implementation and theoretical analysis of the Bee Colony Optimization metaheuristic method. Ph.D. thesis, University of Novi Sad (2017)

15. Jakšić Krüger, T., Davidović, T.: Model convergence properties of the constructive Bee Colony Optimization algorithm. In: Proceedings of 41th Symposium on Operational Research, SYM-OP-IS 2014, pp. 340–345 (2014)

16. Jakšić Krüger, T., Davidović, T., Teodorović, D., Šelmić, M.: The Bee Colony Optimization algorithm and its convergence. Int. J. Bio-Inspired Comput. **8**(5), 340–354 (2016)

17. Jiang, M., Luo, Y., Yang, S.: Stochastic convergence analysis and parameter selection of the standard Particle Swarm Optimization algorithm. Inf. Process. Lett. **102**(1), 8–16 (2007)

18. Karaboga, D., Gorkemli, B., Ozturk, C., Karaboga, N.: A comprehensive survey: Artificial Bee Colony (ABC) algorithm and applications. Artif. Intell. Rev. **42**, 21–57 (2014)

19. Kötzing, T., Neumann, F., Röglin, H., Witt, C.: Theoretical analysis of two ACO approaches for the Traveling Salesman Problem. Swarm Intell. **6**(1), 1–21 (2012)

20. Liu, H., Abraham, A., Snásel, V.: Convergence analysis of swarm algorithm. In: NaBIC, pp. 1714–1719 (2009)

21. Parpinelli, R.S., Lopes, H.S.: New inspirations in swarm intelligence: a survey. Int. J. Bio-Inspired Comput. **3**(1), 1–16 (2011)

22. Pintér, J.: Convergence properties of stochastic optimization procedures. Optimization **15**(3), 405–427 (1984)

23. Solis, F.J., Wets, R.J.B.: Minimization by random search techniques. Math. Oper. Res. **6**(1), 19–30 (1981)

24. Stützle, T., Dorigo, M.: A short convergence proof for a class of Ant Colony Optimization algorithms. IEEE Trans. Evol. Comput. **6**(4), 358–365 (2002)

25. Talbi, E.G.: Metaheuristics: From Design to Implementation. Wiley, New York (2009)
26. Trelea, I.C.: The Particle Swarm Optimization algorithm: convergence analysis and parameter selection. Inf. Process. Lett. **85**(6), 317–325 (2003)
27. Van Den Bergh, F.: An analysis of Particle Swarm Optimizers. Ph.D. thesis, University of Pretoria (2006)
28. Yang, X.-S.: Metaheuristic optimization: algorithm analysis and open problems. In: Pardalos, P.M., Rebennack, S. (eds.) SEA 2011. LNCS, vol. 6630, pp. 21–32. Springer, Heidelberg (2011). https://doi.org/10.1007/978-3-642-20662-7_2
29. Zeng, J.C., Cui, Z.H.: A guaranteed global convergence Particle Swarm Optimizer. J. Comput. Res. Dev. **8**, 1333–1338 (2004)
30. Zlochin, M., Birattari, M., Meuleau, N., Dorigo, M.: Model-based search for combinatorial optimization: a critical survey. Ann. Oper. Res. **131**(1–4), 373–395 (2004)
31. Zlochin, M., Dorigo, M.: Model-based search for combinatorial optimization: a comparative study. In: Guervós, J.J.M., Adamidis, P., Beyer, H.-G., Schwefel, H.-P., Fernández-Villacañas, J.-L. (eds.) PPSN 2002. LNCS, vol. 2439, pp. 651–661. Springer, Heidelberg (2002). https://doi.org/10.1007/3-540-45712-7_63

An Optimization Model for Empty Tank Cars Movement at Railway Petroleum Logistics Market

Ivan A. Davydov[1,2(✉)]

[1] Sobolev Institute of Mathematics, Novosibirsk, Russia
[2] Novosibirsk State University, Novosibirsk, Russia
idavydov@math.nsc.ru

Abstract. We consider the problem of empty tank cars relocation in a railway petroleum transportation. Biggest petroleum carrier in Kazakhstan railway network provides transportation service to the customers. The majority of its expenses consist of an empty run of the cars, as they move to the load stations. Given the demand for empty tank cars, the company seeks to reduce total costs to satisfy all the demand within a planning period. We provide a mathematical model for this problem in terms of integer programming and perform an experimental study with LP solver. Numerical calculations show that new model allows to significantly reduce empty run of the tank cars on the railway petroleum logistics market lowering the total expenses of the company by more than 10% level.

Keywords: Railroads · Transportation · Tank cars · Assignment
Integer linear programming

1 Introduction

The problem of reassignment of empty cargo tank cars is among the most challenging ones in a railway freight transportation. Numerous initial data should be taken into consideration, i.e. demand in empty cars on the loading stations, availability of the cars, their locations and maintenance schedule, compatibility of different cars and types of cargo. A number of studies is devoted to this kind of problems. In [7] the authors consider an empty freight cargo management at Union Pacific Railroad. In the presented model authors seek to reduce transportation costs, and improve delivery time and customer satisfaction while assigning empty freight cars based on demand. They provide a MILP formulation of the problem, mainly based on the transportation problem and propose an efficient solution approach. The UP has introduced the proposed model to the decision-making system resulting in an ROI of 35%. Another modeling approach suggested in [4]. The authors propose a novel empty rail car distribution problem that considers realistic technical and business requirements while assigning

© Springer International Publishing AG, part of Springer Nature 2018
A. Eremeev et al. (Eds.): OPTA 2018, CCIS 871, pp. 267–277, 2018.
https://doi.org/10.1007/978-3-319-93800-4_21

empty cars to customer demands. They provide a path-based multi-commodity capacitated network flow model for this problem. The model takes car substitution, network capacities and classification yard operation complexities into consideration while dissociating the car routing and car distribution decisions. Those two decisions are separated from each other and they are usually made by different departments in railroads. The result of a numerical example showed an 18% improvement, comparing with the non capacitated model. In this work we consider a petroleum logistics market taking its particular properties into account. Petroleum carrier company owns a fleet of tank cars and provides a transportation service on the railway network. The service is generally provided to deliver the petroleum products from refinery plants to retail stations. Some retailers, in turn, generate their own delivery orders in order to supply smaller consumers. Both refinery plants and retailers generate a demand for an empty cars that have to be provided for loading. The demand information includes the number of cars to be provided, due date and also specifies the type of product to be carried. The demand information is updated daily - predicted demand is substituted by explicit data, some orders may be cancelled, new ones may appear. In this work, we present an optimization model for a Kazakhstan Railway Network oil carrier. The majority of its expenses consist of an empty run of the cars, as they move to the load stations. Given the demand for empty tank cars, the company seeks to reduce total costs to satisfy all the demand within a planning period.

2   Problem Formulation

The aim of this work is devoted to the optimization of the empty tank car movement of the biggest oil carrier on the railway network of Kazakhstan Republic. The carrier company owns a fleet of tank cars of different types and provide a transportation service to the customers. The majority of movements are devoted to the transportation of the oil products from three refinery plants to the consumers, or retailer stations. A number of retailers, in turn, make their own orders for empty tank cars thus the demand is not concentrated only on refinery stations. We assume that the movements of the loaded cars are well defined by the set of orders and are made according to schedule. Together with known running times for loaded cars, this data provides us an information about the appearance of unloaded cars on their destination stations. This information is used as an input data. Running times for the empty cars are also assumed to be known as a result of statistical measurements. The set of all stations can be separated into fours subsets in the following way. The first set consists of consumer stations, i.e. those where only unload operations are performed. The demand for empty cars is assumed to be zero, so there is no need to send empty cars to this stations. The second set consist of retailer stations. This stations mostly consume the production of the refinery plants, but also provide retail service to the smaller customers providing a nonzero demand for empty cars by their own. The third set represents the refinery plants. There is a number of refinery plants in the network which supplies a number of different oil products and provides most of the demand for empty cars on the market. This demand is assumed to be known a priori along the whole planning horizon. Due to the extremely high costs for breaking the production process, it is required to ensure that the amount of empty cars on this stations is enough for three consecutive days of operation.

Each refinery has an associated Washing Station which performs the preparation (washing) process. Each car should be washed on such station before load. Washing procedure cost depend on the type of the car to be washed, previous cargo and the cargo to be prepared for. Even though it is possible to prepare a car, used to carry asphalt for a gasoline transportation, the cost of such preparation would be times higher than gasoline-gasoline washing. Together with price levels, each washing station has a given capacity, so only a limited amount of preparations can be performed. The last, fourth set of stations represent the dead ends. This stations used to store the cars, while they are not in demand. The storage is also possible at the refinery stations but is not preferable due to the limited capacity and higher storage price.

Although it is highly desirable to cover all the demand for empty cars, we assume that the lack is possible and leads to a penalty. This penalty represent additional expenses caused by the shortage. The order still can be fulfilled, as it is possible to rent additional cars from the third party. For the sake of simplicity, this aspect is not included in the model.

The company aims to minimize total expenses, which consist of the following terms; mainly it is the cost of an empty run of the tank cars, washing and preparation process, payments for storage and penalty for the shortage of tank cars.

In the following section, we provide a model for this problem in terms of linear integer programming.

3   Mathematical Model

We consider the following formulation of the empty tank cars transportation over the railway network. Sets:

$g \in G$ - types of tank cars;
$c \in C$ - types of cargo;
$i \in I_1$ - stations with only unload operations allowed (consumers);
$i \in I_2$ - stations with both load and unload operations allowed (advanced customers);
$i \in I_3$ - stations connected with factory;
$i \in I_4$ - storage stations;
$i \in I = I_1 \cup I_2 \cup I_3 \cup I_4$ - all stations;
$t \in T = 1..t^{max}$ - days in planning horizon;
$p \in P$ - price level for different types of cargo;

Parameters:

q_{ict}^g - amount of tank cars of type g prepared for transportation of cargo c become available at station i on day t;

V_i -capacity of station i;

τ_{ij} - time to travel from station i to station j;

d_{ijp} - price to travel from station i to station j price level p;

r_{ict} - amount of empty tank cars, prepared for cargo c needed at station i on day t;

$P_{ic_1c_2g}$ - price to prepare(wash) tank car of type g from cargo c_1 to cargo c_2 at station i;

P_i^{max} - upper bound on washing costs on day i;

F^f - fine level induced for insufficient amount of empty tank cars at factory $i \in I_3$;

F^{St} - fine level induced for insufficient amount of empty tank cars at loading station $i \in I_2$;

H_i^{St} - cost for unnecessary storage of the tank car on station $i \in I_2$ per day;

H_i^d - storage cost per tank car per day on the storage station $i \in I_4$;

H_i^f - storage cost per tank car per day on the factory $i \in I_3$;

L^w - duration of the washing process, days;

L^{St} - minimum duration of stay on a storage station, days;

Variables:

$x_{ijcgt} \geq 0$ - amount of empty tank cars of type g prepared for cargo c sent from i to j on day t;

$y_{ic_1c_2gt} \geq 0$ - amount of empty tank cars of type g prepared for cargo c_1 sent to wash for cargo c_2 on station i on day t;

Q_{ict} - number of tank cars prepared for cargo c available on station i on day t;

Q_{ict}^g - number of tank cars of type g prepared for cargo c available on station i on day t;

$\bar{Q}_{ict} \geq 0$ - lack of tank cars for cargo c on station i on day t;

$\bar{Q}_{ict}^g \geq 0$ - lack of tank cars of type g for cargo c on station i on day t;

r_{ict}^g - demand on station i tank cars g cargo c on day t;

S^{store} - total cost of storage of tank cars on the planning horizon;

S^{pen} - total fine, induced for insufficient amount of tank cars on planning horizon;

S^{wash} - total cost for washing on planning horizon;

S^{run} - total cost for empty tank cars run;

With the introduced variables the problem can be formulated as follows. Goal function

$$\min S^{run} + S^{wash} + S^{pen} + S^{store} \qquad (1)$$

subject to

$$\sum_{t' \leq t} q_{icgt'} = \sum_{j \in I} \sum_{t' \leq t} x_{ijcgt'} \quad i \in I_1, c \in C, g \in G, t \in T \qquad (2)$$

$$Q_{ict}^g = \sum_{t' \le t} q_{ict'}^g + \sum_{j \in I} \left(\sum_{t' \le t - \tau_{ji}} x_{jicgt'} - \sum_{t' \le t} x_{ijcgt'} \right)$$
$$- \sum_{t' \le t} r_{ict'}^g + \sum_{t' \le t} \bar{Q}_{ict'}^g \quad i \in I_2, c \in C, g \in G, t \in T \tag{3}$$

$$r_{ict} = \sum_{g \in G} r_{ict}^g \quad i \in I_2 \cup I_3, c \in C, t \in T \tag{4}$$

$$\bar{Q}_{ict} = \sum_{g \in G} \bar{Q}_{ict}^g \quad i \in I_2, c \in C, t \in T \tag{5}$$

$$\bar{Q}_{ict}^g \le r_{ict}^g \quad i \in I_2, c \in C, g \in G, t \in T \tag{6}$$

$$\sum_{t' \le t} q_{ict'}^g + \sum_{j \in I} \left(\sum_{t' \le t - \tau_{ji}} x_{jicgt'} - \sum_{t' \le t} x_{ijcgt'} \right) \ge 0 \quad i \in I_2, c \in C, g \in G, t \in T \tag{7}$$

$$\sum_{g \in G} Q_{ict}^g \le \sum_{t' \ge t} r_{ict} \quad i \in I_2, c \in C, t \in T, \tag{8}$$

$$\sum_{j \in I} \sum_{g \in G} \sum_{t' \le t - \tau_{ji}} x_{jicgt'} \le \sum_{t' \le t} r_{ict'} - \sum_{g \in G} \sum_{t' \le t} q_{ict'}^g \quad i \in I_2, c \in C, t \in T \tag{9}$$

$$Q_{ict}^g = \sum_{t' \le t} q_{ict'}^g + \sum_{j \in I} \left(\sum_{t' \le t - \tau_{ji}} x_{jicgt'} - \sum_{t' \le t} x_{ijcgt'} \right) - \sum_{t' \le t+2} r_{ict'}^g$$
$$+ \sum_{c_1 \in C} \left(\sum_{t' \le t - L^w} y_{ic_1cgt'} - \sum_{t' \le t} y_{icc_1gt'} \right) \quad i \in I_3, c \in C, g \in G, t \in T \tag{10}$$

$$Q_{ict} = \sum_{g \in G} Q_{ict}^g \quad i \in I, c \in C, t \in T, \tag{11}$$

$$\bar{Q}_{ict}^g \ge -Q_{ict}^g \quad i \in I_3, c \in C, g \in G, t \in T \tag{12}$$

$$\sum_{c \in C} Q_{ict} + \sum_{c_1 \in C} \sum_{c_2 \in C} \sum_{g \in G} \sum_{t - L^w + 1 \le t' \le t} y_{ic_1c_2gt'} + \sum_{c \in C} \sum_{t+1 \le t' \le t+2} r_{ict'}^g \le V_i \tag{13}$$
$$i \in I_3, t \in T$$

$$\sum_{c_1 \in C} \sum_{c_2 \in C} \sum_{g \in G} P_{ic_1c_2g} y_{ic_1c_2gt} \le P_i^{max} \quad i \in I_3, t \in T, \tag{14}$$

$$\sum_{c_1 \in C} \sum_{g \in G} \sum_{t' \le t - L^w} y_{ic_1cgt'} \ge \sum_{t' \le t+2} r_{ict'} + \sum_{g \in G} \sum_{j \in I} \sum_{t' \le t} x_{ijcgt'}$$
$$- \sum_{g \in G} \sum_{t' \le t} \bar{Q}_{ict'}^g \quad i \in I_3, c \in C, t \in T \tag{15}$$

$$\sum_{t' \le t} q_{ict'}^g + \sum_{j \in I} \sum_{t' \le t - \tau_{ji}} x_{jicgt'} \ge \sum_{c_2 \in C} \sum_{t' \le t} y_{icc_2gt'} \quad i \in I_3, c \in C, g \in G, t \in T \tag{16}$$

$$\sum_{t' \le t} q_{ict'}^g + \sum_{j \in I} \sum_{t' \le t^{max} - \tau_{ji}} x_{jicgt'} = \sum_{c_2 \in C} \sum_{t' \in T} y_{icc_2gt'} \quad i \in I_3, c \in C, g \in G \tag{17}$$

$$\bar{Q}^g_{ict-1} - \bar{Q}^g_{ict} \le \sum_{c_2 \in C} y_{ic_2cgt-L^w} \quad i \in I_3, c \in C, g \in G, t \in T \tag{18}$$

$$Q^g_{ict} = \sum_{t' \le t} q^g_{ict'} + \sum_{j \in I} \sum_{t' \le t-\tau_{ji}} x_{jicgt'} - \sum_{j \in I} \sum_{t' \le t} x_{ijcgt'} \quad i \in I_4, c \in C, g \in G, t \in T \tag{19}$$

$$\sum_{c \in C} \sum_{g \in G} Q^g_{ict} \le V_i \quad i \in I_4, t \in T \tag{20}$$

$$S^{run} = \sum_{i \in I} \sum_{j \in I} \sum_{c \in C} \sum_{g \in G} \sum_{t \in T} d_{ijp(c)} x_{ijcgt} \tag{21}$$

$$S^{store} = \sum_{c \in C} \sum_{t \in T} \left(\sum_{i \in I_2} Q_{ict} H^{St}_i + \sum_{i \in I_3} (Q_{ict} + \bar{Q}_{ict}) H^f_i + \sum_{i \in I_4} Q_{ict} H^d_i \right) \tag{22}$$

$$S^{wash} = \sum_{i \in I_3} \sum_{c_1 \in C} \sum_{c_2 \in C} \sum_{g \in G} \sum_{t \in T} y_{ic_1c_2gt} P_{ic_1c_2g} \tag{23}$$

$$S^{fpen} = \sum_{i \in I_3} \sum_{c \in C} \sum_{t \in T} \bar{Q}_{ict} F^f + \sum_{i \in I_2} \sum_{c \in C} \sum_{t \in T} \bar{Q}_{ict} F^{St} \tag{24}$$

$$\sum_{j \in I} \sum_{t' + \tau_{ji} \ge t^{max}} x_{jicgt'} = 0 \quad i \in I_2 \cup I_3, c \in C, g \in G, \tag{25}$$

$$x_{ijcgt} = 0 \quad t \in T, i \in I_4, j \in I_4, c \in C, g \in G \tag{26}$$

$$x_{ijcgt} = 0 \quad t \in T, i \in I_3, j \in I_3, g \in G, c \in C \tag{27}$$

$$x_{jicgt} = 0 \quad g \in G, i \in I_2, j \in I_1 \cup I_2 \cup I_4, c \in C, t \in T \tag{28}$$

The objective function specifies the total costs required to cover the demand for empty tank cars at the planning horizon. The expenses consist of the total cost of the empty tank car run, the cost of all washes, the penalty for the lack of tank cars as well as the cost of storing tank cars at stations and dead ends.

The constraints of the problem can be conditionally divided into several groups, each of which simulates the movement of tank cars on one or another set of stations. Constraint (2) describes the flow balance of emerging and departing tank cars at unload stations. Namely, it requires that all tank cars that appear on the station have to be sent to the destination station on the same day. Next group of constraints (3)–(9) describes the operation of stations with load, i.e. stations with a nonzero demand for empty tank cars. The constraint (3) describes the balance of the tank car flow at such stations: the number of tank cars available at the station is calculated as the total number of tank cars that appeared or arrived at the station at that time, minus empty and loaded tank cars sent to other stations. The last term in (3) represents a lack of tank cars at the station. In case it was not possible to cover all the demand for the station, this term will reflect the total shortage of tank cars on each day. The constraint (4) indicates that the demand for empty cars must be split into different categories of tank cars. Similarly, the shortage of tank cars in the constraint (5) splits into categories. According to (6), the shortage of tank cars cannot exceed the demand for them. Inequality (7) indicates that the number of real tank cars at the station cannot be negative. Constraint (8) does not allow to store more tank cars at the station for a given type of cargo that may be needed in the future for loading. Inequality (9) indicates that the number of empty tank cars which can arrive at the station cannot exceed the number of required to the current moment.

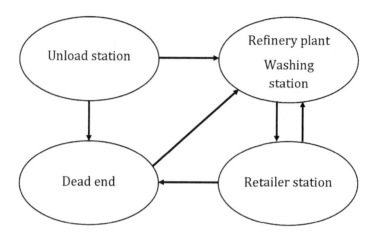

Fig. 1. Scheme of empty movements

The following group of constraints (10)–(18) describes the operation of the refinery plant. The crucial difference between plants and stations with load are an increased priority in meeting the demand and the availability of washing supplies. The first constraint of the group (10) calculates the balance of the tank cars at the refinery. This constraint is constructed similarly to (3). The difference is that washing operations are also included in the balance flow. As a result of washing operations, the tank cars can change the type of cargo associated with them, or simply be prepared for loading. (10) claims to cover the demand for empty tank cars for three days in advance. Constraint (11) calculates the total stock of tank cars for a certain type of cargo. (12) specifies the relationship between the variables for the shortage of tank cars and their number. Constraint (13) counts the total number of tank cars located on the station and claims to satisfy the capacity conditions. Inequality (14) limits the total costs of washing operations at the washing station during the day. Constraint (15) prescribes to wash all tank cars leaving the station in any direction, while (16) and (17) do not allow more tank cars to be used for washing than it was available. Constraint (18) indicates that the tank cars which cover the shortage should also be washed. The group (19)–(20) describes the balance of tank cars in dead ends and satisfies the capacity conditions.

Constraints (21)–(24) calculate the total costs for each group of operations - empty mileage, payment for washes, fines for the shortage, storage fees. Finally (25)–(28) describe special logical constraints, such as the ban on sending tank cars to stations and refineries if the arrival would occur beyond the planning horizon; prohibition of empty cars to pass straightforward between pairs of dead ends, between pairs of oil refineries; dispatch of tank cars to the loading stations is possible only from the refinery plant. The schematic view of possible movements of empty cars is presented in Fig. 1.

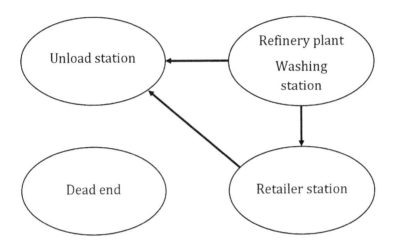

Fig. 2. Scheme of cargo movements

Figure 2 represents the schematic view of the possible cargo movements. It is assumed that these movements are predetermined by the list of orders, thus fixed and represented in the model as an input parameters q_{ict}^g.

4   Experimental Study

The presented model was numerically tested on the real data instances provided by Kazakhstan petroleum carrier. During the study, we applied the GLPK freeware solver to carry out the calculations. The aim of this study was twofold. On the one hand, the desire was to raise the efficiency of the empty car movements and reduce the associated costs. On the other hand, the demand data and other parameters may be updated more than once during the day so the company should be capable to find the solution quickly. The presented model appeared to be successful in both directions. Despite the constraint matrix is not totally unimodular, due to the absence of the parametric coefficients for the variables in the constraints we observed the following property. On every instance, we tested during the experimental study the solution of the linear relaxation of the problem appeared to be fully integer. Thus, the time per instance did not exceed 10 min on a 2,4 Ghz Core i7 PC and an LP solution always provided us with a globally optimal integer solution.

The parameters of the instance are as follows:

6800 tank cars;
3 refinery plants;
638 unload stations;
188 station with demand;
15 dead ends;
Average value for total demand of the refineries is about 500 cars per day;
average travel time is 2.8 days;
A scheme of the Kazakhstan railroads is presented in Fig. 3.

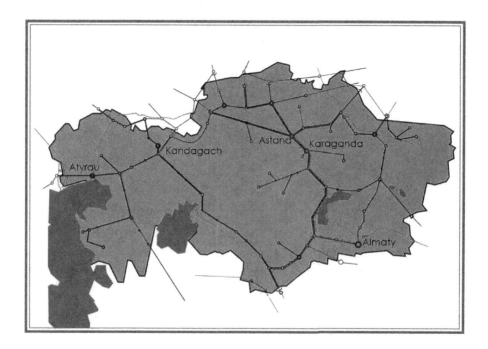

Fig. 3. Map of Kazakhstan railways

During the first experimental study, we conducted a series of runs on a day-by-day basis to compare the results, obtained with the model and real movements of the fleet. As the input data is inexact, especially in a long-term, the study is conducted as follows. We run the model for $[1...T]$ period, obtain an optimal solution and fix the values of the "first-day" variables. Then we run the model for a period $[2...T+1]$, using the solution for day 1 as an input data. I.e. if the value of $x_{i,j,c,g,1} = k$ then we use it on the next step as follows: $q'_{j,c,g,1+\tau_{ij}} = q'_{j,c,g,1+\tau_{ij}} + k$. A series of runs for $T = 30$ times was performed and a period $[1...T]$ was taken into consideration. We obtained the following results: an overall run of the empty cars reduced from 48% to 36% of the total run while the expenses of the company decreased by 10.38% of its real level.

In our second study, we made an attempt to optimize the number of tank cars that the company have to maintain in its fleet. The maintenance cost of each car is a sufficient value, thus the possibility of the shortage of the fleet without losing the quality of service looks quite desirable. During the number of test runs, performed in the same way as before, we compared the empty run and total expenses of the company under different values of the fleet size. We add here another term to the goal function to represent the maintenance cost of the fleet. For fleets big enough this cost is proportional to the size of the fleet. The values for the empty run and total expenses of the company with the real fleet size of 6800 cars, obtained during the previous experiment, were taken as a 100%. The results of the study presented in Table 1.

Table 1. Empty run and expenses.

Fleet size	Empty run (in%)	Total expenses (in%)
5500	110.1	116.3
6000	104.7	98.2
6500	101.3	97.4
6800	100	100
7000	98.1	102.1
7500	91.8	112.4
8000	83.2	123.2

It can be noted from the table that the current fleet of the company is slightly excessive, and on the considered planning horizon it is possible to satisfy all the customers with the smaller fleet. But the company prefers to have a slight reserve in its fleet to be able to manage with any unexpected circumstances. If the fleet is reduced to a size of 5000 cars, the company would not be able to fulfill all orders, so the subsequent fines for unsatisfied orders will raise the expenses. Bigger fleet allows to reduce the empty run, but the maintenance costs for additional cars exceeds the savings.

5   Conclusion and Future Research

This paper studies a new optimization model to minimize the total operational costs of a railway petroleum carrier. We present an integer linear model and apply the GLPK freeware solver. Computational results on the real test instances confirm the efficiency of the approach and allow the company to significantly reduce empty run of the tank cars on the railway petroleum logistics market from 48% to 36% of the total run. Total expenses of the company were also reduced by more than 10%.

As a possible direction for future research, we suggest considering the competition on the market. Similar models for two companies in the field of facility location problem can be found in [1,6]. It is worth to note that such models are harder than well-known NP-complete problems and, in fact, are Σ_p^2-hard [2,5]. The best pricing strategy for each company may be the most intriguing question in such games. Some preliminary results for Stackelberg pricing games can be found in [8–10]. Another direction for future research is the development of the model. Among many the following directions may be considered: Introduce the possibility of car rent. Although, usually the fleet size of the carrier is enough to cover all the demand on the market, due to some emergency situations the company may need additional cars for temporary use. Those may be rented from another carrier (country). The carrier has to pay the rent price for each day, thus it is preferable to use rented cars more intensively than the own ones.

Acknowledgments. The work was supported by the program of fundamental scientific researches of the SB RAS I.5.1., project 0314-2016-0014.

References

1. Alekseeva, E., Kochetov, Yu.: Matheuristics and exact methods for the discrete (r|p)-centroid problem. In: Talbi, El.-G. (ed.) Metaheuristics for Bi-level Optimization. SCI, vol. 482, pp. 189–219. Springer, Heidelberg (2013). https://doi.org/10.1007/978-3-642-37838-6_7
2. Beresnev, V.L., Melnikov, A.A.: A capacitated competitive facility location problem. J. Appl. Ind. Math. **10**(1), 61–68 (2016)
3. Diakova, Z., Kochetov, Yu.: A double VNS heuristic for the facility location and pricing problem. Electron. Not. Discrete Math. **39**, 29–34 (2012)
4. Heydari, R., Melachrinoudis, E.: A path-based capacitated network flow model for empty railcar distribution. Ann. Oper. Res. **253**, 773–798 (2017)
5. Iellamo, S., Alekseeva, E., Chen, L., Coupechoux, M., Kochetov, Yu.: Competitive location in cognitive radio networks. 4OR **13**(1), 81–110 (2015)
6. Lavlinskii, S.M., Panin, A.A., Plyasunov, A.V.: Comparison of models of planning public-private partnership. J. Appl. Ind. Math. **10**(3), 356–369 (2016)
7. Narisetty, A.K., Richard, J.-P.P., Ramcharan, D., Murphy, D., Minks, G., Fuller, J.: An optimization model for empty freight car assignment at union pacific railroad. Interfaces **38**(2), 89–102 (2008)
8. Panin, A.A., Plyasunov, A.V.: On complexity of the bilevel location and pricing problems. J. Appl. Ind. Math. **8**(4), 574–581 (2014)
9. Plyasunov, A.V., Panin, A.A.: The pricing problem. I: Exact and approximate algorithms. J. Appl. Ind. Math. **7**(2), 241–251 (2013)
10. Plyasunov, A.V., Panin, A.A.: The pricing problem. II: Computational complexity. J. Appl. Ind. Math. **7**(3), 420–430 (2013)

Complexity of Bi-objective Buffer Allocation Problem in Systems with Simple Structure

Alexandre B. Dolgui[1], Anton V. Eremeev[2], Mikhail Y. Kovalyov[3], and Vyacheslav S. Sigaev[4(✉)]

[1] IMT Atlantique, Nantes, France
[2] Sobolev Institute of Mathematics, Omsk, Russia
eremeev@ofim.oscsbras.ru
[3] United Institute of Informatics Problems, Minsk, Belarus
[4] Avtomatika-Servis LLC, Omsk, Russia
sigvs@mail.ru

Abstract. We consider a bi-objective optimization problem of choosing the buffers capacity in a production system of parallel tandem lines, each consisting of two machines with a single intermediate buffer. During operation of the system, the equipment stops occur due to failures and these stops are random in the moments when they arise and in their durations. The product is accumulated in an intermediate buffer if the downstream machine is less productive than the upstream machine.

We study the complexity of exact and approximate computations of a Pareto front for the following two bi-objective problem formulations: (i) the expected revenue maximization with minimization of buffers allocation cost and (ii) the expected revenue maximization with minimization of expected inventory costs. The expected revenue is assumed to be an increasing function of the expected throughput of the system.

On the one hand, fully polynomial-time approximation schemes for approximation of Pareto fronts of these problems are proposed and an exact pseudo-polynomial time algorithm is suggested for the first problem in the case of integer buffer capacity costs. On the other hand, we show that both of these problems are intractable even in the case of just one tandem two-machine line.

Keywords: Inventory system · Throughput · Capital costs
Storage costs · Intractability

1 Introduction

Finding the set of Pareto-optimal solutions or a close approximation to it are of great importance in design of automated control and decision support systems. The problem of buffer volume optimization of the volume of buffers arises in

© Springer International Publishing AG, part of Springer Nature 2018
A. Eremeev et al. (Eds.): OPTA 2018, CCIS 871, pp. 278–287, 2018.
https://doi.org/10.1007/978-3-319-93800-4_22

management of such manufacturing systems as automatic lines, flexible production systems and automated assembly lines, where parts are moved from one machine to another using some transport mechanism.

Due to equipment failures, in the process of operation of the line the machine breakdowns occur in random moments and have a random duration. The consequences of failures spread on related operations due to the impossibility to pass an item onto the following operation, or lack of parts coming from the upstream machine. Presence of buffers for storage of parts between the machines allows to reduce the impact of failures on neighboring operations, and to increase the line throughput, i.e. the production rate of the line in the stationary regime. However, installation of buffers is associated with additional capital expenditures and increases the inventory of parts. The problem consists in choosing the volume of buffers based on the throughput of the line, the capital cost of the installation of buffers and the inventory cost.

Significance of solving such problems of optimization of production lines is shown in [28]. Economic effect of implementation of solution methods for such problems in the car production is shown in [26] on the example of PSA Peugeot Citroën.

Analysis of production lines subject to failures is usually conducted using Markov models with discrete or continuous time under the assumption of geometric or exponential distributions of time to fail and time to repair (see [10]). The duration of processing a part can be assumed deterministic or random (typically with geometric, Erlang or the exponential probability distribution). In the case of continuous time and deterministic durations of parts processing, some non-Markov transitions may be approximated by Markov transitions under the assumption of exponential distribution of the corresponding random variables [16,17,23]. At quite natural assumptions thus obtained Markov models have a stationary distribution (see [24], Chap. 2) and the throughput as well as the expected number of parts in each buffer are determined in the stationary regime.

Most of the works in the literature on optimization of buffer volumes are dealing with a single-criterion problem formulations (see [2,18,20]). Other studies consider more than one criterion, but using a weighted sum of criteria [1,12]. In [5], the ant colony algorithm and the evolutionary algorithm of [30] are adapted for multi-criteria buffer allocation problem. Here the optimization criteria are maximization of the throughput of the line, calculated with a simulation algorithm, and minimization of the overall buffers volume. The well-known variant of multi-objective genetic algorithm [11] is adapted in [9] to the bi-criteria buffers allocation problem, where the criteria are the throughput and the capital cost of buffers installation.

In the present paper, we consider three criteria: maximization of average production rate in the steady regime, minimization of capital costs for the installation of buffers and minimization of the average inventory cost for storage of parts in the intermediate buffers.

Exact methods of calculation of the average production rate are known for two-machine tandem lines and, in some special cases, for the three three-machine tandem lines (see e.g., the review [10]). For the general case, one can only apply the approximate decomposition methods,approximate aggregation or simulation methods [10, 16].

In the present paper, we will not assume a specific type of distribution of time to fail and time to repair or processing time of machines. Neither shall we choose a specific method of computing the expected throughput and inventory of a line. Instead of that, we will make two simple monotonicity assumptions which hold in many different versions of the buffers allocation problem (see the details below in Sect. 2).

Suppose that on the set of feasible solutions D, the vector function of criteria $\mathbf{f} = (f_1, f_2)$ is specified with points $\mathbf{f}(x) = (f_1(x), f_2(x)) \in \mathbb{R}^2, x \in D$ in the criteria values space. In our case, f_1 is a maximization criterion and f_2 is a minimization criterion. Let us define the Pareto dominance in the space $\mathbb{R}^2$: vector $\mathbf{f} = \mathbf{f}(x)$, $x \in D$ is Pareto-dominated by vector $\bar{\mathbf{f}} = \mathbf{f}(\bar{x})$, $\bar{x} \in D$, if the inequalities $f_1(x) \geq f_1(\bar{x})$, $f_2(x) \leq f_2(\bar{x})$ hold and there is at least one strict inequality among them. A solution $x \in D$ is dominated by a solution $\bar{x} \in D$, if the vector $\mathbf{f}(x)$ is dominated by vector $\mathbf{f}(\bar{x})$ in the sense of Pareto. The set $\tilde{D}$ of all non-dominated feasible solutions is called the *set of Pareto-optimal solutions*. The *total set of alternatives* is a subset $D^0 \subseteq \tilde{D}$ of a minimum size, such that $f(D^0) = f(\tilde{D})$ [27]. The *Pareto Front* is the set $F := \mathbf{f}(\tilde{D})$. Given $\varepsilon > 0$, the *Pareto set ε-approximation* $\tilde{D}_\varepsilon$ is a set such that for any Pareto optimal solution $\tilde{x} \in \tilde{D}$, there is a solution $x \in \tilde{D}_\varepsilon$ satisfying $f_1(x) \geq (1 - \varepsilon)f_1(\tilde{x})$ and $f_2(x) \leq (1 + \varepsilon)f_2(\tilde{x})$.

In what follows, m denotes the number of machines in a production line, N is the number of intermediate buffers, subject to optimization.

By *system with a simple structure*, we mean a system which consists of N parallel two-machine tandem lines with common input buffer and common output buffer. An example of a system with simple structure is provided in Fig. 1.

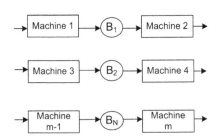

Fig. 1. Example of a series-parallel line with simple structure (N two-machine tandem lines in parallel)

In what follows, we consider the complexity of two bi-objective optimization problems that ask to determine the buffers capacity in a production system with simple structure. The optimization criteria we consider are the same as in [13]: the expected revenue due to line operation, the capital costs for installing buffers, and the expected total inventory cost for intermediate products. The expected revenue is supposed to be an increasing function of the expected throughput of the system.

On the positive side, we propose two fully polynomial-time approximation schemes (FPTASes) for approximation of Pareto front in the following two bi-objective problem formulations: (i) the expected revenue maximization with minimization of capital costs and (ii) the expected revenue maximization with minimization of the expected inventory costs. An exact pseudo-polynomial time algorithm is proposed for computing the Pareto front in the first problem, if the buffers allocation cost is a linear function of buffer sizes with integer coefficients, i.e. assuming integer buffer capacity costs.

On the negative side, we show that the canonical decision problems for the above mentioned bi-objective problems are $\mathcal{NP}$-hard even if the revenue is proportional to the production rate and the buffers allocation cost is linear. In the case of just one tandem two-machine line, both of the problems are complete multiobjective optimization problems in the sense of Emelichev and Perepelitsa [19] and therefore intractable, i.e. their Pareto-front can be of exponential size in the input size. We also show for both of these special cases that if the Pareto front is computable in a polynomial time then $\mathcal{P} = \mathcal{NP}$ holds.

The remainder of the paper is organized as follows. The assumptions of the model of production line and the bi-objective problems formulation are presented in the next section. Section 3 is devoted to the analysis of computational complexity of bi-objective buffer allocation problems on the two-machine tandem lines. This is followed by the analysis of computational complexity and approximability of bi-objective buffer allocation problems for lines of simple structure in Sect. 4. Finally some conclusions are drawn in Sect. 5.

2   Basic Properties and Definitions

2.1   An Illustrative Model of Production Line

Let us consider an illustrative example of a production systems under consideration. Suppose that each machine of the system can be in an operational state or under repair. An operational machine may be blocked and temporarily stopped in case if there is no room in the downstream buffer. An operational machine may be starved if there are no parts to process in the upstream buffer. Otherwise operational machines are working.

A working machine is assumed to have a constant cycle time. It is supposed that machines break down only when they are working. The times to fail and times to repair for each machine are assumed to be mutually independent and exponentially distributed random values. A detailed analysis of steady-state performance of such systems and optimization of its parameters were carried out in a number of works, see e.g. [3, 8, 12–17, 23].

2.2　Optimization Criteria and the Set of Feasible Solutions

Let the buffers in the system be denoted by $B_1, \ldots, B_N$ and let h_j be the capacity of buffer B_j, $j = 1, \ldots, N$, subject to optimization. Denote the vector of decision variables by $H = (h_1, h_2, \ldots, h_N) \in \mathbb{Z}_+^N$, where $\mathbb{Z}_+$ is the set of non-negative integers. Let $D = \{H = (h_1, \ldots, h_N) \in \mathbb{Z}^N \mid 0 \le h_i \le d_i, i = 1, \ldots, N\}$ be the set of feasible solutions, where $d_1, \ldots, d_N$ are the maximal admissible buffer capacities.

The most commonly used optimization criteria are:

- the throughput, i.e. expected number of parts produced by the system per unit of time in the steady state mode (expected steady state production rate) $V(H)$;
- the expected steady state inventory $Q(H) = (q_1(H), \ldots, q_N(H))$, where $q_j(H) \in [0, h_j]$ is the expected steady state number of parts in buffer B_j, $j = 1, \ldots, N$.

Let us introduce the following additional notation, using the symbol $\mathbb{Q}$ for the set of rational numbers:

- $R(V)$ is the revenue related to the production rate V, i.e. $R : \mathbb{Q}_+ \to \mathbb{Q}_+$;
- $B(H)$ is the cost of buffer configuration H, i.e. $B : D \to \mathbb{Q}_+$;
- $C(Q)$ is the cost of expected steady state inventory vector Q, i.e. $C : \mathbb{Q}_+^N \to \mathbb{Q}_+$.

In what follows, $R(V)$ is assumed to be a given non-decreasing function. In the case of lines with simple structure, $V(H)$, $B(H)$ and $C(Q)$ are assumed to be given completely additively separable functions, non-decreasing in each argument. Recall that $f(x_1, \ldots, x_n)$ is called completely additively separable if $f(x) = f(x_1) + \ldots + f_n(x_n)$ for some functions $f_1, \ldots, f_n$, each a function of one variable. We also make two technical assumptions: (i) functions $V(H)$, $B(H)$, $C(Q)$ $Q(H)$ and $R(V)$ are computable in polynomial time, and (ii) denoting any of these functions by $f(\cdot)$, we have the value $|\log f(\cdot)|$ polynomially bounded in the length of the problem input.

The cost function $B(H)$ may be non-linear to model some standard buffer capacities. A stepwise revenue function can be used to model zero revenue in case of an unacceptably low throughput.

2.3　Formulation of the Bi-objective Problems

Let us use the following notation for the problems of finding a Pareto front:

- In (R,B)-PARETO, the criterion f_1 is the expected revenue maximization $R(V(H))$ and f_2 is the minimization of buffer allocation cost $B(H)$.
- In (R,C)-PARETO, the criterion f_1 is the expected revenue maximization $R(V(H))$ and f_2 is the minimization of expected inventory cost $C(Q(H))$.

In accordance with [4,25,29], by *Canonical Decision Problem* for the buffers allocation problem with criteria pairs $f_1(H) \to \max$, $f_2(H) \to \min$, we mean the following decision problem: Given an instance I of buffer allocation problem and a pair $(\alpha, \beta) \in \mathbb{R}^2_+$, decide whether there exists a feasible buffers allocation H', for which $f_1(H') \geq \alpha$, $f_2(H') \leq \beta$.

Let us use the following notation for the canonical decision:

- In (V,linear B)-DEC, the criterion $f_1(H) \equiv V(H)$ and f_2 is the minimization of linear buffers allocation cost $B(H) = \sum_{i=1}^{N} b_i h_i$.
- In (V,linear Q)-DEC the criterion $f_1(H) \equiv V(H)$ and f_2 is the minimization of linear inventory cost $C(Q(H)) = \sum_{i=1}^{N} c_i q_i(H)$.

2.4   Monotonicity Properties

In what follows, we use the following two monotonicity assumptions for each subsystem consisting of two machines, separated by a buffer B_i of size h_i:

- **M1. Monotonicity of expected throughput.** $V(H)$ is an increasing function of h_i, $i = 1, \ldots, N$.
- **M2. Monotonicity of expected inventory.** $q_i(H)$ is an increasing function of h_i, $i = 1, \ldots, N$.

In the case of illustrative model presented in Subsect. 2.1, properties M1 and M2 follow from the analytical solution of the system of Kolmogorov equations describing the two-machine production system [8,17].

3   Computational Complexity of Bi-objective Buffer Allocation Problems on Two-Machine Tandem Line

In the case of $N = 1$, we denote $h := h_1 = H$ and $d := d_1$ for simplicity. Let us consider two increasing functions $V(h)$ and $Q(h)$, defined for $h = 0, 1, \ldots, d$, and taking rational values.

Theorem 1. *(i)   If problem* (R,B)-PARETO *is polynomially solvable in case of $N = 1$, then $\mathcal{P} = \mathcal{NP}$.*

(ii)   If problem (R,C)-PARETO *is polynomially solvable in case of $N = 1$, then $\mathcal{P} = \mathcal{NP}$.*

The proof is similar to the proof of Proposition 1 in [13] and employs an idea of Cheng and Kovalyov [7].

In [19], Emelichev and Perepelitsa give a definition of the *complete multi-objective problem*. A multiobjective optimization problem with k objectives is called *complete* if for any instance I of this problem with a set of feasible solutions D, there exists a vector of criteria $(f_1, \ldots, f_k)$, such that D is the only total set of alternatives w.r.t. $(f_1, \ldots, f_k)$, i.e. $D^0 = D$ holds.

In view of the monotonicity assumptions for $V(h)$ and $Q(h)$, a straightforward verification of the above definition indicates that (R,B)-PARETO and (R,Q)-PARETO are both complete in the sense of Emelichev and Perepelitsa. Note that d is a numerical parameter of the problem and $|D| = d + 1$. Together with the completeness property, this implies that the cardinalities of Pareto fronts of (R,B)-PARETO and (R,Q)-PARETO are not bounded by any polynomial in problem input size and therefore these problems are *intractable*, according to the terminology from [4,21].

4   Computational Complexity of Bi-objective Buffer Allocation Problems on Lines of Simple Structure

Consider the bi-objective buffer allocation problem with criteria of expected revenue $R(V(H))$ maximization and buffers installation cost $B(H)$ minimization on the set D, assuming that $B(H) = \sum_{i=1}^{N} b_i h_i$ where $b_i \in \mathbb{Z}$ for all $i = 1, \ldots, N$.

Proposition 1. *In the case of lines of simple structure, the problem* (V, *linear* B)-DEC *is $\mathcal{NP}$-hard.*

The proof of Proposition 1 is analogous to the proof [15] of $\mathcal{NP}$-hardness of the single-objective buffer allocation problem with the criterion of $B(H)$ minimization, given a lower bound on $V(H)$.

Proposition 2. *In the case of lines of simple structure, the problem* (V,linear Q)-DEC *is $\mathcal{NP}$-hard.*

The proof of Proposition 2 is analogous to the proof [13] of $\mathcal{NP}$-hardness of the single-objective buffer allocation problem with the criterion of $Q(H)$ minimization, given a lower bound on $V(H)$.

Proposition 3. *In the case of lines of simple structure, the Pareto front of buffer allocation problem with the criteria of expected revenue $R(V(H))$ maximization and buffers installation cost $B(H)$ minimization is computable in pseudopolymonial time, assuming integer b_i, $i = 1, \ldots, N$.*

The proof of Proposition 3 is based on the dynamic programming method. It is similar to the proof of pseudo-polynomial solvability of the Integer Knapsack Problem.

Analogous claim for an arbitrary function $B(H)$, increasing in each of its arguments h_i, is problematic. The reason is that the dynamic programming method gives pseudopolymonial solvability only if the cardinality of the set of values, taken by one of the optimization criteria, is polynomially bounded, given polynomially bounded numerical input data. The same difficulty arises in computing the Pareto set of the bi-objective buffer allocation problem with criteria of the expected revenue $R(V(H))$ maximization and the expected inventory cost $C(Q(H))$ minimization.

It is possible, however, to convert the FPTAS for Generalized Knapsack problem [22] into FPTASes for the bi-criteria optimization problems, where expected revenue $R(V(H))$ maximization is combined with buffers installation cost $B(H)$ minimization or with the expected inventory cost $C(Q(H))$ minimization. Recall that a family of algorithms $\{A_\varepsilon\}$ is called a fully polynomial approximation scheme (FPTAS) for a multiobjective optimization problem if for any input instance and any $\varepsilon > 0$, the algorithm A_ε runs in polynomial time w. r. t. the size of the input and $1/\varepsilon$, and outputs a Pareto set ε-approximation.

Theorem 2. *In the case of lines of simple structure, there exist FPTASes for* (R,B)-PARETO *and* (R,C)-PARETO.

The proof is based on the general scheme suggested by Cheng et al. in [6] for the construction of Pareto set ε-approximation of a bi-criteria problem.

5   Conclusions

We have established intractability of several special cases of buffer allocation problem in bi-objective formulation using the proof ideas developed for the analysis of single-criteria formulations of the problem. Our results apply to different particular models of production line, provided that two monotonicity conditions, formulated here, are satisfied. On the positive side, we propose fully polynomial-time approximation schemes for approximation of the Pareto front for two versions of buffers allocation problem and an exact pseudo-polynomial algorithm based on the dynamic programming method.

Acknoeledgement. The work was supported by the program of fundamental scientific research of the SB RAS I.5.1., project 0314-2016-0019.

References

1. Abdul-Kader, W.: Capacity improvement of an unreliable production line - an analytical approach. Comput. Oper. Res. **33**, 1695–1712 (2006)
2. Altiparmak, A., Bugak, A., Dengiz, B.: Optimization of buffer sizes in assembly systems using intelligent techniques. In: Proceedings of the 2002 Winter Simulation Conference, pp. 1157–1162 (2002)
3. Ancelin, B., Semery, A.: Calcul de la productivité d'une ligne integrée de fabrication. RAIRO Autom. Productiq. Inform. Industrielle **21**, 209–238 (1987)
4. Bökler, F.: The multiobjective shortest path problem is NP-hard, or is it? In: Trautmann, H., Rudolph, G., Klamroth, K., Schütze, O., Wiecek, M., Jin, Y., Grimme, C. (eds.) EMO 2017. LNCS, vol. 10173, pp. 77–87. Springer, Cham (2017). https://doi.org/10.1007/978-3-319-54157-0_6
5. Chehade, H., Yalaoui, F., Amodeo, L., De Guglielmo, P.: Optimisation multiobjectif pour le probléme de dimensionnement de buffers. J. Decis. Syst. **18**, 257–287 (2009)
6. Cheng, T.C.E., Janiak, A., Kovalyov, M.Y.: Bicriterion single machine scheduling with resource dependent processing times. SIAM J. Optim. **8**(2), 617–630 (1998)

7. Cheng, T.C.E., Kovalyov, M.Y.: An unconstrained optimization problem is NP-hard given an oracle representation of its objective function: a technical note. Comput. Oper. Res. **29**, 2087–2091 (2002)

8. Coillard, P., Proth, J.M.: Effet des stocks tampons dans une fabrication en ligne. Revue belge de Statistique, d'Informatique et de Recherche Operationnelle **24**(2), 3–27 (1984)

9. Cruz, F.R.B., Van Woensel, T., Smith, J.M.: Buffer and throughput trade-offs in M/G/1/K queuing networks: a bicriteria approach. Int. J. Prod. Econ. **125**, 224–234 (2010)

10. Dallery, Y., Gershwin, S.B.: Manufacturing flow line systems: a review of models and analytical results. Queueing Syst. **12**(1–2), 3–94 (1992)

11. Deb, K., Pratap, A., Agarwal, S., Meyarivan, T.: A fast and elitist multiobjective genetic algorithm: NSGA-II. Proc. IEEE Trans. Evol. Comput. **6**(2), 182–197 (2002)

12. Dolgui, A., Eremeev, A., Kolokolov, A., Sigaev, V.: A genetic algorithm for the allocation of buffer storage capacities in a production line with unreliable machines. J. Math. Model. Algorithms **1**, 89–104 (2002)

13. Dolgui, A., Eremeev, A.V., Kovalyov, M.Y., Sigaev, V.S.: Complexity of buffer capacity allocation problems for production lines with unreliable machines. J. Math. Model. Algorithms Oper. Res. **12**(2), 155–165 (2013)

14. Dolgui, A., Eremeev, A.V., Sigaev, V.S.: Analysis of a multicriterial buffer capacity optimization problem for a production line. Autom. Remote Control **78**(7), 1276–1289 (2017)

15. Dolgui, A., Eremeev, A.V., Sigaev, V.S.: HBBA: hybrid algorithm for buffer allocation in tandem production lines. J. Intell. Manuf. **18**(3), 411–420 (2007)

16. Dolgui, A.B., Svirin, Y.P.: Models of evaluation of probabilistic productivity of automated technological complexes. Vesti Akademii Navuk Belarusi: phisikatechnichnie navuki **1**, 59–67 (1995)

17. Dubois, D., Forestier, J.-P.: Productivité et en cours moyen d'un ensemble de deux machines séparées par une zone de stockage. RAIRO Automat **16**(2), 105–132 (1982)

18. D'Souza, K., Khator, S.: System reconfiguration to avoid deadlocks in automated manufacturing systems. Comput. Industr. Eng. **32**, 445–465 (1997)

19. Emelichev, V.A., Perepelitsa, V.A.: Multiobjective problems on the spanning trees of a graph. Soviet Math. Dokl **37**(1), 114–117 (1999)

20. Hamada, M., Martz, H., Berg, E., Koehler, A.: Optimizing the product-based avaibility of a buffered industrial process. Reliab. Eng. Syst. Saf. **91**, 1039–1048 (2006)

21. Hansen, P.: Bicriterion path problems. In: Fandel, G., Gal, T. (eds.) Multiple Criteria Decision Making Theory and Application. Lecture Notes in Economics and Mathematical Systems, vol. 177, pp. 109–127. Springer, New York (1979). https://doi.org/10.1007/978-3-642-48782-8_9

22. Kovalyov, M.Y.: A rounding technique to construct approximation algorithms for knapsack and partition type problems. Appl. Math. Comput. Sci. **6**(4), 101–113 (1996)

23. Levin, A.A., Pasjko, N.I.: Calculating the output of transfer lines. Stanki i Instrument **8**, 8–10 (1969). (in Russian)

24. Li, J., Meerkov, S.M.: Production Systems Engineering. Springer, New York (2009). https://doi.org/10.1007/978-0-387-75579-3

25. Serafini, P.: Some considerations about computational complexity for multiobjective combinatorial problems. In: Jahn, J., Krabs, W. (eds.) Recent Advances and Historical Development of Vector Optimization. Lecture Notes in Economics and Mathematical Systems, vol. 294, pp. 222–231. Springer, Heidelberg (1986). https:// doi.org/10.1007/978-3-642-46618-2_15

26. Patchong, A., Lemoine, T., Kern, G.: Improving car body production at PSA Peugeot Citroen. Interfaces **33**(1), 36–49 (2003)

27. Perepelitsa, V.A.: Mnogokriterial'nye modeli i metody dlya zadach optimizatsii na grafakh. Lambert Academic Publishing, 330 (2013)

28. Tempelmeier, H.: Practical considerations in the optimization of flow production systems. Int. J. Prod. Res. **41**(1), 149–170 (2003)

29. Warburton, A.: Approximation of pareto optima in multiple-objective shortestpath problems. Oper. Res. **35**(1), 70–79 (1987)

30. Zitzler, E., Laumanns, M., Thiele, L.: SPEA2: Improving the Strength Pareto Evolutionary Algorithm. Technical Report 103, Comput. Engin. Commun. Networks Lab, Swiss Federal Institute Technology, Zurich (2001)

Inventory Policies in Dual Sourcing Systems with Uncertain Yield

Adriana F. Gabor[1] and Andrei Sleptchenko[2(✉)]

[1] University of United Arab Emirates, Al Ain, UAE
adriana.gabor@uaeu.ac.ae
[2] Khalifa University of Science and Technology, Abu Dhabi, UAE
andrei.sleptchenko@ku.ac.ae

Abstract. We study a spare part supply system where both Traditional Manufacturing and Additive Manufacturing (also know as 3D printing) are used for replenishment of the spare parts inventories. Demands for the spare parts occur according to a Poisson process, and the failed parts are immediately replaced from the inventory, if available. If inventory is not available, items are backordered and fulfilled when a spare becomes available (i.e., a replenishment is received from one of the suppliers). Additive Manufacturing offers the advantage of shorter lead times, however, at higher production costs. Moreover, Additive Manufacturing processes often have uncertain yield, leading to the fact that not every produced part will satisfy the quality control and can be used to replenish the inventory. In this paper, we propose a Linear Programming (LP) based optimization problem that decides which of the processes to use for replenishment of the inventory while minimizing the total (holding + backorder) system costs.

Keywords: Spare part logistics · Dual sourcing
Additive Manufacturing · Markov-Decision Processes

1 Introduction

Spare part inventories play an important role in modern life, keeping downtime of advanced capital goods within reasonable limits and ensuring availability, safety and eco-friendliness of different production and service systems, cf. [1]. At the same time, spare parts inventories have low turnover and form a substantial cost item in the production or service budgets. One of the main reasons for the large costs of spare parts inventories is the long replenishment lead-time, in particular for customized parts manufactured using Traditional Manufacturing (TM) technologies like milling, drilling or molding.

Additive manufacturing (AM), referred to as 3D printing, has greatly improved in the last years and is often used as an alternative (sometimes even as the main process) for the production of assembly components and replacement parts. Due to the use of very computerized printers that do not require specialized skills, the production can be moved closer to the inventory location, thus

© Springer International Publishing AG, part of Springer Nature 2018
A. Eremeev et al. (Eds.): OPTA 2018, CCIS 871, pp. 288–295, 2018.
https://doi.org/10.1007/978-3-319-93800-4_23

reducing lead times considerably. However, (AM) often produces items that are not of desired quality and the production costs might be higher than in Traditional Manufacturing (TM).

In this paper, we study inventory policies for the case where both AM and TM may be used for replenishment. That is, depending on the stock level and yield uncertainty, one may decide whether to source a spare part from an AM or/and TM supplier.

This paper is related to the dual sourcing models, where suppliers have different lead times. Motivated by the offshoring phenomenon that became very popular after the 1990s, these models assume an expensive supplier with shorter lead-time and a cheaper supplier, with a longer lead-time. Due to the complexity of the problem, however, the literature mainly focuses on models with certain yield.

For a lead-time difference larger than one, it has been shown in [2] that the optimal policy does not have a simple structure. In the recent years, several close to optimal policies have been proposed in the literature for this problem [3–6]. Our model differs from these papers by taking into account yield uncertainty for the supplier with short lead-times.

In the context of dual sourcing, uncertainty in supply has been considered mainly regarding capacity or yield. [7,8], analyze periodic inventory models with uncertain capacities of the faster supplier. [8] assumes that the fast supplier is the in-house production unit, and is thus cheaper than the outside supplier, who is merely used when in-house capacity is not sufficient. For the case with no ordering costs and capacity restrictions revealed at the moment of order, [8] show that a capacity dependent base stock policy for both suppliers is optimal. A similar problem is studied in [7], however, under the assumption that the capacity of the fast supplier is unknown at the moment of ordering. Both papers assume a lead time difference of at most one.

Uncertain yield is mainly studied in the context of single sourcing [9–12]. For dual sourcing, the most studied form of random yield is when suppliers face a random disruption (either can satisfy the whole order or nothing). [13] looks at the joint pricing and inventory control problem under a finite horizon for a firm with two suppliers, both facing random disruptions and price-sensitive random demands. They show that when both suppliers are unreliable, for a lead-time difference of one and disruptions following a Markov Chain, the optimal inventory policy in each period is a reorder point policy and the optimal price is decreasing in the starting inventory level in that period. To the best of our knowledge, the only paper considering binomial random yield in the context of dual sourcing with arbitrary lead time difference is [14]. The unreliable supplier is the slow, cheaper supplier. They propose an approximate policy by reducing the problem with uncertain yield to a dual sourcing problem with certain yield and modified demand.

In our paper, we assume a dual sourcing model, where the quality of the items delivered by the fast and expensive supplier (the AM supplier) is uncertain. This supplier is used only to compensate for the long lead-times of the regular supplier (the TM supplier), which produces at lower costs, but has longer lead time. In our model, we assume expensive items, that justify the assumption that orders are placed when demands occur and have size at most one for each supplier (similar to the (S−1, S) policy for single sourcing models). Note that when the order size is one, random disruptions and random yield are equivalent, if the disruption becomes known only upon delivery. We assume exponential lead times for both suppliers. This assumption allows us to work with lead times with a difference larger than 1, unlike the previous papers studying random disruptions.

The rest of the paper is organized as follows. In Sect. 2, we describe the problem and provide necessary assumptions and notations. In Sect. 3, we present how the problem can be analyzed using the Continuous Markov Decision Process. Finally, in Sects. 4 and 5, we present some numerical results and initial observations about advantages of using the Additive Manufacturing as a resupply option.

2   The Problem

In this paper, we consider a single-site spare parts supply system. We assume that demands for spare parts occur according to Poisson processes and the failed spare parts are immediately replaced from the inventory, if available. If the replacement parts are not available in the inventory, the failed parts are backordered and fulfilled when a spare of the same type becomes available. When a demand occurs, a replenishment order is sent out immediately, to either the TM or the AM supplier, or to both see Fig. 1. The last option is necessary to compensate for the uncertain yield of the AM supplier whose parts might not pass quality control. When a part from the *AM* supplier does not pass the quality control, it will be discarded. Replenishment times from both suppliers are assumed exponentially distributed and independent of each other.

The decision of which supplier to use (the AM or the TM supplier) is based on the trade-off between the higher purchasing costs (but shorter lead-times) for the AM supplier and the lower purchasing costs (but longer lead-times) for the TM supplier. The model below assumes only one type of items in the system. This assumption, however, can be easily relaxed since the proposed optimization model is fully separable, and the decisions for each item type can be taken independently.

The main optimization challenge is to find an ordering policy that minimizes the average per-period total cost, consisting of the ordering (production) costs, the inventory holding costs, and the stockout (backorder) penalty costs.

2.1   Assumptions and Notations

In the remainder of the paper, we use the following assumptions and notations:

- λ: arrival rate of failed items; the arrival process is assumed to be Poisson.
- μ^{TM} and μ^{AM}: replenishment rates from the TM and AM suppliers;

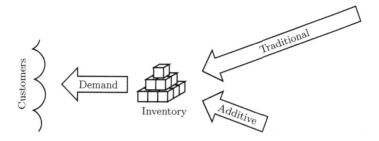

Fig. 1. System diagram

- P^g: the probability of an item from an AM supplier to pass quality control
- C^H: holding costs, paid per time unit per parts available in the inventory (net inventory).
- c^{BO}: Penalty costs (or backorder costs) incurred when the required part is not available; paid per time unit per not available part,
- c^{TM}, c^{AM}: ordering costs from the TM and AM suppliers, correspondingly.

3   The Markov Decision Problem

The behavior of this inventory system can be modeled by a continuous time Markov Chain (MC) with states given by $s = (n^I, m^{TM}, m^{AM})$, where

n^I – items in the inventory (net),
m^{TM} – items in the ordering pipeline at the TM supplier,
m^{AM} – items in the ordering pipeline at the AM supplier.

Let α_s, β_s, and γ_s be binary variables that indicate whether the AM, TM or both suppliers are used in state (n^I, m^{TM}, m^{AM}). We assume that

$$\alpha_s + \beta_s + \gamma_s = 1, \tag{1}$$

i.e., a replenishment order is always placed.

For a given ordering policy $\{(\alpha_s, \beta_s, \gamma_s), s \in \Omega\}$, the transition diagram is as shown in Fig. 2. Please note, that when a part replenished by the AM process does not satisfy the quality control (with probability $1 - P^g$), the total number of parts (inventory + ordered) in the system drops by 1 (lower right node in Fig. 2). Therefore, extra parts might be ordered at the next failure event (lower left node in Fig. 2).

The equilibrium equations for the steady-state probabilities $p_{n^I, m^{TM}, m^{AM}}$ in this Continuous Time Markov Chain system (for all states $s = (n^I, m^{TM}, m^{AM}) \in \Omega$) are as follows:

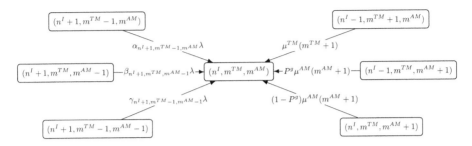

Fig. 2. Markov Chain flow diagram

$$(\lambda + m^{TM}\mu^{TM} + m^{AM}\mu^{AM})p_{n^I,m^{TM},m^{AM}}$$
$$= \alpha_{n^I+1,m^{TM}-1,m^{AM}}\lambda p_{n^I+1,m^{TM}-1,m^{AM}}$$
$$+ \beta_{n^I+1,m^{TM},m^{AM}-1}\lambda p_{n^I+1,m^{TM},m^{AM}-1}$$
$$+ \gamma_{n^I+1,m^{TM}-1,m^{AM}-1}\lambda p_{n^I+1,m^{TM}-1,m^{AM}-1} \qquad (2)$$
$$+ (m^{TM}+1)\mu^{TM}p_{n^I-1,m^{TM}+1,m^{AM}}$$
$$+ P^g(m^{AM}+1)\mu^{AM}p_{n^I-1,m^{TM},m^{AM}+1}$$
$$+ (1-P^g)(m^{AM}+1)\mu^{AM}p_{n^I,m^{TM},m^{AM}+1},$$

Adding the normalization equality

$$\sum_{s\in\Omega} p_s = 1 \qquad (3)$$

we obtain a system of independent linear equations that gives the steady-state probabilities $p_{n^I,m^{TM},m^{AM}}$.

We truncate the state space Ω by limiting the net inventory n^I by the minimum and maximum possible inventory levels (n^I_{min} and n^I_{max}). We also limit the number of items in pipelines, m^{TM} and m^{AM}, by m^{TM}_{max} and m^{AM}_{max}. At the same time, the range $n^I_{max} - n^I_{min}$ should not exceed $\max(m^{TM}_{max}, m^{AM}_{max})$.

The goal of the model is to find an ordering policy $\{(\alpha_s, \beta_s, \gamma_s), s \in \Omega\}$ that minimizes the average per-period total cost consisting of the ordering (production) costs, the inventory holding costs, and the stockout (backorder) penalty costs:

$$\min_{\alpha,\beta} \sum_{s\in\Omega} \left[c^{TM}\lambda(\alpha_s + \gamma_s) + c^{AM}\lambda(\beta_s + \gamma_s) + c^H(n^I_s)^+ - c^{BO}(n^I_s)^- \right]p_s \qquad (4)$$

where a^+ and a^- denote $\max(0, a)$ and $\min(0, a)$, and n^I_s represents the first component of state s.

The optimization problem consists of the objective (4), balance Eqs. (2–3) and relation (1). Due to the multiplication of the variables α, β and γ by the steady-state probabilities, the optimization problem is not linear. However, the problem can be shown to be equivalent to a linear program, whose solutions are integer (see for example [15,16]).

3.1   Linear Programming Formulation

To create the LP formulation, we relax first the integrality condition for $(\alpha_s, \beta_s, \gamma_s)$. As a result, these variables will present in fact probabilistic decisions to order from TM or AM suppliers: in state s, a part will be ordered from the TM supplier with probability α_s, from the AM supplier with probability β_s and from both with probability γ_s.

Next, we introduce new decision variables:

$$y_s^\alpha = \alpha_s p_s,$$
$$y_s^\beta = \beta_s p_s.$$

These new variables allow us to the rewrite the optimization problem (1–4) as the following linear program:

$$\min_{\alpha,\beta} \sum_{s \in \Omega} \left[c^{TM} \lambda (\alpha_s + \gamma_s) + c^{AM} \lambda (\beta_s + \gamma_s) + c^H (n_s^I)^+ - c^{BO}(n_s^I)^- \right] p_s \quad (5)$$

$$\text{s.t. } (\lambda + m^{TM} \mu^{TM} + m^{AM} \mu^{AM}) p_{n^I, m^{TM}, m^{AM}}$$
$$= \lambda \big[y_{n^I+1, m^{TM}-1, m^{AM}}^\alpha + y_{n^I+1, m^{TM}, m^{AM}-1}^\beta$$
$$+ p_{n^I+1, m^{TM}-1, m^{AM}-1}$$
$$- y_{n^I+1, m^{TM}-1, m^{AM}-1}^\alpha - y_{n^I+1, m^{TM}-1, m^{AM}-1}^\beta \big] \quad (6)$$
$$+ (m^{TM} + 1) \mu^{TM} p_{n^I-1, m^{TM}+1, m^{AM}}$$
$$+ P^g (m^{AM} + 1) \mu^{AM} p_{n^I-1, m^{TM}, m^{AM}+1}$$
$$+ (1 - P^g)(m^{AM} + 1) \mu^{AM} p_{n^I, m^{TM}, m^{AM}+1}$$

$$\sum_{s \in \Omega} p_s = 1 \quad (7)$$

$$y_s^\alpha + y_s^\beta \le p_s \quad (8)$$

$$y_s^\alpha, y_s^\beta, p_s \ge 0 \quad (9)$$

Note here that γ_s is substituted by $1 - \alpha_s - \beta_s$, hence $\gamma_s p_s = p_s - y_s^\alpha - y_s^\beta$.

For this problems, it is possible to show (see for example [15]) that the optimal y_s^α and y_s^β are always equal either p_s or 0. That is, the optimal α_s and β_s will always be integer. Observe that when $\alpha_s = \beta_s = 0$ corresponds to the situation when we order from both suppliers.

4   Numerical Experiments

In this section, we present few numerical experiments to demonstrate advantages of using the Additive Manufacturing as an extra supply option.

In the presented experiments, we use the following settings. The cost parameters and the probability that an AM supplied part is good are: $c^H = 1.0$,

$c^{TM} = 10.0$, $c^{AM} = 100.0$, $c^{BO} = 1000.0$, $P^g = 0.9$. The maximum and minimum inventory levels (n^I_{max} and n^I_{min}) are set to 10 and -20, correspondingly. For the failure and replenishments rates we following values:

$$\lambda = 0.1, 0.2, 0.3 \quad \text{– failure rates per year}$$
$$\mu^{TM} = 2, 3, 4 \quad \text{– lead times of 6, 4, and 3 month}$$
$$\mu^{AM} = 30, 50, 100 \quad \text{– lead times of 12, 7.5, and 3.6 days}$$

The combinations of these failure and replenishments rates produce 27 experiments, the results of which can be found in Table 1.

Table 1. The total cost and inventory cost reductions when additive manufacturing is used as a secondary supply option.

Total cost reduction					Inventory cost reduction				
		λ					λ		
μ^{TM}	μ^{AM}	0.1	0.2	0.3	μ^{TM}	μ^{AM}	0.1	0.2	0.3
2	30	4.65%	0.53%	1.50%	2	30	33.83%	0.13%	0.03%
	50	6.66%	0.81%	2.03%		50	33.83%	0.13%	0.03%
	100	8.10%	1.54%	2.47%		100	33.83%	0.16%	0.03%
3	30	0.49%	0.01%	0.02%	3	30	0.16%	0.00%	0.00%
	50	1.97%	0.02%	0.02%		50	0.16%	0.00%	0.00%
	100	3.14%	0.06%	0.15%		100	0.16%	0.00%	-0.01%
4	30	0.01%	0.01%	0.00%	4	30	0.00%	0.00%	0.00%
	50	0.09%	0.02%	0.01%		50	0.00%	0.00%	0.00%
	100	0.10%	0.02%	0.01%		100	0.00%	0.00%	0.00%

The results presented in Table 1 indicate that:

1. Additive Manufacturing can help reduce the total system cost despite its higher costs and yield uncertainty.
2. The main reduction in cost is due to inventory cost reduction. Reducing inventory costs plays an important role, due to the high costs of spare parts, in particular of advanced equipment.
3. The cost reduction depends very much on the problem parameters, and the larger savings are achieved when the TM supplier is slow.
4. The optimal policy is state dependent, making harder to derive simple ordering rules.

The obtained results are rather limited and are just to demonstrate the advantages of the Additive Manufacturing in spare parts supply. At the same time, the obtained optimization problem is linear and can be easily solved for relatively large problem sizes using any commercial LP solver.

5   Conclusions

In this paper, we studied the spare parts supply system where both AM and TM methods are used for replenishment of the spare parts inventories. One of the main aspects of our model is that we assume yield uncertainty in the replenishments from the AM source. The obtained results indicate that the Additive Manufacturing can be an interesting option for the replenishment, despite its obvious drawbacks (price and yield uncertainty).

References

1. Alrabghi, A., Tiwari, A.: State of the art in simulation-based optimisation for maintenance systems. Comput. Ind. Eng. **82**, 167–182 (2015)
2. Whittemore, A.S., Saunders, S.: Optimal inventory under stochastic demand with two supply options. SIAM J. Appl. Math. **32**(2), 293–305 (1977)
3. Tagaras, G., Vlachos, D.: A periodic review inventory system with emergency replenishments. Manag. Sci. **47**(3), 415–429 (2001)
4. Veeraraghavan, S., Scheller-Wolf, A.: Now or later: a simple policy for effective dual sourcing in capacitated systems. Oper. Res. **56**(4), 850–864 (2008)
5. Allon, G., Van Mieghem, J.: Global dual sourcing: tailored base-surge allocation to near- and offshore production. Manag. Sci. **56**(1), 110–124 (2010)
6. Boute, R.N., Van Mieghem, J.A.: Global dual sourcing and order smoothing: the impact of capacity and lead times. Manag. Sci. **61**(9), 2080–2099 (2014)
7. Jakšič, M., Fransoo, J.: Dual sourcing in the age of near-shoring: trading off stochastic capacity limitations and long lead times. Eur. J. Oper. Res. **267**(1), 150–161 (2018)
8. Yang, J., Qi, X., Xia, Y.: A production-inventory system with Markovian capacity and outsourcing option. Oper. Res. **53**(2), 328–349 (2005)
9. Bollapragada, S., Morton, T.E.: Myopic heuristics for the random yield problem. Oper. Res. **47**(5), 713–722 (1999)
10. Henig, M., Gerchak, Y.: The structure of periodic review policies in the presence of random yield. Oper. Res. **38**(4), 634–643 (1990)
11. Inderfurth, K., Transchel, S.: Technical note on 'Myopic heuristics for the random yield problem'. Oper. Res. **55**(6), 1183–1186 (2007)
12. Inderfurth, K., Kiesmüller, G.P.: Exact and heuristic linear-inflation policies for an inventory model with random yield and arbitrary lead times. Eur. J. Oper. Res. **245**(1), 109–120 (2015)
13. Gong, X., Chao, X., Zheng, S.: Dynamic pricing and inventory management with dual suppliers of different lead times and disruption risks. Prod. Oper. Manag. **23**(12), 2058–2074 (2014)
14. Ju, W., Gabor, A.F., van Ommeren, J.C.: An approximate policy for a dual-sourcing inventory model with positive lead times and binomial yield. Eur. J. Oper. Res. **244**(2), 490–497 (2015)
15. Sleptchenko, A., Johnson, M.E.: Maintaining secure and reliable distributed control systems. INFORMS J. Comput. **27**(1), 103–117 (2015)
16. Knofius, N., van der Heijden, M.C., Sleptchenko, A., Zijm, W.H.: Improving effectiveness of spare part supply by additive manufacturing as dual sourcing option. Beta working paper series (530) (2017, submitted for publication)

A Genetic Algorithm for the Pooling-Inventory-Capacity Problem in Spare Part Supply Systems

Hasan Hüseyin Turan[1], Andrei Sleptchenko[2(✉)], and Fuat Kosanoglu[3]

[1] School of Engineering and Information Technology,
University of New South Wales, Canberra, Australia
h.turan@adfa.edu.au
[2] Department of Industrial and Systems Engineering,
Khalifa University of Science and Technology, Abu Dhabi, UAE
andrei.sleptchenko@ku.ac.ae
[3] Department of Industrial Engineering, University of Yalova, Yalova, Turkey
fuat.kosanoglu@yalova.edu.tr

Abstract. We study a pooling-inventory-capacity problem that arises in the design of repair shops for repairable spare part logistic systems. We formulate the problem as a stochastic nonlinear integer programming model and propose a two-stage sequential solution algorithm. At the first stage, a genetic algorithm (GA) generates a set of feasible pooled repair shop design schemes. A pooled design can be viewed and modeled as the union of mutually exclusive and total exhaustive multi-class multi-server queueing systems. Thus, we exploit this fact and optimize each queueing system separately. In the second stage, optimal inventory and capacity levels for each independent system are calculated by using a queueing approximation technique and a local greedy heuristic. Finally, the performed numerical experiments show that proposed two-stage approach achieves high-quality solutions in reasonable time.

Keywords: Spare part logistics · Repair shop · Genetic algorithm
Queueing

1 Introduction

Maintenance plays a very important role in the modern life ensuring availability, safety, and eco-friendliness of different production and service systems, cf. [1]. At the same time, maintenance costs form a substantial cost item in the production or service budgets. In production systems, maintenance related costs can be typically anywhere between 15% and 70% of the total production expenses [2].

In our research, we study supply systems for repairable spare parts, where malfunctioned parts or components are immediately replaced by ready-for-use spares. The failed components are sent to a repairshop, and once the repair is finished, they are forwarded back to stock as good as new. The repair shop has

A. Eremeev et al. (Eds.): OPTA 2018, CCIS 871, pp. 296–308, 2018.
https://doi.org/10.1007/978-3-319-93800-4_24

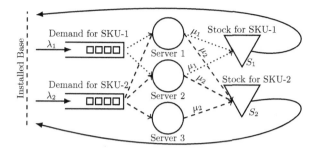

Fig. 1. Repair shop architecture for 2 failure types and 3 Cross-Trained Servers.

several multi-skilled parallel servers (technicians) that are capable of handling certain types of spares, see Fig. 1.

Previously (see [3–5]), different simulation based evolutionary heuristics were applied to optimize the skill assignments in the spare parts supply systems described above. There was shown that the optimal assignments of skills could reduce the total system cost by almost 30% compared to the cases where servers can handle all failures. The presented heuristics used the simulation models for objective function evaluation due to the lack of exact methods for analysis of queueing systems with cross-training.

The simulation-based optimization methods, however, are very time-consuming. Therefore, one can utilize the existing queueing theory models to speed up the optimization of the skill-server assignments, in order to be able to analyze bigger queueing systems. One of such models is the model for a multi-class multi-server queueing system [6], where all incoming requests can be processed by all servers, i.e. with "full" cross-training. The idea of pooled design was analyzed in [7,8], where it was shown, that the best pooled design is very often within the 10% range from the optimal solution.

In this paper, we propose a new Genetic Algorithm (GA) based optimization heuristic where the search space is limited only to the pooled designs. This allows us to use the existing exact queueing models and to speedup the evaluation of the objective function.

The rest of the paper is organized as follows. Section 2 describes the assumptions and the optimization model of the studied problem. In Sect. 3, the proposed optimization heuristic is presented. Numerical experiments comparing the proposed heuristic with other optimization models are presented in Sect. 4. Conclusions and future research directions are discussed in Sect. 5.

2   Problem Definition and Optimization Model

We study a single echelon spare part supply system with one repair facility and multiple repairable stock keeping units (SKUs) at stock points. The repair facility may have several parallel multi-skilled servers with different amount of cross-training as shown in Fig. 2. We focus on a design scheme known as pooling in which repairables are clustered by some measure of similarity. A repair shop has a pooled structure if the spare parts are divided into several independent and non-overlapping groups (clusters) and each cluster has their group of servers as depicted in Fig. 2(b).

When a part fails in the system, it is immediately replaced by a ready-for-use part of the same SKU type from the stock point, and the failed part is sent to the repair shop (Fig. 1). If the spare is not available in the stock, the request is backordered. In this case, the system goes down, and a downtime cost occurs until the requested ready-for-use part is delivered. The failures of parts obey a Poisson process with constant rates, and they are mutually independent of each other. We assume the repair times are exponentially distributed and mutually independent. We also assume the expected repair times depend on the SKU type and are independent of the processing server providing that the server has that skill to repair the failed part. We use first come first served (FCFS) queueing discipline. This discipline implies that whenever a server gets idle, it picks the failed part that has the longest waiting time in the queue, as long as the server has the needed skill. These are the most commonly used assumptions in the repairable spare part supply systems [9].

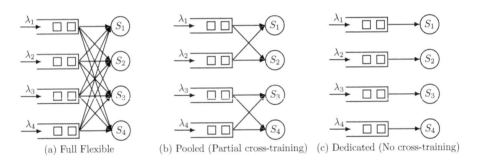

(a) Full Flexible   (b) Pooled (Partial cross-training)   (c) Dedicated (No cross-training)

Fig. 2. An example of possible different repair shop designs.

In addition to the common assumptions, we model each cluster inside the repair shop as a Markovian multi-class multi-server $M/M/k$ queueing system with dedicated queues, i.e., every server inside a cluster can repair all SKUs that are assigned to that cluster. Also, clusters inside the repair shop are mutually exclusive (disjoint) and collectively exhaustive, i.e., a particular failed SKU can be repaired at exactly one cluster, and all SKUs are assigned to exactly one cluster. The rest of the assumptions can be found in [5].

We use the similar modeling approach developed in [8]. The parameters and decision variables used for the development of stochastic model and solution procedures are presented below.

Sets:

N: Set of distinct types of repairables (SKUs)

Problem parameters:

λ_i: Failure rate of SKU type i $(i = 1, \ldots, |N|)$

μ_i: Service rate of SKU type i $(i = 1, \ldots, |N|)$

h_i: Inventory holding cost of SKU type i per unit time per part $(i = 1, \ldots, |N|)$

b: Penalty cost for each backordered demand per unit time, which is equivalent to the cost of downtime of the system

f: Operation cost of a server per unit time

c_i: Cost of having the skill to repair SKU type i per unit time per server $(i = 1, \ldots, |N|)$

ϵ: Very small positive real number

Decision variables:

S_i: Initial inventory quantity (basestock level) kept in stock for SKU type i $(i = 1, \ldots, |N|)$ and $\mathbf{S} = (S_1, \ldots, S_{|N|})$

z_j: Number of the operational servers in the cluster j $(j = 1, \ldots, |N|)$

x_{ij}: Binary variable indicating that whether the cluster j has the skill to repair SKU type i $(i = 1, \ldots, |N|)$ or not

The objective function (1) shows the sum of costs included in the model. The first term represents the cost of acquiring capacity (servers), whereas the second term represents the cross-training cost of the servers in each cluster, the third summation term shows the cost of holding spare parts inventories, and backorder cost is represented by the last term. The backorder cost term is calculated using the penalty cost b and the expected total number of backordered parts $\mathbb{EBO}_i[S_i, \mathbf{X}, \mathbf{Z}]$ for each SKU type i in the steady-state; under the given initial inventory level S_i, pooling scheme of the repair shop $\mathbf{X}$ and the server assignment policy $\mathbf{Z}$. The variable $\mathbf{X}$ represents the $(|N| \times |N|)$–matrix of the binary decision variables x_{ij} denoting how SKUs are pooled in the repair shop. The variable $\mathbf{Z}$ represents a $1 \times |N|-$ row matrix of integer decision variables z_j denoting the number of servers in each cluster of the repair shop. Constraints (2) guarantee that an SKU is assigned to the exactly one cluster. The overall utilization rate of a particular cluster k $(\sum_{i=1}^{N} x_{ik} \lambda_i / \mu_i)$ must be strictly smaller than capacity (total number of servers in the cluster z_k) of that cluster, which is ensured by Constraints (3). Constraints (4) restrict x_{ij} to be binary so that clusters in any pooling scheme $\mathbf{X}$ become mutually exclusive and totally exhaustive.

The remaining constraint sets (5) and (6) are required the decision variables to be integers and non-negative.

$$\min_{\mathbf{S},\ \mathbf{X},\ \mathbf{Z}} \sum_{j=1}^{|N|} f z_j + \sum_{j=1}^{|N|} z_j \left(\sum_{i=1}^{|N|} c_i x_{ij} \right) + \sum_{i=1}^{|N|} h_i S_i + b \sum_{i=1}^{|N|} \mathbb{EBO}_i \left[S_i, \mathbf{X}, \mathbf{Z} \right] \quad (1)$$

$$\text{Subject to:} \sum_{j=1}^{|N|} x_{ij} = 1 \qquad\qquad i \in \{1, \ldots, |N|\} \qquad (2)$$

$$\sum_{i=1}^{|N|} x_{ij} \frac{\lambda_i}{\mu_i} \leq (1 - \epsilon) z_j \qquad j \in \{1, \ldots, |N|\} \qquad (3)$$

$$x_{ij} \in \{0, 1\} \qquad\qquad i, j \in \{1, \ldots, |N|\} \qquad (4)$$

$$z_j \in \{0\} \cup \mathbb{Z}^+ \qquad\qquad j \in \{1, \ldots, |N|\} \qquad (5)$$

$$S_i \in \mathbb{N}_0 \qquad\qquad i \in \{1, \ldots, |N|\} \qquad (6)$$

3   A Two-Stage Solution Heuristic

An arbitrary pooling policy $\mathbf{X}$ is a feasible solution to the above model iff it corresponds to a partition of the set of SKUs, N. The number of ways a set of $|N|$ elements can be partitioned into non-empty subsets is called a Bell number. The number of possible pooling schemes $\mathbf{X}$; i.e., partitions, increases exponentially for increasing number of SKUs in the system. That is, the presented optimization problem is most probably NP-hard, as many other clustering problems. Therefore, we search for the optimal pooled repair shop design policy, $\mathbf{X}$, by utilizing a genetic algorithm (GA) based heuristic. This GA based heuristic, generates, first, a set of feasible pooled repair shop design policies. Afterwards, these candidate feasible solutions (policies) are passed through fitness evaluation function to find optimal values of server assignment policy $\mathbf{Z}$ and inventory levels of spares $\mathbf{S}$. Figure 3 presents details of the proposed solution algorithm. The details of the algorithm are provided in the following subsections.

3.1   Pooling Policy Generation via GA

The GA is a stochastic optimization technique that depends on a random-based searching mechanism, and it has been successfully adapted in many areas to solve a large number of optimization problems [10]. GAs are inspired by natural selection and biological evolutionary philosophy. A population of individuals is represented by a chromosome, a string of information which is randomly generated [4]. Figure 4 shows the chromosome encoding procedure chosen in this study. Each chromosome corresponds to a particular repair shop design policy, $\mathbf{X}$. Every chromosome in the population has $|N|$ genes. The value of the gene i indicates the cluster that SKU type i is assigned into. Each chromosome also

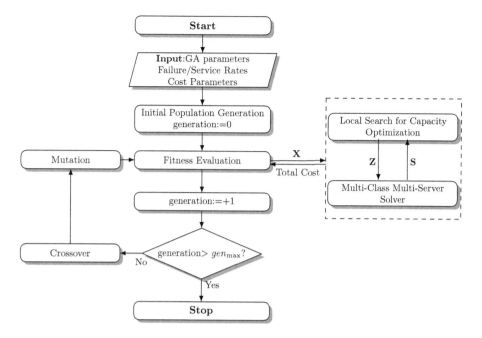

Fig. 3. The proposed GA-integrated two-stage sequential solution algorithm.

Fig. 4. An illustrative chromosome representation for a pooled design

carries information about the number of clusters exist in the repair shop. The total number of distinct integer in the chromosome represents the number of clusters. For example in Fig. 4, there are five clusters and SKUs 1, 5 and 10 are in the same cluster.

The algorithm starts by creating an initial population of individuals (chromosome); i.e., feasible pooling policies. The value of genes is generated by assigning a random integer from 1 to $|N|$ with equal probability. Then, individuals in the population are evaluated for fitness; i.e., total cost/objective function in Eq. (1). The details of fitness evaluation are discussed in the Subsect. 3.2.

Individuals that have better fitness are used to produce new offspring solutions which have a greater chance of being superior to the previous population of individuals. Less successful individuals from the previous generation are consequently deleted [4]. Selection processes in GA determine which solutions are to be preserved and allowed to reproduce and which ones deserve to die out. There are different techniques to implement selection in GAs. We chose tournament selection with tournament size 10. In tournament selection, several tournaments are played among a few individuals. The individuals are chosen at random from the population. The winner of each tournament is selected for next generation.

Crossover is an operation where two chromosomes (parents) partially contribute characteristics to a new chromosome (child). We use a uniform random one-point crossover technique to generate new feasible pooling policies. After crossover, we apply mutation operation to randomly selected individuals. Mutation operation is used to avoid trapping in local optima and to explore new pooling policies. We develop and try two different mutation operations as illustrated in Fig. 5.

(a) switch-mutation: The procedure in which a random gene is chosen and switched to the value of another gene in the chromosome with equal probability.

(b) swap-mutation: The procedure in which two genes are selected at random, and then they are swapped by value.

Table 1 documents GA parameters and functions used in the solution algorithm development.

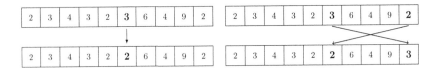

Fig. 5. An illustrative (a) switch-mutation (b) swap-mutation operations

Table 1. Summary of genetic algorithm parameters and details.

Parameter	Value
Population size	100
Number of generation	25
Crossover rate	0.8
Chromosome mutation rate	0.4
Switch-mutation probability	0.5
Swap-mutation probability	0.5
Operation	Detail
Initial population	Uniform random integer $\in [1, |N|]$
Fitness evaluation	Queuing approximation and local search
Selection	Tournament selection
Crossover	One-point uniform
Mutation(s)	Switch
	Swap

3.2   Fitness Evaluation

After the generation of the pooling policies $\mathbf{X}$ by above described GA procedures, multi-class multi-server solver and local search algorithm are called to evaluate the fitness (total cost) of each policy in the population as shown in Fig. 3.

Each cluster in $\mathbf{X}$ can be analyzed and optimized separately due to the clusters being mutually exclusive and independent from each other. For each cluster, the local search takes into account trade-off between adding an extra server to the cluster and decreasing the spare inventories in that cluster. The local search algorithm is basically a greedy search to find the optimal number of servers z_k in the cluster k by solving newsboy-like subproblems (see [11] for details).

To solve above-mentioned newsboy-like subproblems, the probability distribution of the number of failed SKU type i at the steady-state, $p_i(q)$, is needed to find the value of $\mathbb{EBO}_i[S_i, \mathbf{Z}, \mathbf{X}]$. In this direction, each cluster k in the repair shop for a given number of servers z_k is modeled as a multi-class multi-server $M/M/z_k$ queueing system. The method proposed by [6] is deployed to analyze multi-class multi-server $M/M/z_k$ and to derive the probability distribution of the number of failed SKU type i. When the number of SKU types and number of servers are reasonably large in the cluster, this method may become computationally expensive in terms of running time. Thus, we utilized queueing approximation algorithms discussed in [12,13]. By this approximation, marginal probability distribution (and several performance characteristics) of the SKU type i in the cluster k is derived by aggregating all other SKUs in the cluster k into a single SKU type (class). To obtain the remaining probability distributions for the other SKUs in the cluster, the procedure is repeated [8,11].

Figure 6 visualizes how N-class $M/M/z_k$ system is decomposed into N independent 3-class $M/M/z_k$ for approximation, where Λ_A and $\Lambda_{A'}$ denote the arrival rates of aggregated classes.

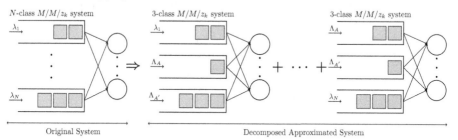

Fig. 6. Approximation of an N-class $M/M/z_k$ queueing system with decomposition into N 3-class $M/M/z_k$ subsystem

We can find the optimal value of inventory levels $\mathbf{S}$ for each SKU by using the approximated distributions, $\tilde{p}_i(q)$. The smallest S_i value that satisfies the following equation is the optimal inventory level for given $\mathbf{X}$ and $\mathbf{Z}$ (see [14] for details).

$$\sum_{q=0}^{S_i} \tilde{p}_i(q) \geq \frac{b - h_i}{b} \, i = 1, \ldots, |N| \tag{7}$$

4   Numerical Study

In this section, first, we describe the experiment testbed used for the numerical study in Subsect. 4.1. Then, we test the proposed solution algorithm under different input settings. In this direction, Subsect. 4.2 presents the total system cost reductions achieved by the proposed algorithm in comparison with fully flexible and dedicated repair shop designs, and Subsect. 4.3 documents the run-times of the proposed optimization algorithms under different factors.

4.1   Testbed

For benchmarking, we use the same testbed of instances as in [5,8]. In this dataset, a full factorial design of experiment (DoE) with seven factors and two levels per factor is used to generate a total of 128 instances. The number of SKUs, N, and the initial total number of servers, M, are the first two DoE factors with levels 10 and 20 for the numbers of SKUs, and 5 and 10 for the initial numbers of servers. The failure rates and the service rates are generated based on the system (repair shop) utilization rate with the assumption that all SKUs are processed on all servers, i.e., a repair shop design with one cluster and fully flexible servers. The system utilization rate, ρ, is the third design factor with levels 0.65 and 0.80. For the chosen utilization rate, we randomly generate two sets of parameters:

(a) the failure rates λ_i, such that $\sum_{i=1}^{N} \lambda_i = 1$, and

(b) workload percentages δ_i, such that $\sum_{i=1}^{N} \delta_i = 1$.

Using the generated λ_i and δ_i, we produce the service rates μ_i as $\mu_i = \frac{\lambda_i}{\delta_i \rho M}$, where $\delta_i \rho M$ is the total workload of SKU type i. The pattern of the holding costs, h_i, is the fourth design factor with two variants (levels): (i) IND: completely randomly (independent) within a range $[h_{min}, h_{max}]$, and (ii) HPB: hyperbolically related to the workloads $w_i = \lambda_i/\mu_i = \delta_i \rho M$:

$$h_i = \frac{h_{max} - h_{min} + 10}{9 \frac{w_i - w_{min}}{w_{max} - w_{min}} + 1} - 10 + h_{min} + \xi_i$$

where

$$\xi_i \in U[-\frac{h_{max} - h_{min}}{20}, \frac{h_{max} - h_{min}}{20}], w_{min} = \min_{i=1,...,N} w_i$$

$$and \ w_{max} = \max_{i=1,...,N} w_i$$

The parameters of the hyperbolic relation are chosen such that it replicates some of the real-life scenarios where more expensive repairables are repaired less frequently. The minimum holding cost, h_{min}, is the fifth factor with levels 1 and 100. The maximum holding cost is fixed at 1,000. The server cost, f, and the skill cost, c_i, are the last two factors in our DoE. The server cost levels are set as

10,000 and 100,000 ($10h_{max}$ and $100h_{max}$). The skill cost is assumed as 1% or 10% of the chosen server cost for all SKUs. The penalty cost, b, is set as fifty-fold of the average holding cost so that about 98% of requests can be met from spare stocks. That means the probability of backorder is only 0.02. The overview of all factors and levels are presented in Table 2.

Table 2. Problem parameter variants for test bed.

Factors	Levels
No. of SKUs (N)	10, 20
No. of initial servers (M)	5, 10
Utilization rate (ρ)	0.65, 0.80
Minimum holding cost (h_{min})	1, 100
Maximum holding cost (h_{max})	1000
Holding cost/Workload relation	IND, HPB
Server cost (f)	$10h_{max}$, $100h_{max}$
Cross-training cost (c_i)	$0.01f$, $0.10f$
Penalty cost (b)	$50\frac{\sum_{i=1}^{N} \lambda_i h_i}{\sum_{i=1}^{N} \lambda_i}$

4.2 Cost Reduction

Using the proposed pooling-based GA heuristic, we find the optimal pooling scheme together with optimal capacity and inventory levels of spares for the cases described in the previous subsection. We compare the total system cost with the costs obtained from fully flexible (as in Fig. 2(a)) and dedicated (as in Fig. 2(c)) designs.

We define two metrics $\Delta_{GA}^{Flexible}$ and $\Delta_{GA}^{Dedicated}$. The metric $\Delta_{GA}^{Flexible}$ represents the relative percentage difference between the total cost of the pooled design found by pooling based GA heuristic and the total cost of the fully flexible design. In the same manner, $\Delta_{GA}^{Dedicated}$ denotes the relative percentage gap between the total cost of the pooled and the dedicated designs.

Figure 7 shows pair-wise total cost comparisons for different problem factors. When all the test cases are considered, the pooled designs found by proposed heuristic can yield ~44% and ~21% cost savings in comparison with dedicated and fully flexible designs, respectively. In only 11% of the cases (15 out 128) in the testbed, the fully flexible repair shop designs outperform the pooled designs in terms of the total system cost. Figure 7 also indicates that dedicated repair shop designs are not cost-effective because of to the excess idle server capacity and high total capacity acquisition cost. Dedicated designs are outperformed by proposed heuristic in all of the test cases, and in 25% of cases (32 out of 128), dedicated designs achieve a better (lower) total system cost compared to fully flexible systems.

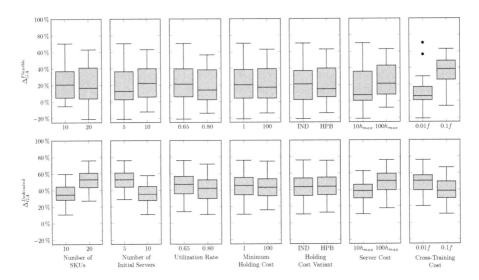

Fig. 7. Pair-wise total cost comparison with fully flexible and dedicated systems

4.3   Run-Time Performance

We implement the algorithm and run all the experiments on a computer with 16 GB RAM and 2.8 GHz i7 CPU. The run-time distributions of solution algorithm under different factors are provided in Fig. 8. On an average, the presented pooling-based GA heuristic converges the best solution in 1456 CPU seconds. The worst case performance is around 4000 CPU seconds, which is acceptable for tactical and operational level decisions in real-life spare part supply systems.

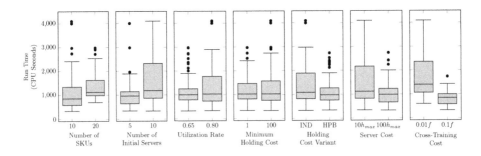

Fig. 8. Run-time performance of solution algorithms under varying factors

The computational study shows that the increasing problem size (number of SKUs and the number of initial servers) has a negative impact on the run-time of the algorithm due to the increasing effort to solve the multi-class multi-server queueing system approximation. Interestingly, the cross-training cost factor indirectly affects the size of multi-class multi-server queueing problem that has to be solved several times during execution of the algorithm, which results in longer run-times.

5 Conclusions

An effective spare part supply system planning is essential to achieve a high capital asset availability. In this direction, a two-stage sequential solution heuristic is developed to solve repair shop design problems in the spare parts supply systems. The developed algorithm uses a GA and queueing approximation as its subroutines to find and evaluate possible pooled repair shop designs. The proposed methodology achieves lower total system costs in comparison to dedicated and fully flexible designs within reasonable run-times.

As future research, to test the applicability of the methodology with real-life cases (with larger problem sizes; i.e., a larger number of SKUs) would be an invaluable contribution. Improving the performance of the GA by developing new mutation and crossover operations would be worthwhile. We also plan to use other metaheuristics such as simulated annealing and particle swarm optimization rather than GA to generate feasible pooling schemes. Integrating pooling decision with static and dynamic routing and prioritization rules in the part repair processes would provide interesting managerial insights.

References

1. Alrabghi, A., Tiwari, A.: State of the art in simulation-based optimisation for maintenance systems. Comput. Ind. Eng. **82**, 167–182 (2015)
2. Wang, L., Chu, J., Mao, W.: An optimum condition-based replacement and spare provisioning policy based on Markov chains. J. Qual. Maint. Eng. **14**(4), 387–401 (2008)
3. Sleptchenko, A., ElMekkawy, T.Y., Turan, H.H., Pokharel, S.: Simulation based particle swarm optimization of cross-training policies in spare parts supply systems. In: The Ninth International Conference on Advanced Computational Intelligence (ICACI 2017), pp. 60–65 (2017)
4. Turan, H.H., Pokharel, S., Sleptchenko, A., ElMekkawy, T.Y.: Integrated optimization for stock levels and cross-training schemes with simulation-based genetic algorithm. In: 2016 International Conference on Computational Science and Computational Intelligence, pp. 1158–1163 (2016)
5. Sleptchenko, A., Turan, H.H., Pokharel, S., ElMekkawy, T.Y.: Cross-training policies for repair shops with spare part inventories. Int. J. Prod. Econ. (2017)
6. Van Harten, A., Sleptchenko, A.: On markovian multi-class, multi-server queueing. Queueing Syst. **43**(4), 307–328 (2003)
7. Al-Khatib, M., Turan, H.H., Sleptchenko, A.: Optimal skill assignment with modular architecture in spare parts supply systems. In: 4th International Conference on Industrial Engineering and Applications (ICIEA) 2017, pp. 136–140. IEEE (2017)

8. Turan, H.H., Pokharel, S., Sleptchenko, A.A., ElMekkawy, T.Y.T., Al-Khatib, M.: A pooling strategy for flexible repair shop designs. In: Proceedings of the 7th International Conference on Operations Research and Enterprise Systems (ICORES 2018), pp. 272–278 (2018)
9. Sherbrooke, C.C.: Optimal Inventory Modeling of Systems: Multi-echelon Techniques, vol. 72. Springer, New York (2006). https://doi.org/10.1007/b109856
10. Hiassat, A., Diabat, A., Rahwan, I.: A genetic algorithm approach for location-inventory-routing problem with perishable products. J. Manuf. Syst. **42**, 93–103 (2017)
11. Turan, H.H., Sleptchenko, A., Pokharel, S., ElMekkawy, T.Y.: A clustering-based repair shop design for repairable spare part supply systems (2018, submitted for publication)
12. Altiok, T.: On the phase-type approximations of general distributions. IIE Trans. **17**(2), 110–116 (1985)
13. Van Der Heijden, M., Van Harten, A., Sleptchenko, A.: Approximations for markovian multi-class queues with preemptive priorities. Oper. Res. Lett. **32**(3), 273–282 (2004)
14. Turan, H.H., Sleptchenko, A., ElMekkawy, T.Y.T., Pokharel, S., Al-Khatib, M.: Pooled repair shop designs for repairable spare part supply systems (2018, submitted for publication)

A Core Heuristic and the Branch-and-Price Method for a Bin Packing Problem with a Color Constraint

Artem Kondakov[1,2] and Yury Kochetov[1,2(✉)]

[1] Novosibirsk State University, Novosibirsk, Russia
tyxtyxyc@gmail.com
[2] Sobolev Institute of Mathematics, Novosibirsk, Russia
jkochet@math.nsc.ru

Abstract. We study a new bin packing problem with a color constraint. A finite set of items and an unlimited number of identical bins are given. Each item has a set of colors. Each bin has a color capacity. The set of colors for a bin is the union of colors for its items and its cardinality can not exceed the bin capacity. We need to pack all items into the minimal number of bins. For this NP-hard problem, we design the core heuristic based on the column generation approach for the large-scale formulation. A hybrid VNS matheuristic with large neighborhoods is used for solving the pricing problem. We use our core heuristic in the exact branch-and-price method. Computational experiments illustrate the ability of the core heuristic to produce optimal solutions for randomly generated instances with the number of items up to 250. High-quality solutions on difficult instances with regular structure are found.

Keywords: Branch-and-price · Column generation · Matheuristic

1 Introduction

In the classical bin packing problem, a set of weighted items must be packed into the minimal number of identical bins so that the sum of weights of items in each bin does not exceed the bin's capacity. In this paper, we continue to study a new variant of the bin packing problem with a color constraint [13]. We assume that each item has some colors rather than a weight. The bin capacity limits the total number of colors for its items. The goal is to pack all items into the minimal number of identical bins so that the total number of colors in each bin does not exceed the bin capacity. It is a NP-hard problem in the strong sense and the classical bin packing problem can be reduced to it [16].

The bin packing problem [9] with color constraints (BPC) is a recent line of research in combinatorial optimization. One of the applications of the problem is in the beverage package printing [16]. In [19] the bin packing problem with classes of items (colors) is used to model video-on-demand applications.

© Springer International Publishing AG, part of Springer Nature 2018
A. Eremeev et al. (Eds.): OPTA 2018, CCIS 871, pp. 309–320, 2018.
https://doi.org/10.1007/978-3-319-93800-4_25

A biclique covering (biclustering), which is a straight reformulation of the BPC, is used to model protein-protein interactions [5].

There are some versions of the problem with online and offline settings [3,17]. In the colored bin packing [4], each bin has a maximum color capacity, i.e. a limit on the number of items of a particular color. This version is originated in the production planning of a steel plant. In a generalized version of the color bin packing problem [10,15], a conflict graph describes some constraints for the items in the same bin. In the black and white bin packing problem with alternation constraints [2], two items with the same color cannot be packed adjacently to each other.

In this paper, we continue to research a new version of the color bin packing problem which is a special case of the co-printing problem [13,16]. We assume that each item has zero weight and a set of colors. We design the branch-and-price method for this problem based on the large-scale reformulation. Following this framework, column generation procedure is implemented and additional stabilization constraints are added to speed up convergence of the procedure [7]. The VNS matheuristic with large neighborhoods is used for solving the pricing problem [13]. To obtain upper bounds, we consider the set of generated columns as a core of this problem and apply commercial solver (GUROBI) to this core. A similar idea was suggested by Avella et al. [1] for solving large-scale p-median problem by Lagrangian relaxations. It is interesting to note that the solver is very efficient in the case of core subproblem, although we spend a lot of efforts to perform the initial column generation procedure.

In our computational experiments, we observe that such core heuristic provides strong results even for a subcore when we terminate the column generation at an intermediate step for large instances. We illustrate this useful idea in computational experiments for instances with the number of items up to 500. The core heuristic has found optimal solutions for all test instances with the number of items up to 250 if each item has a random subset of colors. Moreover, integrality gap is 0 for this class of instances. A similar effect is known for the classical bin packing problem. Lower bound coincide with upper bound in the root of branching tree and such random instances are easy for our approach. Therefore, we design a set of difficult instances based on the idea from [12]. We create the random instances by the following rule: all items have the same number of colors and each color is used in the same number of items. For such regular instances, the integrality gap is positive, up to 20%, the branching tree is non-trivial and the running time increases rapidly when dimension grows. Nevertheless, the core heuristic produces optimal or near optimal solutions with minimal deviation from the optimal value.

The paper is organized as follows. In Sect. 2 we introduce notations and the mathematical model for the BPC with a large number of variables. In Sect. 3 we outline the branch-and-price scheme. In Sect. 4 we describe column generation procedure and its usage during the branch-and-price method. To reduce the number of iterations of the column generation method, we introduce deep dual-optimal inequalities to the restricted master problem. In Sect. 5, we design our core heuristic. Finally, Sect. 6 presents computational experiments for the random and difficult test instances with large integrality gap.

2   Mathematical Model

Let us introduce the following notations:

$I = \{1, \ldots, n\}$ is the set of items;
$J = \{1, \ldots, m\}$ is the set of colors;
$K_i \subset J$ is the set of colors for item i;
b is the upper bound for number of different colors in each bin;
$p = (p_1, \ldots, p_i, \ldots, p_n)$ is a bin pattern, or bin for shot, where $p_i \in \{0, 1\}$ denotes whether item i is in the bin or not;
$P = \{p : | \cup_{i \in I}(K_i : p_i = 1) | \leq b\}$ is the set of all feasible bins.

Decision variables:
$y_p = 1$ if bin p is used in the solution and $y_p = 0$ otherwise.
Now we can write the BPC problem as follows:

$$\min \Big\{ \sum_{p \in P} y_p \; : \; \sum_{p \in P} p_i y_p \geq 1, i \in I, \; y_p \in \{0, 1\} \Big\}. \tag{1}$$

We use this large-scale formulation for our methods for several reasons. First of all, the large-scale formulation allows us to avoid symmetries which usually accompanies compact bin packing representations [14]. Second, large-scale formulations, as a rule, have small integrality gap even for the classical bin packing problem [11]. Despite the tremendous size of the resulting problem, it's relaxation can be solved with the column generation technique and the optimal solution can be found via the branch-and-price method.

To present column generation approach described in Sect. 4, we need to introduce relaxed model with continuous variables $\bar{y}_p, p \in P$ to which we will also refer as the master problem:

$$\min \Big\{ \sum_{p \in P} \bar{y}_p \; : \; \sum_{p \in P} p_i \bar{y}_p \geq 1, i \in I, \; 0 \leq \bar{y}_p \leq 1 \Big\}. \tag{2}$$

The optimal solution of this model gives us lower bound. These bounds are actively used in the branch-and-price method described in Sect. 3.

Another interesting formulation of the BPC problem can be done in terms of bipartite clique covering. Let $G(V_1, V_2, E)$ be the bipartite graph with parts of vertices V_1, V_2 and set of edges $E \subset V_1 \times V_2$. A biclique K_{st} in G is a complete bipartite subgraph with s vertices from V_1 and t vertices from V_2. We wish to cover vertices from V_1 by a minimal number of bicliques for large $t, t \geq D$ for given threshold D. It is easy to see that the BPC problem is equivalent to this biclique covering problem. We put $V_1 = I$, $V_2 = J$ and an edge (ij) belongs to E if $j \notin K_i$. Each feasible bin in the BPC problem is a biclique in the covering problem and vice versa.

3   Branch-and-Price Method

The branch-and-price technique utilizes the column generation approach for solving linear programming relaxation. This relaxation generates tight lower bounds, which reduces the total amount of iterations required to find global optima.

Branching scheme corresponds to the one described in [18], referred as "branching based on a set of bounds on the components of q". A similar scheme was used in [16]. Two types of branches are applied:

1. Applied whenever the sum of optimal LP variables $\bar{y}$ corresponding to bins containing item i is not integer: $\sum_{p_i > 0} \bar{y}_p = \alpha$ is fractional. Then the two nodes are added into the branching tree:

$$\sum_{p_i > 0} \bar{y}_p \geq \lceil \alpha \rceil \ ; \quad \sum_{p_i > 0} \bar{y}_p \leq \lfloor \alpha \rfloor. \tag{3}$$

2. Applied if the sum is integer, but some of the variables are fractional. These branch inequalities forces (or forbids) usage of particular item combination in a bin. Let $q = (q_1, ..., q_n)$ be the so-called partial bin, with at least two different items. Then the branching constraints are defined as follows:

$$\sum_{p \geq q} \bar{y}_p = 1 \ ; \quad \sum_{p \geq q} \bar{y}_p = 0, \tag{4}$$

where $p \geq q$ means $p_i \geq q_i$ for all $i \in I$.

Inequalities (3) and (4) are added to Model (2) at each node of the search tree. As shown for a general column generation procedure in [18], branching scheme described above can eliminate all fractional solutions. In Sect. 4 we'll demonstrate corresponding changes in pricing problem.

Now a general overview of a procedure can be given:

1. Find the optimal solution for the relaxed master problem (2): the result will be an initial lower bound.
2. Obtain upper bound for the integer problem with a core heuristic, described in Sect. 5.
3. If upper bound does not equal to lower bound, determine an item i for which a fractional column exists.
4. If the sum of the columns containing item i is fractional, branch according to the first type, else determine a partial bin and branch according to the second type.
5. Re-optimize an LP, apply core heuristic and repeat from step 3.

Partial bin for the second type of branching selected as follows. An item i which is combinable with the minimum number of other items is selected. Then, the partial bin is extended with the minimum number of other items such that sum $\sum_{p \geq q} \bar{y}_p$ is fractional. If at some previous level of branching tree a partial bin is already opened, we extend that bin instead of opening new bin.

As we'll demonstrate in Sect. 6, for random examples optimal value for Model (2) doesn't differs significantly from optimal value for Model (1). In other words, integrality gap of Model (1) is relatively small: usually lesser than one bin for random examples. However, calculation of lower bound requires solving Model (2), which can be challenging due to a huge number of columns. A column generation approach, which will be described in the next section, is applied to solve this problem.

4 Lower Bounds

According to the classical column generation approach, we restrict ourselves to a small subset $P' \subset P$, initially generated by some heuristic. Then at each iteration, a pricing problem is solved, and new columns with negative reduced cost are generated. We add all such columns to the restricted master problem. Now we consider the dual linear programming problem to the master problem:

$$\max\left\{ \sum_{i \in I} w_i : \sum_{i \in I} p_i w_i \leq 1, p \in P', \quad w_i \geq 0 \right\}, \tag{5}$$

where the dual variable w_i can be considered as a price for item i. To enlarge the subset P' or terminate the method, we should solve the following pricing problem with optimal values w_i^* of the dual variables.

Let us introduce additional variables:

$x_i = 1$ if item i is placed in a bin and $x_i = 0$ otherwise.

$z_j = 1$ if a bin contains item with color j and $z_j = 0$ otherwise.

Let us also introduce additional parameters and variables which is required to properly account branching constraints.

1. For each constraint of type $\sum_{p_i > 0} \bar{y}_p \geq \lceil \alpha \rceil$ we add variable ρ_i to (5), let ρ_i^* be an optimal value;
2. For each constraint of type $\sum_{p_i > 0} \bar{y}_p \leq \lfloor \alpha \rfloor$ we add variable σ_i to (5), let σ_i^* be an optimal value;
3. For each constraint of type $\sum_{p \geq q} \bar{y}_p = 1$ we add new item t with color set $K_t = \cup_{i:q_i>0} K_i$. Such color set forces any newly generated bin, which contains item t, to contain all items that lying in partial bin q, and vice versa. We aggregate all such items in set T;
4. For each constraint of type $\sum_{p \geq q} \bar{y}_p = 0$ we add corresponding partial bin q to a set Q. These partial bins induce new constraints in pricing problem, modifying search space for new columns in a way that every bin which contains q will not be generated.

Now the pricing problem can be stated as follows:

$$\min(1 - \sum_{i \in I \cup T} (w_i^* + \rho_i^* + \sigma_i^*)x_i)$$

$$\text{s.t.} \sum_{j \in J} z_j \leq b;$$

$$x_i \leq z_j, \quad j \in K_i, i \in I \cup T;$$

$$\sum_{i:q_i=1} x_i \leq |q| - 1, \quad q \in Q;$$

$$x_i, z_j \in \{0,1\}, \quad i \in I \cup T, j \in J.$$

The objective function minimizes the reduced cost. The first constraint controls the total number of colors for a bin. The second constraint shows the relations between items and colors. The third constraint induced by branching, as we've described above. It is easy to see that a knapsack problem can be reduced to the pricing problem, and as a result, the pricing problem is NP-hard. A block diagram for the column generation procedure is presented in Fig. 1.

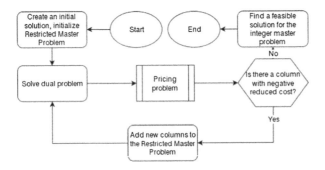

Fig. 1. Block diagram of the column generation method

To reduce the number of iterations of the column generation method, we introduce *deep dual-optimal inequalities* [7] to the restricted master problem. It is a set of restrictions, which, if added to the dual problem, does not cut off at least one optimal solution. The following proposition establishes a foundation for our inequalities:

Proposition 1. Let I_0, I' be two disjoint sets of items such that $\cup_{i \in I_0} K_i \supseteq K_{i'}$ for all $i' \in I'$. Then there exist a dual optimal solution w^* for the restricted master problem such that $\sum_{i \in I_0} w_i^* \geq \sum_{i' \in I'} w_{i'}^*$.

Proof. It is easy to verify that the set of constraints in model (1) has the (s,t) exchange property, with $s_i = 1$ if $i \in I_0$, $t_i = 1$ if $i \in I'$ and 0 otherwise, as defined in [7]. Indeed, set of items I_0 can be replaced with all items from I' without breaking the capacity constraint. Therefore, using Proposition 2 from [7], the cut $\sum_{i \in I_0} w_i^* \geq \sum_{i' \in I'} w_{i'}^*$ is a dual-optimal inequality. □

At each iteration, we identify a set of items I_0 for which an inequality

$$\sum_{i \in I_0} w_i^* < \sum_{i \in \{i \mid K_i \subseteq \cup_{i' \in I_0} K_{i'}\} \setminus I_0} w_i^* \tag{6}$$

holds, and add it to the master problem. We will demonstrate the efficiency of those inequalities in Sect. 6.

To solve pricing problem and generate new column effectively, we apply previously designed the hybrid VNS matheuristic [13]. It can be applied into the branch-and-price method with minor modifications only:

- Branching constraints of type 1 modify only cost of items in the pricing problem;
- Each branching constraint of type $\sum_{p \geq q} \bar{y}_p = 1$ induces one new item with the same structure as any other item;
- Each branching constraint of type $\sum_{p \geq q} \bar{y}_p = 0$ induces new constraint. This constraint can be accounted by GUROBI [8] while obtaining any solution from any neighborhood described in [13].

When optimal solution for Model (2) is found, we can apply the core heuristic described in the next section to obtain a feasible solution for the BPC.

5   Core Heuristic

Three heuristics for the BPC are presented in [13]. Two of them, the well-known FFD heuristic for the classical bin packing problem and so-called FillBin heuristic produce weak solutions, but they do not require the optimal solution to the Model (2). Third heuristic, LP heuristic, produce much better solutions, but it can be used only at the last iterations of the column generation method.

Now we introduce a new core heuristic which produces better solutions than previously mentioned ones. Although it also requires the optimal solution to the Model 2, this heuristic works faster and produce better solutions than LP heuristic.

Formal description of the core heuristic is the following: after the column generation procedure terminates, we find optimal integer solution for the restricted master problem. To get a solution at some level of the branching tree, we only need to find the optimal solution that satisfies the branching constraints. We use commercial solver GUROBI to this end. The idea was inspired by [1] for the large-scale p-median problem. But instead of Lagrangian relaxation to the compact problem representation, we use the column generation method to the large-scale problem formulation. Moreover, as we will see late, we can apply a truncated version of this approach for large-scale instances as well.

According to our computational experiments, the integrality gap in the restricted master problem is close to zero for randomly generated instances. As a rule, at least one optimal solution belongs to the problem's core. For this reason, we can find optimal integer solution with GUROBI easily. Nevertheless, there are some difficult instances with a large integrality gap, as we will see in Sect. 6. For these instances, running time of the core heuristic can increase dramatically.

6   Computational Experiments

We conduct our computational experiments for the randomly generated test instances in 10 sets of parameters n, m and b. For each set, we generate 10 instances. Thus, the total amount of tests is equal to 100. The color set K_i for each item is generated by the following rule. We uniformly choose an integer l from 1 to b and uniformly choose a 0–1 vector with exactly l ones.

Table 1. Computational results for three variants of the column generation method

n	m	b	$v1$			$v2$			$v3$		
			It	C	T	It	C	T	It	C	T
150	55	30	87	261	2256	110	313	2601	89	270	2893
175	55	30	122	294	2691	143	339	3055	124	302	2914
200	55	30	98	316	2772	132	342	3258	105	312	3151
225	55	30	119	322	3024	148	361	3406	120	331	3383
250	55	30	141	347	4017	170	405	4249	141	341	4373
150	90	40	101	277	2318	128	320	2913	105	282	2766
175	90	40	125	305	2652	151	334	3122	124	317	3078
200	90	40	136	329	2910	154	357	3289	136	335	3166
225	90	40	156	362	3573	183	398	4170	157	367	3884
250	90	40	202	481	5837	257	530	6714	210	486	6526

Table 1 presents the results of our experiments for different variants of the column generation method. Column It is the averaged amount of column generation iterations among 10 instances, column C is average amount of generated columns, and column T is average time in seconds for PC ASUS x550L Intel Core i5 2.300 MHz. Block of columns $v1$ shows the performance of the method with all features that have been described earlier; $v2$ shows the performance of the same method without stabilization inequalities (2); and $v3$ shows the performance of the method with a simple variant of the VNS matheuristic for the pricing problem. According to [13], items with large color sets are the most inconvenient for this heuristic. Thus, we can select all large items and solve the pricing problem for each of them separately. Then we apply the VNS matheuristic for the remaining items and the best found solution is returned as a result of the modified VNS heuristic. This modified variant is applied in $v1$ to contrast $v3$. As we can see in Table 1, the modified VNS heuristic and stabilization inequalities are useful for the column generation method and improve its performance. For all instances, lower bounds coincide with upper bounds and we have got optimal solutions.

The last iterations of the column generation method are the most time consuming. If the VNS heuristic cannot find a new column with negative reduced cost, we need to apply exact method (GUROBI solver). But the core heuristic can be used in intermediate iterations of the column generation method as well. We use this truncated heuristic to find near optimal solutions for the large-scale instances. The accuracy of those solutions can be estimated with intermediate lower bound [6]: given an optimal cost Z^*_{RMP} for the restricted master problem, an optimal solution w for the dual-master problem, an optimal solution x^* for the pricing problem, then the following inequality holds:

$$\frac{Z^*_{RMP}}{w^\top x^*} \leq Z^*_{MP} \tag{7}$$

where Z^*_{MP} is an optimal cost for the full master problem. We terminate the column generation method when the VNS heuristic for the pricing problem can't find a solution with negative reduced cost. We apply the core heuristic and solve the pricing problem by GUROBI to estimate lower bound from (7). We observe that this intermediate lower bound is rough: if we run the column generation method without interruptions, the core solution improves just a little, while the lower bound improves significantly.

Table 2. Computational results for the truncated heuristic

n	m	b	It	C	T_{CG}	T_{CH}	Acc_{LB}	Acc_{CG}
150	90	40	28	232	416	2.12	8.03%	0%
175	90	40	41	257	621	2.61	8.1%	0.25%
200	90	40	65	281	1683	2.58	9.25%	1.33%
225	90	40	86	314	2258	3.13	11.38%	1.2%
250	90	40	93	384	3104	3.6	12.6%	1.56%
300	90	40	97	426	4841	4.21	15.33%	-
350	90	40	95	485	6103	4.88	20.6%	-
400	90	40	98	515	7057	5.37	21.14%	-
450	90	40	101	648	8508	6.16	23.78%	-
500	90	40	103	810	10736	7.71	35.53%	-

Table 2 presents the results for that truncated heuristic. Column Acc_{LB} shows the average percentage deviation of heuristic solutions from the intermediate lower bound, column Acc_{CG} shows the average percentage deviation from the optimum. Columns T_{CG} and T_{CH} show the running time of column generation procedure and the core heuristic respectively. Empty cells in the table show the cases when the running time of the column generation is too high. Note that optimal integer solution for the restricted master problem has been found easily.

Table 3. The branch-and-price performance

n	m	b	T_0	T	Branches	Type1	Type2	RelativeGap	OptIt
30	30	10	11.43	20.234	6	5	1	0.2	2
30	30	10	10.293	18.305	5	5	0	0.2	0
30	30	10	13.818	21.004	8	7	1	0.16	0
42	42	14	45.942	64.942	17	14	3	0.167	3
42	42	14	50.595	77.152	21	16	5	0.153	0
42	42	14	48.87	78.526	21	15	6	0.181	2
54	54	18	159.204	218.569	31	22	9	0.188	5
54	54	18	157.707	202.651	29	23	6	0.133	4
54	54	18	159.204	249.569	35	25	10	0.2	5
66	66	22	323.748	430.730	68	47	21	0.2	8
66	66	22	356.235	512.399	76	50	26	0.15	11
66	66	22	328.748	456.730	71	49	21	0.15	9
78	78	6	772.654	985.148	105	75	30	0.188	14
78	78	26	721.515	906.148	96	70	26	0.12	13
90	90	30	1602.083	2172.342	153	98	55	0.13	12
90	90	30	1684.211	2328.487	160	101	59	0.153	18

We note that the performance of the method strongly depends on the integrality gap and the bin capacity. For instances without the gap, our approach shows good performance in running time. It seems that the uniformly generated instances are quite *easy* for the method. Many of them have not the integrality gap in the large-scale formulation. If the integrality gap is positive, we need a lot of efforts to solve the linear programming relaxation. Nevertheless, we can apply the core heuristic in such a case as well even for large bin capacity and before the termination of the column generation method.

In our last experiment, we try to create the most difficult test instances with large integrality gap. We use the idea from [12] for the difficult instances for the facility location problems. We generate the random instances by the following rule: each item has exactly $\lceil b/2 \rceil$ colors and every color is used in the same number of items. Table 3 shows the performance of our branch-and-price method for the difficult instances. We consider 2–3 instances for each set of parameters n, m and b. Column T reflects the total computational time in seconds, column T_0 reports the amount of time required to find initial lower and upper bounds. Total amount of branching nodes demonstrated in column *Branches*. Column *Type1* stands for the first type of branches, and *Type2* stands for the second type of branches. Column *RelativeGap* shows the relative integrality gap in this instance between optimal solution and lower bound obtained by the column generation method. The gap is calculated as $(z^* - z_{LB})/z^*$, where z^* is the optimal solution of the BPC, and z_{LB} is the optimal solution to linear programming relaxation of the BPC. Column *OptIt* shows the iteration when the optimal solutions were discovered.

As we can be seen in Table 3, the gap is quite large, from 12% to 20%. The initial lower and upper bounds are the most time consuming. In non root nodes of the branching tree, we already have a lot of generated columns. Thus, we need to generate a few additional columns only. In fact, we have got enough information to find good solutions in root node and the optimal solution arises pretty fast (see column *OptIt*).

7   Conclusions

In this paper, we study a new variant of the bin packing problem with a color constraint. We have designed the exact branch-and-price method and the core heuristic which is very efficient for the random instances without integrality gap and produces near optimal solutions for difficult test instances with large gap. It is interesting to note that for the classical bin packing problem we still cannot find any test instances with a large integrality gap [11]. We guess that the approximability properties of the new and the classical bin packing problems are different. We know some polynomial time approximation algorithms for the classical problem. But similar results for the new problem are still unknown. It is a new line for future research.

Acknowledgements. The research in Sects. 1, 2, 3 and 4 was supported by RFBR grant 18-07-00599. The research in Sects. 5 and 6 was supported by the program of fundamental scientific researches of the SB RAS I.5.1., project 0314-2016-0014.

References

1. Avella, P., Boccia, M., Salerno, S., Vasilyev, I.: An aggregation heuristic for large scale p-median problem. Comput. Oper. Res. **39**(7), 1625–1632 (2012)
2. Balogh, J., Békési, J., Dosa, G., Kellerer, H., Tuza, Z.: Black and white bin packing. In: Erlebach, T., Persiano, G. (eds.) WAOA 2012. LNCS, vol. 7846, pp. 131–144. Springer, Heidelberg (2013). https://doi.org/10.1007/978-3-642-38016-7_12
3. Böhm, M., Sgall, J., Veselý, P.: Online colored bin packing. In: Bampis, E., Svensson, O. (eds.) WAOA 2014. LNCS, vol. 8952, pp. 35–46. Springer, Cham (2015). https://doi.org/10.1007/978-3-319-18263-6_4
4. Dawande, M., Kalagnanam, J., Sethuraman, J.: Variable sized bin packing with color constraints. Electron. Not. Discrete Math. **7**, 154–157 (2001)
5. Ding C., Zhang Y., Li T., Holbrook S.R.: Biclustering protein complex interactions with a biclique finding algorithm. In: Sixth International Conference on Data Mining (ICDM 2006), pp. 178–187. IEEE (2006)
6. Farley, A.A.: A note on bounding a class of linear programming problems, including cutting stock problems. Oper. Res. **38**(5), 922–923 (1990)
7. Gschwind, T., Irnich, S.: Dual inequalities for stabilized column generation revisited. INFORMS J. Comput. **28**(1), 175–194 (2016)
8. Gurobi Optimization, Inc.: Gurobi optimizer reference manual (2015)

 9. Jansen, K., Porkolab, L.: Preemptive parallel task scheduling in $O(n) + \text{Poly}(m)$ time. In: Goos, G., Hartmanis, J., van Leeuwen, J., Lee, D.T., Teng, S.-H. (eds.) ISAAC 2000. LNCS, vol. 1969, pp. 398–409. Springer, Heidelberg (2000). https://doi.org/10.1007/3-540-40996-3_34

10. Jansen, K.: An approximation scheme for bin packing with conflicts. J. Comb. Optim. **3**(4), 363–377 (1999)

11. Kartak, V.M., Ripatti, A.V., Scheithauer, G., Kurz, S.: Minimal proper non-irup instances of the one-dimensional cutting stock problem. Discrete Appl. Math. **187**, 120–129 (2015)

12. Kochetov, Yu., Ivanenko, D.: Computationally difficult instances for the uncapacitated facility location problem. In: Ibaraki, T., Nonobe, K., Yagiura, M. (eds.) Metaheuristics: Progress as Real Problem Solvers, vol. 32, pp. 351–367. Springer, Boston (2005). https://doi.org/10.1007/0-387-25383-1_16

13. Kochetov, Yu., Kondakov, A.: VNS matheuristic for a bin packing problem with a color constraint. Electron. Not. Discrete Math. **58**, 39–46 (2017)

14. Margot, F.: Symmetry in integer linear programming. In: Jünger, M., Liebling, T.M., Naddef, D., Nemhauser, G.L., Pulleyblank, W.R., Reinelt, G., Rinaldi, G., Wolsey, L.A. (eds.) 50 Years of Integer Programming 1958–2008, pp. 647–686. Springer, Heidelberg (2010). https://doi.org/10.1007/978-3-540-68279-0_17

15. Muritiba, A.E.F., Iori, M., Malaguti, E., Toth, P.: Algorithms for the bin packing problem with conflicts. Informs J. Comput. **22**(3), 401–415 (2010)

16. Peeters, M., Degraeve, Z.: The co-printing problem: a packing problem with a color constraint. Oper. Res. **52**(4), 623–638 (2004)

17. Shachnai, H., Tamir, T.: Polynomial time approximation schemes for class-constrained packing problems. J. Sched. **4**(6), 313–338 (2001)

18. Vanderbeck, F.: On Dantzig-Wolfe decomposition in integer programming and ways to perform branching in a branch-and-price algorithm. Oper. Res. **48**(1), 111–128 (2000)

19. Xavier, E.C., Miyazawa, F.K.: The class constrained bin packing problem with applications to video-on-demand. Theoret. Comput. Sci. **393**(1), 240–259 (2008)

On Calculation and Estimation of Flow Transmission Probability in a Communication Network

Alexey S. Rodionov, Olga A. Yadykina, and Denis A. Migov[✉]

Institute of Computational Mathematics and Mathematical Geophysics of SB RAS,
prospekt Akademika Lavrentieva 6, 630090 Novosibirsk, Russia
alrod@sscc.ru, artjuha@gmail.com, mdinka@rav.sscc.ru
http://www.sscc.ru/

Abstract. We study the problem of estimating a probability that a flow of a given capacity may be transferred in a communication network. Network is represented by a random graph with absolutely reliable nodes and unreliable links with given operational probabilities and capacities. The algorithm for fast decision making whether a network is reliable enough for transmission of a given flow is proposed. Case studies show applicability of the proposed approach.

Keywords: Communication network · Transport network
Connectivity · Random graph · Network reliability · Flow network
Flow transmission

1 Introduction

Problems relating to analysis of flow stochastic network have been subject of considerable research (see, for example [1–6]). In this paper, we consider such important index of network reliability as a probability of transmission of a flow with given capacity between two terminal nodes. This characteristic evaluates the ability to transfer a predetermined amount of a resource (it can be information, vehicles, natural resources, etc.) between nodes, even in the case of partial network failure. We use random graph with absolutely reliable vertices and unreliable edges which fail independently as a network's model, which is quite common choice [7].

To calculate and evaluate the network reliability, we use cumulative updating of its bounds. This approach was first considered for estimating the probabilistic connectivity of random graphs [8–10]. Further, such methods have been proposed for other reliability indices: the average pairwise network reliability [11], the diameter constrained network reliability [12,13], the mathematical expectation of a size of a connected subgraph that contains some special node [14], and

Supported by Russian Foundation for Basic Research under grants 17-07-00775, 18-07-00460.

the wireless sensor network reliability [15]. A profound survey of techniques mentioned above can be found in [16].

The basic idea lays in incremental updating exact lower (LB) and upper (UB) reliability boundaries and comparing them with a predetermined reliability threshold R_0. As a result, if $LB \geq R_0$, then a network is reliable, and if $UB < R_0$, then a network is unreliable one. Since problems related to the reliability analysis of networks are mostly NP-hard [17,18], such approach makes it possible to avoid the exhaustive search. For calculating reliability boundaries we use the well-known factoring method [19], and for verifying suitability of obtained particular realizations of a network, we use the Ford-Fulkerson method of maximum flow searching [20]. The main advantage of combining these methods is a possibility for obtaining an answer to the question: "if the network is reliable enough?" in a reasonable time.

2   Cumulative Updating of Network Reliability

Recent research [8] considered problem of determination whether a network is reliable enough in terms of network probabilistic connectivity. The idea of the proposed method is to check if a network is feasible without exact calculating a value of network reliability. For this purpose authors defined so called threshold R_0 which is a requirement of the network reliability. Let us denote by LU and BU the lower and upper bounds of a reliability index respectively. For $R(G)$, we use original notations RL and RU for the lower and upper bounds respectively, and initialize them by 0 and 1. These bounds are updated in such a way that on i-th iteration $\bar{RL}_i \geq \bar{RL}_{i-1}$ and $\bar{RU}_i \leq \bar{RU}_{i-1}$. Decision process stops when either RL_l exceeds R_0 or R_0 exceeds RU_l. In the first case the network is supposed to be reliable and in the second one the network is unreliable.

Let us assume that during recursive factoring procedure we obtain L final graphs $G_1, G_2, \ldots, G_L$, for which the reliability can be easily calculated. Let P_l for $1 \leq l \leq L$ be the probability to have G_l. Thus, $\sum_{l=1}^{L} P_l = 1$ and the following inequality holds for any $1 \leq k \leq L$ [8]:

$$\sum_{l=1}^{k} P_l R(G_l) \leq R(G) \leq 1 - \sum_{l=1}^{k} P_l(1 - R(G_l)). \tag{1}$$

This inequality gives the algorithm for cumulative updating of the lower and upper bounds of $R(G)$. Every time whenever reliability of some G_l for any $1 \leq l \leq L$ is calculated, we can update RL_l and RU_l in the following way:

$$\begin{aligned} RL_l &= RL_{l-1} + P_l R(G_l) \\ RU_l &= RU_{l-1} - P_l(1 - R(G_l)). \end{aligned} \tag{2}$$

RL_l and RU_l approach exact $G(R)$ value every time when l increases. Once either RL_l or RU_l reaches R_0, the proposed algorithm concludes the feasibility of G: if RL_l reaches R_0, G is feasible; if RU_l passes R_0, G is infeasible. Thus, we can set any acceptable value of R_0 in order to stop the method during execution without performing exact calculating of the network reliability.

Later, this approach was applied for other reliability indices [16]. Present study proposes how cumulative updating can be used for reliability analysis of flow stochastic networks.

3 Problem Statement

We consider a flow random network with unreliable edges which is represented by graph $G(V, E)$. For each edge $e_{ij} \in E$ its capacity c_{ij} and its reliability (probability of existence) p_{ij} are given. Two terminal nodes are given: source S and sink T.

As a measure of reliability, let us consider the probability of transmission of a flow with given capacity f_0 between terminal nodes. Further we call it reliability and denote it by $FP(G, S, T, f_0)$. FP_0 is the predefined reliability threshold. If it turns out that the reliability is less than the threshold, we decide that the network is unreliable.

The problem is to find out whether the $FP(G, S, T, f_0)$ is greater than or equal to FP_0.

4 The Algorithm

We propose to calculate $FP(G, S, T, f_0)$ by the well-known factoring method [19], which is suitable for reliability calculation of various reliability indices. For a random graph, two procedures are performed in the factorization process for the chosen (pivot) edge: removing the edge or making it absolutely reliable. An edge is absolutely reliable if its reliability is equal to 1.

When an edge became absolutely reliable, the lower bound changes, and when an edge is removed, the upper bound changes:

$$LB = LB + P(H)I(S), \tag{1.1}$$

$$UP = UP - P(H)(1 - I(S)), \tag{1.2}$$

where $P(H)$ is a probability of a graph realization obtained during factoring process at the current step. $I(S)$ is the Boolean function equals to one if the specified flow threshold f_0 is no greater than the maximum flow in the graph, and equals to zero otherwise:

$$I(S) = \begin{cases} 1, & \text{if } f_{max} \geq f_0, \\ 0, & \text{if } f_{max} < f_0. \end{cases} \tag{1.3}$$

The maximum flow f_{max} is calculated either for a graph with removed edge, or for a graph consisting only of absolutely reliable edges. For this purpose the Ford-Fulkerson method of maximum flow searching is used.

After each update of reliability bound by expressions (1.1), (1.2) we check whether the network is reliable enough and can stop the calculation process with the following results: if $FP_0 > UB$ then the network is unreliable, if $FP_0 \leq LB$ then the network is reliable.

Thus, we obtain the following algorithm

1. Choose an edge e_{ij} in the graph G.
2. Remove the pivot edge, having received the graph $H = G \backslash \{e_{ij}\}$.
   2.1. IF the flow can not be passed through the received graph H, THEN change the upper reliability bound (UB).
   2.2. IF $FP_0 > UB$ THEN the network is unreliable.
   2.3. ELSE GOTO 1. assuming $G = H$.
3. Make e_{ij} absolutely reliable. Let $H = G/\{e_{ij}\}$ be a corresponding graph
   3.1. IF the flow can be transferred via absolutely reliable edges, THEN change the lower reliability bound (LB).
   3.2. IF $FP_0 \leq LB$ THEN the network is reliable.
   3.2. ELSE GOTO 1. assuming $G = H$.

Therefore, the algorithm is a decision tree. The initial graph is at the root node (assigned to the input) with all its attributes and parameters. On the right branch we make a pivot edge absolutely reliable, on the left branch we remove a pivot edge. The branching process ends in the following cases:

- The flow can be transferred via absolutely reliable edges in the received graph H, which is obtained by making a pivot edge absolutely reliable.
- The flow cannot be passed in the received graph, which is obtained by removing a pivot edge, that is:
  - the capacity of the remaining edges is not sufficient for transmitting the flow
  - the obtained graph is not connected.

4.1   Improvements of the Algorithm

To improve the algorithm performance, the factoring can be performed not with an arbitrary edge, but along the edges from S–T paths. We offer to find all the complementary paths from source to sink in the process of maximum flow searching by the Ford-Fulkerson method. The advantage of the approach is that after each edge removal the Ford-Fulkerson algorithm is launched for finding all complementary S–T paths, which can be stored without any loss of computational speed.

However, set of paths mentioned above does not coincide with the set of all paths from source to sink. So after each removal of an edge, this set must be changed. Therefore, there is no possibility to use some optimization techniques related to the analysis of a state of all paths from S to T.

To choose a next pivot edge in the proposed algorithm, we use two different approaches ("pessimistic" and "optimistic"). In the first strategy we choose pivot edges from each path starting from the source in a row and first check removing branch, thus trying destroy all paths and thus bring closer a moment of $I(S) = 0$. On the contrary, in the second strategy we choose pivot edges from one path and only then go to a next path and first check branch in which the pivot edge became absolutely reliable, thus bring closer a moment of $I(S) = 1$.

Note that we can accelerate calculations if calculate one bound only: lower by "optimistic" strategy or upper by "pessimistic" one, as we know that both bounds goes to an exact solution. Choice of a strategy depends on our a-priory certainty in result.

In addition, after each edge removal, to accelerate the maximum flow search, hanging vertices are deleted, except the case when a hanging vertice is the source or the sink.

4.2   The Algorithm's Pseudocode

1. Obtain set L of $S - T$ paths, $l_i = \{e_{ij}\}$, $n = 0$, $m = 0$. Calculate maximum flow f_{max} in G by Ford-Fulkerson method.
2. IF $f_{max} < f_0$, THEN G is unreliable.
3. **Removing e_m from G.**
   3.1. IF e_m is absolutely reliable and the last in l_n THEN
      – $m = 0$, $n = n + 1$ – choose the first edge in the next chain
      – GOTO 3.3.
   3.2. ELSE
      – $m = m + 1$, $n = n$ – choose the next edge in the current chain
      – GOTO 3.3.
   3.3. Remove e_m in G, $G\backslash\{e_m\} = G_i$.
   3.4. Obtain set L_i of $S - T$ paths and calculate maximum flow f_{max} in G_i.
   3.5. $P(G_i) = P(G)p_{ij}$.
   3.6. IF flow f_0 can be transferred in G_i GOTO 3. assuming:
      – $G = G_i$
      – $P(G) = P(G_i)$
      – $G_{trust} = G_{trust}$
      – $L = L_i$
      – $n = 0$
      – $m = 0$
   3.7. $UB- = P(G_i)$
   3.8. IF $FP_0 > UB$ THEN G is unreliable.
4. **Making e_m absolutely reliable in G**
   4.0 Return the edge e_m in G
   4.1 IF e_m is absolutely reliable and is the last in the chain l_n THEN $m = 0$, $n = n + 1$.
   4.2. ELSE if e_m is absolutely reliable THEN $m = m + 1$, $n = n$.
   4.3. Make e_{ij} absolutely reliable, $G/\{e_{ij}\} = G_i$.
   4.4. Update G_{trust}.
   4.5. Calculate maximum flow f_{max} in G_{trust}.
   4.6. $P(G_i) = P(G)p_{ij}$.
   4.7. IF flow f_0 can be transferred in G_{trust} GOTO 3. assuming:

- $G = G_i$
- $P(G) = P(G_i)$
- IF e_m is the last in l_n THEN
  - $m = 0$
  - $n = n + 1$
- ELSE
  - $m = m + 1$
  - $n = n$

4.8. ELSE $LB+ = P(G_i)$

4.9. IF $FP_0 \leq LB$ THEN Graph is reliable.

In this algorithm we can swap steps 3 and 4. That is, we can swap sequence of the delete function and the function for making an edge absolutely reliable. As it is shown in the next section, the program runtime strongly depends on the sequence choice.

5   Case Studies

We consider two topologies: $4 * 4$ grid (G_1, 16 nodes, 24 edges), and $4 * 5$ grid (G_2, 20 nodes, 31 edges) in assumption that all edges have equal reliabilities $p = 0.99$ and capacities $c = 15$. The source is placed in the upper left corner, and the sink is placed in the lower right corner. For each topology we have computed exact reliabilities values for the various FP_0 values and also have demonstrated the decision making process for the different threshold values. The obtained results and the computational are present below.

Table 1. Results of numerical experiments for G_1

Network requirements		FP	Runtime of the exact algorithm (s)	Runtime of the cumulative updating algorithm (s)		Result
f_0	FP_0			Delete branch	Contract branch	
19	0.995>	0.96018	4.23	0.038	0.451	False
	0960<	0.96018	4.23	0.585	0.431	True
	0.900<	0.96018	4.23	1.509	0.51	True
14	0.9999>	0.9998	40.496	0.048	0.277	False
	0.990<	0.9998	40.496	0.457	0.354	True
	0.900<	0.9998	40.496	0.504	0.034	True

The decision making algorithm regardless of order of the functions (first delete or contract) shows a great runtime advantage over the exact algorithm. If we compare the runtime values in the Table 1 with the runtime values in the

Table 2. Results of numerical experiments for G_2

Network requirements		FP	Runtime of the exact algorithm (s)	Runtime of the cumulative updating algorithm (s)		Result
f_0	FP_0			Delete branch	Contract branch	
19	0.995>	0.96017	250.627	0.045	0.241	False
	0.960<	0.96017	250.627	2.489	2.226	True
	0.900<	0.96017	250.627	2.643	0.179	True
14	0.9999>	0.9997	>1 h	0.43	1.757	False
	0.998<	0.9997	>1 h	2.530	1.805	True
	0.900<	0.9997	>1 h	2.017	0.67	True

Table 2, we can conclude that this advantage increases with increasing dimension of a network.

As it was stated above, there is time difference between variants of factorization. If a graph is reliable then the algorithm by deleting works slower. On the contrary, this algorithm works faster for an unreliable graph. This is due to the peculiarity of comparing the threshold value of reliability and calculated boundaries.

When making an edge absolutely reliable, only the lower limit changes, which tends from below to the exact value. On the other hand, when removing a edge, the upper limit is lowered, tending to the exact value from the top.

So, if the graph is reliable, then to complete the algorithm the condition $LB \geq FP_0$ must be fulfilled, which is attainable with the making an edge absolutely reliable.

And if the graph is unreliable, then it is necessary to obtain the inequality $UB < FP_0$, which is achieved when removing edges.

5.1 Convergence of Reliability Boundaries

Let us show the convergence of the boundaries of LB and UB to the exact reliability value. We consider $3 * 3$ grid topology in assumption that all edges have equal reliabilities $p = 0.65$ and capacities $c = 20$, $f_0 = 20$. The runtime is less than 1 s.

The figure below shows the changing of the lower and upper bounds of the reliability as a function of step of the algorithm N (the iteration number) (Fig. 1).

In the next figure we show how bounds are changing for $4 * 4$ grid topology in assumption that reliability of each is between 0.5 and 1 with a uniform distribution, and $f_0 = 20$. The total calculation time is about 41 s (Fig. 2).

Due to the fact that the calculation of UB and LB is based on the exact factorization method, these boundaries converge to the exact reliability value, as figures show.

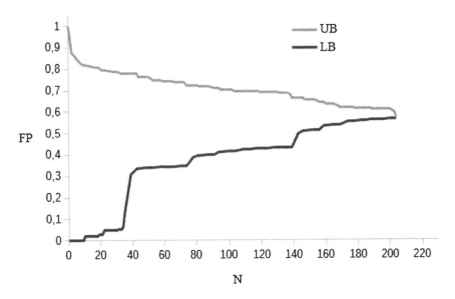

Fig. 1. Convergence of bounds to the exact solution for $3 * 3$ grid topology

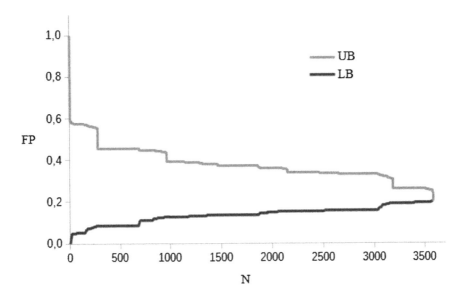

Fig. 2. Convergence of bounds to the exact solution for $4 * 4$ grid topology

6   Conclusion

As a result, we have obtained cumulative estimation of the probability of transmission a flow of a given value between a selected pair of nodes in a random graph to make a decision about its reliability by this criterion.

Experiments show the efficiency of using the developed algorithm in comparison with the exact factorization algorithm.

The aim of further research is to develop a parallel realization of the proposed algorithm.

In addition, optimization of the algorithms, both sequential and parallel, is required in order to take into account the structural features of a graph. For example, a graph may contain chains, bridges, points of articulation, or have others structural features.

References

1. Evans, J.R.: Maximum flow in probabilistic graphs-discrete case. Networks **6**, 161–183 (1976)
2. Lin, Y.K.: MC-based algorithm for a telecommunication network under node and budget constraints. Appl. Math. Comput. **190**, 1540–1550 (2007)
3. Lin, Y.K.: Reliability of k separate minimal paths under both time and budget constraints. IEEE Trans. Reliab. **59**(1), 183–190 (2010)
4. Forghani-Elahabad, M., Mahdavi-Amiri, N.: An efficient algorithm for the multi-state two separate minimal paths reliability problem with budget constraint. Reliab. Eng. Syst. Saf. **142**, 472–481 (2015)
5. Todinov, M.T.: Topology optimization of repairable flow networks and reliability networks. Int. J. Simul. Syst. Sci. Technol. **11**(3), 75–84 (2010)
6. Wu, W.W., Ning, A., Ning, X.X.: Evaluation of the reliability of transport networks based on the stochastic flow of moving objects. Reliab. Eng. Syst. Saf. **93**, 838–844 (2008)
7. Colbourn, Ch.J.: The Combinatorics of Network Reliability. Oxford University Press, New York (1987)
8. Won, J.-M., Karray, F.: Cumulative update of all-terminal reliability for faster feasibility decision. IEEE Trans. Reliab. **59**(3), 551–562 (2010)
9. Won, J.-M., Karray, F.: A greedy algorithm for faster feasibility evaluation of all-terminal-reliable networks. IEEE Trans. Syst. Man Cybern. Part B Cybern. **41**(6), 1600–1611 (2011)
10. Rodionov, A.S., Migov, D.A., Rodionova, O.K.: Improvements in the efficiency of cumulative updating of all-terminal network reliability. IEEE Trans. Reliab. **61**(2), 460–465 (2012)
11. Rodionov, A.S., Rodionova, O.K.: Exact bounds for average pairwise network reliability. In: the 7th ACM International Conference on Ubiquitous Information Management and Communication (Kota Kinabalu, Malaysia), Article no. 45. ACM New York (2013)
12. Migov, D.A., Nesterov, S.N.: Methods of speeding up of diameter constrained network reliability calculation. In: Gervasi, O., Murgante, B., Misra, S., Gavrilova, M.L., Rocha, A.M.A.C., Torre, C., Taniar, D., Apduhan, B.O. (eds.) ICCSA 2015. LNCS, vol. 9156, pp. 121–133. Springer, Cham (2015). https://doi.org/10.1007/978-3-319-21407-8_9

13. Migov, D.A., Nechunaeva, K.A., Nesterov, S.N., Rodionov, A.S.: Cumulative updating of network reliability with diameter constraint and network topology optimization. In: Gervasi, O., Murgante, B., Misra, S., Rocha, A.M.A.C., Torre, C., Taniar, D., Apduhan, B.O., Stankova, E., Wang, S. (eds.) ICCSA 2016. LNCS, vol. 9786, pp. 141–152. Springer, Cham (2016). https://doi.org/10.1007/978-3-319-42085-1_11

14. Rodionov, A.S.: Cumulative estimated values of structural networks reliability indices and their usage. In: IEEE Conference on Dynamics of Systems, Mechanisms and Machines (Omsk, Russia), pp. 1–4. IEEE Press, New York (2016)

15. Migov, D.A.: Evaluation of wireless sensor network reliability with use of reliability bounds cumulative updating. In: IEEE International Forum on Strategic Technology (Ulsan, Korea), pp. 120–124. IEEE Press, New York (2017)

16. Rodionov, A.S., Migov D.A.: Obtaining and using cumulative bounds of network reliability. In: Volosencu, C. (ed.) System Reliability, pp. 93–112. InTech, Rijeka, Croatia (2017). Chap. 5

17. Ball, M.O.: Computational complexity of network reliability analysis: an overview. IEEE Trans. Reliab. **35**, 230–239 (1986)

18. Canale, E., Cancela, H., Robledo, F., Romero, P., Sartor, P.: Full complexity analysis of the diameter-constrained reliability. Int. Trans. Oper. Res. **22**(5), 811–821 (2015)

19. Page, L.B., Perry, J.E.: A practical implementation of the factoring theorem for network reliability. IEEE Trans. Reliab. **37**(3), 259–267 (1998)

20. Ford, L.R., Fulkerson, D.R.: Maximal flow through a network. Can. J. Math. **8**, 399–404 (1956)

Profit Maximization and Vehicle Fleet Planning for a Harbor Logistics Company

Natalia B. Shamray[1] and Nina A. Kochetova[2($\boxtimes$)]

[1] Institute for Automation and Control Sciences, Vladivostok, Russia
shamray@dvo.ru
[2] Sobolev Institute of Mathematics, Novosibirsk, Russia
nkochet@math.nsc.ru

Abstract. We present a new optimization model to maximize the total operating profit of a harbor logistics company on a finite time horizon. Some local providers supply the company with a scrap-metal materials of different qualities. The materials are reprocessed into the high-quality product and exported to abroad by different types of ships. The company has to cover the purchase cost for the materials, the transportation cost to deliver the materials, the reprocessing and storage cost in a warehouse, shipping cost, and payment for international declarations. To find the best strategy for the company we present a mixed integer nonlinear model. We linearize the objective function and aggregate the set of providers in order to apply CPLEX software efficiently. We conduct computational experiments on real test instances and discuss how to use the model for planning fleet of vehicles, a capacity of the warehouse, and price strategy for the company.

Keywords: Supply chain · Heterogeneous fleet · Vehicle
Reprocessing

1 Introduction

In the last few years, international sea freight transportation showed an outstanding growth causing to harbors to reach their maximum capacity. Operations are nowadays unthinkable without effective usage of information technology and optimization methods [1]. In this paper, we present a mathematical model for a logistics company to maximize its profit in the harbor supply chain system (see Fig. 1). Some local providers have scrap metal of different types. The company gets it by known price and delivers to reprocessing center by own or rented vehicles. Each vehicle has a capacity. The company has some own vehicles and can rent the limited number of vehicles with small, medium, or large capacity. The working time of company drivers is bounded. In the reprocessing center, the scrap metal is transformed into the exported final product and stored in the harbor warehouse. The reprocessing center has a capacity which limits the total volume of materials per day. In this center, the metal is cut and all impurities are

© Springer International Publishing AG, part of Springer Nature 2018
A. Eremeev et al. (Eds.): OPTA 2018, CCIS 871, pp. 331–342, 2018.
https://doi.org/10.1007/978-3-319-93800-4_27

removed. The company has to pay for reprocessing and storage in the warehouse for each ton of the final product. According to the export contracts, the company must supply the known volume of the product to the foreign partners by ships within a given time horizon. Some ships with different capacities are available in the harbor by different prices. The company has to cover the expenses for ship registration and sea freight. The goal is to find the best logistic strategy to maximize the total profit of the company or minimize the total cost to realize the export contracts and determine

- the flow of each type of scrap metal from providers to reprocessing center;
- the fleet of vehicles for delivering these flows and the route for each vehicle;
- the storage plan for the warehouse;
- the schedule of ships within the time horizon.

To solve this optimization problem, we create a mixed integer nonlinear program and linearize the objective function in order to apply the CPLEX software. To reduce the dimension of the problem we aggregate the set of providers and introduce a joint provider in each local town. We conduct computational experiments on real test instances for a harbor near Vladivostok city and show how to apply the model for planning the own fleet of vehicles of the company, modify the capacity of the reprocessing center and the price strategy for local providers.

The paper is organized as follows. In Sect. 2 we introduce notations and detailed mathematical model. In Sect. 3 we rewrite the model as the linear mixed integer program and present an approximation method. In Sect. 4 we discuss the computational experiments and finally, Sect. 5 concludes the paper.

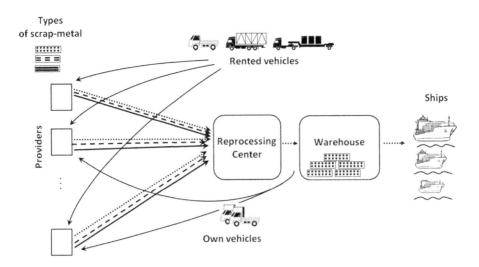

Fig. 1. Supply chain framework

2   Mathematical Model

Let us introduce the following notations. We will use the following sets:

J is the set of local providers;
I is the set of scrap types;
K^0 is the set of vehicles of company;
K is the set of vehicles available for renting;
S is the set of ships;
T is the time horizon.

Parameters of the Model:

s_{ij} is the total amount of scrap type i (or scrap i for short) for provider j;
r_i is the cost of reprocessing the scrap i into the final product per ton;
ζ_i is the percent of impurities for scrap i;
c_{kj}^0 is the transportation cost for one trip to provider j by own vehicle k;
c_{kj} is the transportation cost for one trip to provider j by rented vehicle k;
p_{ij} is the price function for provider j for scrap i;
v_k is the capacity of vehicle k;
$\bar{\tau}$ is the duration of driver's shift for company vehicles;
τ_j is the duration of trip to provider j and back;
v is the capacity of the reprocessing center per day;
c is the storage cost of ton of final product per day;
v^0 is the initial amount of final product in warehouse;
f_s is the registration fee for ship s;
f is the sea freight of final product per ton;
$\bar{v}_s$ is the capacity of ship s;
d is the total amount of final product that must be sent to foreign partners.

Variables of the Model:

$V_{ijt} \geq 0$ is the amount of scrap i which is purchased from provider j in day t;

$V_{ij} \geq 0$ is the total amount of scrap i which is purchased from provider j;

$Y_t \geq 0$ is the total amount of final product in warehouse at the end of day t;

$W_t \geq 0$ is the amount of final product which is sent by a ship in day t;

$Q_{st} \in \{0,1\}$ is the ship scheduling; $Q_{st} = 1$ iff ship s departures in day t;

Q_{kjt}^0 is the number of trips for company's vehicle k to provider j in day t;

Q_{kjt} is the number of trips for rented vehicle k to provider j in day t.

The profit of the company is the difference between the revenue R which is a constant defined by the contracts, and the total operational cost that consists of the following items:

– payment to local providers:

$$C_1(V) = \sum_{i \in I} \sum_{j \in J} p_{ij}(V_{ij});$$

– reprocessing and storage cost:

$$C_2(V, Y) = \sum_{i \in I} \sum_{j \in J} r_i V_{ij} + c \sum_{t \in T} Y_t;$$

– transportation cost:

$$C_3(Q) = \sum_{t \in T} \sum_{j \in J} \left(\sum_{k \in K^0} c_{kj}^0 Q_{kjt}^0 + \sum_{k \in K} c_{kj} Q_{kjt} \right);$$

– ship registration and sea freight:

$$C_4(Q, W) = \sum_{t \in T} \sum_{s \in S} f_s Q_{st} + f \sum_{t \in T} W_t.$$

The problem is to maximize the total profit

$$P(V, W, Y, Q) = R - C_1(V) - C_2(V, Y) - C_3(Q) - C_4(Q, W) \qquad (1)$$

subject to the following constraints:

- rented and own vehicles can deliver all scrap to reprocessing center

$$\sum_{i \in I} (1 + \zeta_i) V_{ijt} \leq \sum_{k \in K^0} v_k Q_{kjt}^0 + \sum_{k \in K} v_k Q_{kjt}, \quad j \in J, t \in T; \qquad (2)$$

- each provider has limited store of each type of scrap

$$V_{ij} = \sum_{t \in T} V_{ijt} \leq s_{ij}, \quad i \in I, j \in J; \qquad (3)$$

- working time of the company drivers is bounded

$$\sum_{j \in J} Q_{kjt} \tau_j \leq \bar{\tau}, \quad k \in K^0, t \in T; \qquad (4)$$

- the capacity constraint for reprocessing center

$$\sum_{i \in I} \sum_{j \in J} (1 + \zeta_i) V_{ijt} \leq v, \quad t \in T; \qquad (5)$$

- the initial amount of the product at the warehouse

$$Y_0 = v^0; \qquad (6)$$

- the balance constraint at the warehouse for each day

$$Y_t = Y_{t-1} + \sum_{i \in I} \sum_{j \in J} V_{ijt} - W_t, \quad t \in T; \tag{7}$$

- the export capacity of the warehouse

$$W_t \leq Y_{t-1}, \quad t \in T; \tag{8}$$

- the ship capacity constraints

$$W_t \leq \sum_{s \in S} Q_{st} \bar{v}_s, \quad t \in T; \tag{9}$$

- the assignment constraint for ships

$$\sum_{s \in S} Q_{st} \leq 1, \quad t \in T; \tag{10}$$

- the contracts demand for foreign partners

$$\sum_{t \in T} W_t = d; \tag{11}$$

- variables constraints

$$Q^0_{kjt}, Q_{kjt} \geq 0 \text{ integer}, \quad Q_{st} \in \{0,1\}, \quad W_t, V_{ijt}, V_{ij}, Y_t \geq 0. \tag{12}$$

The problem (1)–(12) has the linear constraints and nonlinear objective function. We assume that all functions $c_{ij} = c_{ij}(V_{ij})$ are piecewise convex linear functions. In Sect. 3 we reduce this nonlinear problem to the linear one. Nevertheless, the problem is NP-hard as the knapsack problem with Q_{st} variables is a special case of the problem (1)–(12).

3   Approximation Method

Let us assume that each function $p_{ij} = p_{ij}(V_{ij}), i \in I, j \in J$ is continuous piecewise convex linear functions (see Fig. 2a) and can be presented as follows:

$$p_{ij}(V_{ij}) = \begin{cases} p^0_{ij} V_{ij}, & \text{if } 0 \leq V_{ij} \leq \delta^1_{ij}, \\ (p^0_{ij} + p^1_{ij})V_{ij} - \xi^1_{ij}, & \text{if } \delta^1_{ij} \leq V_{ij} \leq \delta^2_{ij}, \\ \quad \vdots \\ (p^0_{ij} + \cdots + p^q_{ij})V_{ij} - \xi^q_{ij}, & \text{if } \delta^{q-1}_{ij} \leq V_{ij} \leq s_{ij} \end{cases}$$

and it has q linear segments with positive coefficients $p^l_{ij}, \xi^l_{ij}, l \leq q$.

In fact, coefficients p_{ij}^l are additional bonuses for local provider for large supplies. To get the linear objective function, we replace each function $p_{ij}(V_{ij})$ by the sum of linear functions $p_{ij}^l(V_{ij}^l)$ with new nonnegative variables:

$$p_{ij}(V_{ij}) = \sum_{l=0}^{q} p_{ij}^l V_{ij}^l, \quad i \in I, j \in J$$

subject to additional constraints $V_{ij}^l \geq V_{ij} - \delta_{ij}^l, l = 1, \ldots, q$ (see Fig. 2b).

It is easy to see that $V_{ij}^l = \max\{0, V_{ij} - \delta_{ij}^l\}, l = 1, \ldots, q$ and this linearization approach is correct. Thus we can apply the classical branch and bound method for the mixed integer linear program (CPLEX solver).

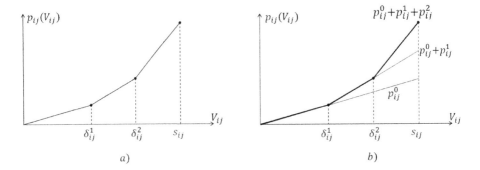

Fig. 2. Price function $p_{ij}(V_{ij})$ and its linearization

Unfortunately, the dimension of the real-world instances is too large for this method and we need additional efforts to reduce the number of variables. To this end, we aggregate the set of providers [7] and replace all providers in each town by a joint provider. From the point of view the objective function (1) it is incorrect. Joint providers can get an additional bonus for common supplies function. Thus, the aggregated model can get an approximate solution only. But we can try to improve the obtained solution by re-assignment of the total supply of joint provider by its local providers.

On the other hand, the concept of joint provider can improve the model and reduce the transportation cost. The variables Q_{kjt}^0 and Q_{kjt} mean the number of vehicles which go to provider j on day t. According to constraints (2), the split delivery is not allowed. For each joint provider, we can do it implicitly. The savings in distance and the number of vehicles that can be achieved by using split deliveries can by large [2,3]. Nevertheless, we consider the simples trips only to simplify the model. Otherwise, we need to modify the model for vehicle routing and design special type sophisticated algorithms to optimize the transportation cost [4–6].

4   Computational Experiments

We conduct our computational experiments on real test instance of Russian logistics company which is located in a harbor near Vladivostok city. The company has 96 local providers in 26 towns and villages (see Fig. 3).

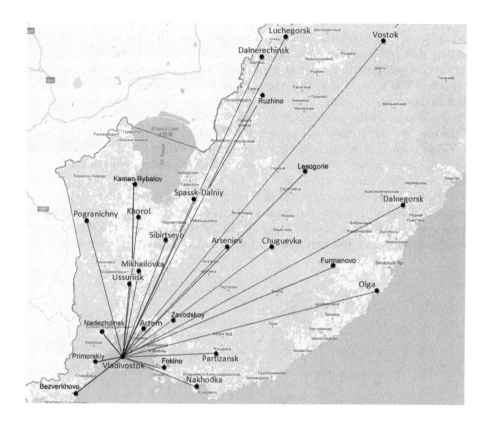

Fig. 3. Regional providers

Three types of scrap metal (3A, 5A, 12A) are supplied and two of them (5A and 12A) are reprocessed into the type 3A as the final product for the contracts. Time horizon is one month, $|T| = 30$, the total amount of the final product for export $d = 8000$ ton. Five types of ships are available for rent with the appropriate parameters (Table 1).

The sea freight f is 850 rubles per ton. Three types of vehicles can be rented:

- 10 ton at most 20 vehicles,
- 15 ton at most 15 vehicles,
- 30 ton at most 10 vehicles

Table 1. The ship parameters

	$s=1$	$s=2$	$s=3$	$s=4$	$s=5$	$s=6$
capacity $\bar{v}_s$	2500	3000	3500	4000	4500	5000
regist. fee f_s	25000	30000	35000	40000	45000	50000

and only one 5 ton vehicle belongs to the company, $|K^0| = 1$; $|K| = 45$; the working shift is 7 hours, transportation cost matrix $\{c_{ij}\}$ includes elements from 2500 rubles to 5500 rubles per trip to providers. The capacity of reprocessing center $v = 300$ ton per day, the storage cost for warehouse $c = 2, v^0 = 0$. The price function $p_{ij}(V_{ij})$ has three linear segments and grows from 7500 to 9750 per ton identically for providers from the same town.

In our first experiment, we aggregate the set of providers and get 26 joint providers instead of 96 ones. We find near optimal solution with deviation from the optimum at most $0,1\%$ in one minute by CPLEX software. Figure 4 depicts providers in the solution obtained.

Figure 5 shows the supply schedule for three types of scrap materials from the providers. Figure 6 presents the total amount of final product in the warehouse at the end of each day, variables $Y_t, t \in T$. To get the feasible solution to original problem instance, we re-distribute the total supply of each joint provider by its

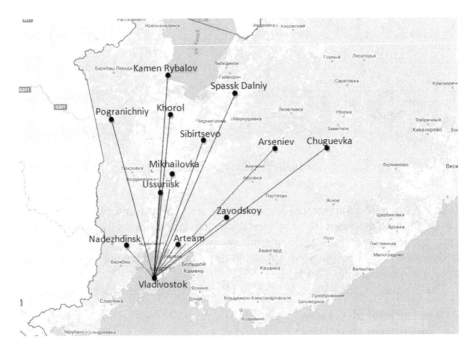

Fig. 4. Regional providers in the solution

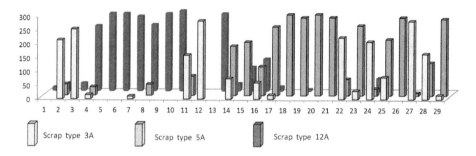

Fig. 5. Supply scheduling for three types of scrap materials from all providers

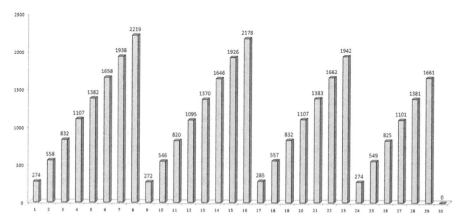

Fig. 6. The total amount of final product in warehouse at the end of each day

providers and compare the results. The total profit of the company increases but at most 1% only. We conclude that our aggregation approach is efficient and can produce feasible solutions with small deviations from the optimum.

In the second experiment, we modify the price functions p_{ij} in order to understand the influence of the bonus coefficients $p_{ij}^l, l \leq q$. We compare the structure of the solutions obtained for the bonus coefficients $\beta p_{ij}^l, l \leq q$ for $\beta = 0, 1, 2, \ldots, 5$. In case $\beta = 0$, we can improve the total profit. In case $\beta > 1$, the total profit decreases. But the structure of the solutions obtained is identical for these cases. The number of joint providers in final solutions increases from 13 to 15 only for $\beta = 5$. We guess that the optimal value has a weak dependence on the bonuses. Optimization model tries to smoothen the influence of the bonuses by set of providers in the final solution. Thus, the convexity of the price function is important for marketing purposes. It stimulated local providers to store enough materials for the company. But the model can smooth its influence. As a result, the company can use large bonuses p_{ij}^l without large reduction of the total profit.

In the third experiment, we study the influence of the company fleet of vehicles on the total profit. We conduct several experiments with different sets K^0, $|K^0| = 1, 2, \ldots, 10$ with different vehicle's capacities and observe slight improving the total profit. This improvement easy to understand because of $c_{kj}^0 \leq c_{kj}$ for identical vehicles. By the current prices for renting vehicles, the including new vehicles into the company fleet can be useful for large time horizon only, 4–5 years. Thus, we conclude that the total cost for delivering all scrap materials to the reprocessing center is a small part of the total operational cost. For improving the total profit, we need to enlarge the revenue instead of reducing the operational cost.

In our final experiment, we replace the Eq. (10) by inequality

$$\sum_{t \in T} W_t \geq d \tag{13}$$

and determine the maximal profit for this time horizon. In this case, the company can send 8400 ton of final product instead of 8000 (according to contracts) and improve the profit from 28 568 000 to 29 816 000.

To find the bottleneck of the system, we modify the capacity of reprocessing center and solve the problem (1)–(10), (12), (13) with $v = 300, 350, 400, \ldots, 650$. The results of this experiment are shown in Fig. 7. We can see the fast-growing the profit and an increase the amount of final product for export $\sum_{t \in T} W_t$ from 8000 to 15600. The detailed analysis of the experimental results can help us to find the best capacity of reprocessing center if we know the payment for enlarging the center and the total demand for export contracts in a future.

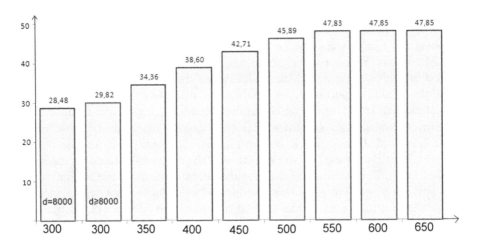

Fig. 7. The total profit for different capacity of reprocessing center

5   Conclusions

This paper studied a new optimization model to maximize the total operating profit of a harbor logistics company. We present a mixed integer nonlinear model, linearize the objective function, and aggregate the set of local providers to apply CPLEX solver efficiently. Computational results for real test instances confirm the efficiency of the approach and allow the company to improve the structure of supply chain system. Specifically, we observe that the capacity of the reprocessing center is a bottleneck for the system.

For future research, it is interesting to consider some competitive logistics companies. Each company has own contracts, warehouse, and reprocessing center. But the set of local providers is the same for all companies. Similar models for two companies in the field of facility location can be found in [8–11].

It is worth to note that such models are harder than well-known NP-complete problems and, in fact, are Σ_2^P–hard [12–14]. The best pricing strategy for each company may be the most intriguing question in such games. Some preliminary results for Stackelberg pricing games can be found in [15].

Acknowledgements. The research in Sects. 1, 2 and 3 was supported by RFBR grant 18-07-00599. The research in Sects. 4 and 5 was supported by the program of fundamental scientific researches of the SB RAS I.5.1., project 0314-2016-0014.

References

1. Grazia Speranza, M.: Trends in transportation and logistics. Eur. J. Oper. Res. **264**(3), 830–836 (2018)
2. Chen, S., Golden, B.: The split delivery vehicle routing problem: applications, algorithms, test problems, and computational results. Networks **49**(4), 318–329 (2007)
3. Dror, M., Trudeau, P.: Saving by split delivery routing. Transp. Sci. **23**(2), 141–145 (1989)
4. Kochetov, Y.A., Khmelev, A.V.: A hybrid algorithm of local search for the heterogeneous fixed fleet vehicle routing problem. J. Appl. Ind. Math. **9**(4), 503–518 (2015)
5. Khmelev, A., Kochetov, Y.: A hybrid local search for the split delivery vehicle routing problem. Int. J. Artif. Intell. **13**(1), 147–164 (2015)
6. Vidal, T., Crainic, T.G., Gendreau, M., Prins, C.: Heuristics for multi-attribute vehicle routing problems: a survey and synthesis. Eur. J. Oper. Res. **231**(1), 1–21 (2013)
7. Irawan, C.A., Salhi, S., Scaparra, M.P.: An adaptive multiphase approach for large unconditional and conditional p-median problems. Eur. J. Oper. Res. **237**(2), 590–605 (2014)
8. Alekseeva, E., Kochetov, Y., Plyasunov, A.: An exact method for the discrete $(r|p)$-centroid problem. J. Glob. Optim. **63**(3), 445–460 (2015)
9. Alekseeva, E., Kochetov, Y.: Matheuristics and exact methods for the discrete $(r|p)$-centroid problem. In: Talbi, E.G. (ed.) Metaheuristics for Bi-level Optimization. SCI, vol. 482, pp. 189–219. Springer, Berlin (2013). https://doi.org/10.1007/978-3-642-37838-6_7

10. Davydov, I.A., Kochetov, Y.A., Carrizosa, E.: A local search heuristic for the $(r|p)$-centroid problem in the plane. Comput. Oper. Res. **52**, 334–340 (2014)
11. Lavlinskii, S.M., Panin, A.A., Plyasunov, A.V.: Comparison of models of planning public-private partnership. J. Appl. Ind. Math. **10**(3), 356–369 (2016)
12. Iellamo, S., Alekseeva, E., Chen, L., Coupechoux, M., Kochetov, Y.: Competitive location in cognitive radio networks. 4OR **13**(1), 81–110 (2015)
13. Davydov, I., Kochetov, Y., Plyasunov, A.: On the complexity of the $(r|p)$-centroid problem in the plane. TOP **22**(2), 614–623 (2014)
14. Panin, A.A., Plyasunov, A.V.: On complexity of the bilevel location and pricing problems. J. Appl. Ind. Math. **8**(4), 574–581 (2014)
15. Diakova, Z., Kochetov, Y.: A double VNS heuristic for the facility location and pricing problem. Electron. Notes Discret. Math. **39**, 29–34 (2012)

Author Index

Adukova, Natalia V. 207
Ageev, Alexander 45
Al-Ameri, Jehan Mohammed 137

Borisovsky, Pavel 56

Davidović, Tatjana 251
Davydov, Ivan A. 267
Dolgui, Alexandre B. 278

Eremeev, Anton V. 278

Gabor, Adriana F. 288
Gimadi, Edward Kh. 131
Gorban, Alexander N. 137

Il'ev, Victor 3
Il'eva, Svetlana 3

Jakšić Krüger, Tatjana 251

Kazaeva, K. E. 193
Kel'manov, Alexander 109, 120
Khachay, Michael 68
Khamidullin, Sergey 120
Khandeev, Vladimir 109, 120
Kochetov, Yury 309
Kochetova, Nina A. 331
Kondakov, Artem 309
Kononov, Aleksandr V. 16, 78
Kosanoglu, Fuat 296
Kovalenko, Yulia V. 93
Kovalyov, Mikhail Y. 278
Kudryavtsev, Konstantin N. 207

Lavlinskii, Sergey 220
Levesley, Jeremy 137

Memar, Julia 78
Migov, Denis A. 321

Neznakhina, Katherine 68
Nikolaev, Andrei V. 155

Panasenko, Anna 109
Panin, Artem A. 16, 220
Plyasunov, Aleksandr V. 16, 220
Popov, Leonid D. 170
Pyatkin, Artem 120

Rodionov, Alexey S. 321
Rykov, Ivan A. 131

Seliverstov, Alexandr V. 183
Shamardin, Yury V. 120, 131
Shamray, Natalia B. 331
Shenmaier, Vladimir 120
Sigaev, Vyacheslav S. 278
Sleptchenko, Andrei 288, 296

Terekhov, Valery A. 137
Turan, Hasan Hüseyin 296
Tyukin, Ivan Y. 137

Vasil'ev, Valery A. 235
Veremchuk, Natalia S. 29

Yadykina, Olga A. 321

Zabotin, I. Ya. 193
Zabudsky, Gennady G. 29
Zakharov, Aleksey O. 93
Zinder, Yakov 78

Printed in the United States
By Bookmasters